OCCUPY THE EARTH: GLOBAL ENVIRONMENTAL MOVEMENTS

ADVANCES IN SUSTAINABILITY AND ENVIRONMENTAL JUSTICE

Previously ADVANCES IN ECOPOLITICS
Series Editor: Liam Leonard

PUBLISHED UNDER SERIES TITLE '**ADVANCES IN ECOPOLITICS**'

Community Campaigns for Sustainable Living: Health, Waste & Protest in Civil Society Volume 7
Edited by Liam Leonard

Sustainable Politics and the Crisis of the Peripheries: Ireland and Greece Volume 8
Edited by Liam Leonard & Iosif Botetzagias

Enterprising Communities: Grassroots Sustainability Innovations Volume 9
Edited by Anna Davies

Transnational Migration, Gender and Rights Volume 10
Volume Editor: Ragnhild Sollund
Series Editor: Liam Leonard

PUBLISHED UNDER SERIES TITLE '**ADVANCES IN SUSTAINABILITY AND ENVIRONMENTAL JUSTICE**'

International Business, Sustainability and Corporate Social Responsibility Volume 11
Edited by Maria Alejandra Gonzalez-Perez and Liam Leonard

Principles and Strategies to Balance Ethical, Social and Environmental Concerns with Corporate Requirements Volume 12
Edited by Liam Leonard and Maria Alejandra Gonzalez-Perez

Environmental Philosophy: The Art of Life in a World of Limits Volume 13
Edited by Liam Leonard, John Barry, Marius De Geus, Peter Doran and Graham Parkes

The Sustainability of Restorative Justice Volume 14
Edited by Paula Kenny and Liam Leonard

ADVANCES IN SUSTAINABILITY AND ENVIRONMENTAL JUSTICE VOLUME 15

OCCUPY THE EARTH: GLOBAL ENVIRONMENTAL MOVEMENTS

EDITED BY

LIAM LEONARD
West Virginia University, Morgantown, WV, USA

SYA BURYN KEDZIOR
Towson University, Towson, MD, USA

Emerald

United Kingdom – North America – Japan
India – Malaysia – China

Emerald Group Publishing Limited
Howard House, Wagon Lane, Bingley BD16 1WA, UK

First edition 2014

British Library Cataloguing in Publication Data
A catalogue record for this book is available from the British Library

ISBN: 978-1-78350-697-2
ISSN: 2051-5030 (Series)

CONTENTS

ACKNOWLEDGEMENTS

The editors would like to thank their friends and families for their support during the making of this book. We would also like to thank all of the contributors, and Commissioning Editor Aimee Nixon and the Emerald Group team.

This book is dedicated to our children, in the hope that they may see a better future.

LIST OF CONTRIBUTORS

Saleh Ahmed	Utah State University, Logan, UT, USA
Keith Bettinger	University of Hawai'i, Honolulu, HI, USA
Michael Briguglio	University of Malta, Msida, Malta
Micah Fisher	University of Hawai'i, Honolulu, HI, USA
Benny Furst	Ministry of Environmental Protection, Jerusalem, Israel
Nan Marie Greer	University of Hawai'i, Honolulu, HI, USA
Sya Buryn Kedzior	Towson University, Towson, MD, USA
Liam Leonard	West Virginia University, Morgantown, WV, USA
Charles Z. Levkoe	Wilfrid Laurier University, Ontario, Canada
Robert J. Mason	Temple University, Philadelphia, PA, USA
Darren McCauley	University of St. Andrews, St. Andrews, UK
Wendy Miles	University of Hawai'i, Honolulu, HI, USA
Milos Milicevic	Central European University, Budapest, Hungary
Rebecca Schiff	Lakehead University, Ontario, Canada
Leslie E. Sponsel	University of Hawai'i, Honolulu, HI, USA
Tamara Steger	Central European University, Budapest, Hungary

INTRODUCTION

The rise of global environmental movements has occurred in response to the threats posed by large scale industrialisation and pollution across the world. Starting with the conservation movements of the 1800s, the modern environmental movement emerged from the Anti-Nuclear protests of the Campaign for Nuclear Disarmament (CND) in the early 1960s. Rachel Carson's bestseller *Silent Spring* was published in 1962. Environmental protest formed part of the Counterculture of the 1960s, alongside campaigns for Civil Rights, Women's Liberation and the Peace Movement. Minds became more focused on ecology with the first 'Earth Day' in 1970, with 20 million Americans taking part, and led to the foundation of the Environmental Protection Agency (EPA). The 1990 Earth Day event was followed by United Nations Earth Summit in Rio de Janeiro.

On the streets of Seattle in 1999, the growing anti-globalisation movement protested the rise of corporate power. These protests against the World Trade Agreement and the injustices caused by globalised corporate activity gave rise to a worldwide movement which was supported by the rise of social media. Regional environmental movements began to link up through internet technologies as part of a phenomenon termed 'The Network Society' by sociologist Manuel Castells. Pro-democracy protests and campaigns in support of emerging and developing nations became a feature of the 2000s. Writers such as Noam Chomsky provided intellectual support for this movement. The economic crash which came at the end of that decade also gave rise to the 'Occupy' movement, and gave rise to the phrase 'Another World is Possible'. This movement is a coalition which stands for economic equality, the rights of emerging nations and oppressed communities, labour rights, environmental justice and human rights. Ultimately, it remains to be seen if these movements, both historical and contemporary, will lead to a world of greater human rights and environmental justice. In the words of Arundhati Roy, 'Another world is not only possible, she is on her way. On a quiet day, I can hear her breathing'.

This book, *Occupy the Earth: Global Environmental Movements*, is the latest in the Advances in Sustainability and Environmental Justice series with Emerald Publishing. The book contains contributions on movements from across the globe, and as such provides an interesting and crucial insight into the state of the world's contemporary environmental movements. The onset of global concerns about environmental risks, climate change and threats to the planet from industry has focused the minds of a generation. Throughout the world, new movements are emerging in an attempt to challenge those who would put profits before the planet. This volume brings together global contributions that represent the cutting edge of research in the area of global environmental movements.

Contributions include chapters on the spatial impacts of environmental groups in Israel, the work of Greenpeace in Brazil, environmental activism in Ireland, animal rights and anti-hunt activism in Malta, the global de-growth movement, environmental movement mobilisation in China, and anti-pollution activism in India. The scope and breath of this research indicates the emergence of both a global grassroots environmental mobilisation in addition to analysis and documentation of these responses by researchers world-wide. With increased threats from climatic change and ecological degradation being highlighted as a threat to much of the world's population in the coming century, this activism and ensuing research becomes all the more significant.

Liam Leonard
Sya Buryn Kedzior
Editors

ONE GLOBAL MOVEMENT, MANY LOCAL VOICES: DISCOURSE(S) OF THE GLOBAL ANTI-FRACKING MOVEMENT

Tamara Steger and Milos Milicevic

ABSTRACT

In this chapter, we "occupy the earth" with an overview of the anti-fracking discourse(s) of diverse local initiatives converging as a global movement opposed to fracking. By mapping the discourse(s) of the anti-fracking movement, the articulation of the problems and solutions associated with fracking raise questions not only about the environment but draw attention to a crisis of democracy and the critical need for social and environmental justice. With the help of a multiple theoretical framework we draw on insights about environmental movements and their democratizing potential; conceptualizations about power and (counter) discourse; and depictions of the environmental justice movements in the United States. Toward this end, we analyze the framing of the anti-fracking movement: the many local voices engaging in political struggles to sustain their communities, places and ways of life, and the global movements' forum for collective solidarity, recognition, and civic action.

Occupy the Earth: Global Environmental Movements
Advances in Sustainability and Environmental Justice, Volume 15, 1–35

ISSN: 2051-5030/doi:10.1108/S2051-503020140000015002

Shedding light on the multiple frames employed by movement members, we discuss the implications and potential embodied in this widening debate.

Keywords: Social movement; environmental movement; hydraulic fracturing or "fracking"; political ecology

INTRODUCTION

In this chapter, we "occupy the earth" with an overview of the anti-fracking discourse(s) of diverse local initiatives converging as a global movement opposed to fracking. By mapping the discourse(s) of the anti-fracking movement, the articulation of the problems and solutions associated with fracking raise questions not only about the environment but also draw attention to a crisis of democracy and the critical need for social and environmental justice. With the help of Arturo Escobar's insights about environmental movements and their role in articulating "alternative imaginaries"; Michel Foucault's conceptualization of power and (counter) discourse; and Giovanna Di Chiro's depictions of the environmental justice movements in the United States, we analyze the framing of the anti-fracking movement: framing across the many local voices engaging in political struggles to sustain their communities, places and ways of life, and the global movement's forum for collective solidarity, recognition, and civic action. In this process, three important spheres of the movement are identified: (1) what constitutes knowledge about fracking; (2) communities struggling in and across places; and (3) the explicit and powerful articulation of a resonant, alternative political and economic imaginary based on democracy and social equality.

OVERVIEW OF FRACKING

Hydraulic fracturing, also referred to as "fracking" (or "fracing") – a controversial high pressure, water (and often, chemical) based, horizontal drilling technique for removing oil and gas largely from shale – is engaged more and more across the globe. While the science and technology on fracking has been advancing for decades, it has only reached a point of recognition in this second decade of the 21st century as an energy "game changer" and the advent of a "New Gas Age" shifting the geopolitics of global energy. Fracking is currently widespread in the media and on the political agenda of major countries around the world, where the gas and oil industry and many governments are hailing it as: an economic necessity or even a boon ("Gas Rush"); an opportunity for energy security and independence; and a clean "transition fuel" to mitigate climate change. It has raised energy production capacity in the United States (US), where fracking is most widespread, to the point where the US is anticipating a role as a new major gas exporter. China, Russia, parts of South America, and the United Kingdom and others are joining its leagues as they encounter great potential in their own territories.

As the gas and oil industry and governments get excited about the further prospects of fracking, however, local and global communities have another story to tell. Fracking, they assert, damages their communities, consumes and/or pollutes precious water, air, and soil resources, causes earthquakes, and creates noise pollution. Fracking, they say, will furthermore not solve unemployment problems and will further exacerbate climate change and deter the much-needed attention to alternative, more sustainable, energy developments. Especially important, however, in the anti-fracking movement is the quest for (a real) democracy, less corruption, and the articulation of a better life and a better world.

Hundreds of local anti-fracking initiatives have appeared all over the world, and as these initiatives are increasingly creating links between one another, they are generating a truly global movement against fracking. Protests on the streets and in camps are popping up all over the world, as social media are flourishing with solidarity across movement activists working to organize their own communities and join and support other communities in their anti-fracking efforts. Many localities such as New York and Pittsburgh, Philadelphia (see foodandwaterwatch.org for an exhaustive list) and countries (e.g., France and Bulgaria) are winning bans and moratoriums against fracking.

Such wins have been paramount in the global movement against fracking. Global Frackdown, Camp Frack, and others have generated global forums for local and state anti-fracking initiatives to share information, find and generate solidarity, and record the strategies and successes to date. For example, according to the Global Frackdown web page, their first global effort in 2012 accounted for 200 actions across 20 countries. The organization Food and Water Watch quantifies moratoriums, bans, delays and other political successes, and report 416 measures (to date) to prevent or stop fracking in the US (Food and Water Watch, 2014a, 2014b). Thus, while the global movements coalesce anti-fracking efforts to promote solidarity, exchange, and widespread recognition, it is in the everyday of the people of the local initiatives that struggle for water, air, food, and land in the after-waves of a severe financial crisis and an unprecedented gap in social inequality where the thrust of the movement gains resonance.

In these places and across these communities, people are not just struggling for the subsistence of their everyday lives, they have found themselves in the middle of a crisis of democracy where "truth" and "power" are one and the same; and where their "truths" about fracking confront and repoliticize a dominant global discourse on energy politics that is largely orchestrated and perpetrated by the gas and oil industry and various political entities at local, state, federal, and international levels. In this context, the movement is challenged to originate an "alternative imaginary" in which a different path, an energy path included, can be articulated to guide the movement as a force for change. In this chapter, we explore the counter-discourse of the anti-fracking movement from the perspective of local and global anti-fracking efforts. Through anti-fracking frames, we discover the heart of the movement and shed light on "the power of the people" in a time of emerging hyperglobalization and consolidation of industry and governmental interests.

ANALYTICAL FRAMEWORK

Much of social movement theory has sought to understand how and why people engage in collective action. Prominent theories along these lines include crowd theory, political opportunity structures, resource mobilization theory, identity formation, framing, and others. Each has contributed to the knowledge base for understanding how people get together to assert change or maintain the status quo when confronted with change. Further work has delineated environmental justice movements centered around, among others, climate justice, water justice, defense of the commons, extractivism, and degrowth movements (Martinez-Alier et al., 2014).

For our purposes, we engage the work of Arturo Escobar on environmental social movements. This analytical framework is further complemented by an understanding of power and discourse based largely on the works of Michel Foucault. Finally, we consider the work of Giovanna Di Chiro on environmental justice movements. It is our assertion that this multi-theoretical framework is the most appropriate for making sense of local anti-fracking initiatives and the global anti-fracking movement. We are thus particularly interested in the political strategies of these efforts manifested operationally through power and discourse and their implications for environmental and social justice, particularly through movement framing. And, true to the title of this chapter, we further explore some of the distinctions across local and global initiatives.

Articulating Place and Power through Social Movement Mobilization

For Escobar, environmental social movements constitute particular relationships between nature and culture, or society characterized by political strategies that imagine alternative worlds (Escobar, 2001). While Escobar takes on the issue and politics of biodiversity, we similarly focus on (anti-) fracking and raise the question of whether or to what extent the anti-fracking movement can be categorically considered as an environmental movement (as it often is), and how (un)helpful it is to talk about the anti-fracking movement in this way. We are largely concerned with the articulation of a (counter)-discourse that brings (anti-)fracking into meaning through talk and texts, networks, institutions, and practices.

Central to understanding environmental and social justice movements is the articulation of the relationship between nature and culture or nature and society. The case of anti-fracking is especially illuminated by these relationships. In this context, the negotiation of "local" and "global" worlds is further articulated. For while the dominant global articulations of the nature-society relationship tend to produce socially homogenizing views that obscure local (cultural, historical, ecological, economic, and political) differences; it is exactly these differences that are often at the forefront of local articulations. This is why Escobar, drawing on the work of Edward Casey, asserts the need of "getting back into place." "We are, in short, placelings" (Escobar, 2001, p. 143) and further referring to Casey, "To live is to live locally, and to know is first of all to know the places one is in" (Casey, 1996, p. 18). Humanity is wrapped up in a world of places that act as contexts for human experience and are thus bound up with personal and cultural identity. Place has everything to do with the everyday which is embedded in institutions and cultural practices. It is for this reason that "emplacement of all cultural practices" is often highlighted by anthropologists who see this emplacement as stemming from the fact that "culture is carried into places by bodies" and that "place, body, and environment integrate with each other" (Escobar, 2001, p. 143).

The relationship between nature and society is further debated in the literature on environmental justice, particularly at the empirical level in the analysis of environmental justice movements. Di Chiro, for example, engages the environmental justice movement in the US in which "nature" and the "environment" are those "places and sets of relationships that sustain a local community's way of life" (Di Chiro, 1998, p. 300). Hence, environmental justice mobilization and activism can be seen as efforts to sustain places and relationships necessary for everyday life. Unlike mainstream environmentalism which emphasizes the protection of non-human "surroundings," environmental justice activists see this separation of humans and nature as misleading.

Di Chiro further explains how the construction of environment as separate from society is associated with the need to control and dominate ("tame") the environment. Alternatively, the environmental justice movement is engaged in countering this conception of environment and, as a result, changing the way we understand and relate to each other and the places within which we reside and sustain our lives. This reconceptualization, she argues, is the foundation for innovative political organization and action at the community level with regard to environmental threats that is "diverse, egalitarian, and non-hierarchical" as compared to mainstream environmentalism which posits a "technocratic rationality and top-down managerialism" approach (Di Chiro, 1998, p. 306). As we will discuss later in our analysis, local anti-fracking initiatives are the ethnographic articulation of place in which people reinterpret their experiences and assert new meanings in the context of fracking which relate largely to everyday living, particularly with regard to the quality of water, air, and food, but also sociocultural life more broadly.

Yet, as much as place may have "ontological priority in the generation of life and the real" (Escobar, 2001, p. 143), the hyper-connectedness of people and places in our 21st century world requires us to consider wider contexts and how places and identities are themselves, at least partly, structured by these "global connections." In other words, while people and places are very much "local," they are also interwoven in the fabric of globality. We find in these spheres of existence a bifurcated construction of social reproduction characterized by notions of locality and a dominant discourse of (neoliberal) capitalism, in which a market ideology characterized by hyperconsumerism, deregulation, state austerity, and privatization conceptually define the day.

At the global scale, we find fracking located in a discourse of "energy security," "energy independence," "clean transition fuel," – a "shale gas revolution." From an anti-fracking perspective, the global fracking world looks more like "an obstacle to renewable energy development," or as Food and Water Watch, an international non-profit non-governmental organization that works to protect and promote the safety of food, water, and fish, put it, "Fracking takes a huge toll on affected communities, generates massive volumes of toxic waste, creates hazardous air pollution problems, poses long-term risks to vital drinking water resources and threatens to lock in catastrophic changes to our climate" (Food and Water Watch, 2014a, 2014b).

A sense of and defense of place emerge especially noteworthy in the face of environmental threats such as mining. Not only are shale mining and fracking sites popping up in localities across the globe but these activities are also led by a very limited number of (multinational) global enterprises. This defense of place that is at once both local and global emerges in an evolving discourse that can be powered in part by the political strategies of social movements. In the context of anti-fracking social mobilization, what is contested is the energy production convening a localized extractive technology with global capital(ism). Anti-fracking movements intensify the politicization of energy production, and perhaps unlike climate change, assert (and experience) a clear connection between this struggle and everyday life.

Alternative Imaginaries and Power Struggles

Escobar points out that social movements create the possibility for the articulation of alternative imaginaries (Escobar, 1992). Through a plurality of people constructing an intensification of direct democracy, transformation becomes possible. And, this transformation is evident in the discourse and its inherent relationship to power. Social movements are thus constituted by discourses. Yet, while social movements do require attention to "resource mobilization" and issues of identity, they do not simply erupt (like volcanoes) nor do they necessarily fit the causal models or methodologies of measurement. They are waves that swell and gain force, and can rally forward the momentum of everyday life onto previously unimagined shores. Further, counter to some common perceptions of social movements, they do not end when people stop hitting the streets with signs or when the media loses interest, rather they reverberate in their inevitable remolding of the discourse which may or may not be immediately evident in a policymaking framework. And even if corporate and/or governmental interests find a way to ultimately engage in further widespread fracking as the case may be (see, e.g., Weile, 2014) or further implement policies in their own interests (see Gillens & Page, 2014) albeit even counter to the articulations of social movements, the social imaginary does not revert back, but rather persists in having reshaped or in reshaping the discourse.

While Escobar critically discusses different theories of social movements (i.e., Tourraine, Laclau, Mouffe, Melucci) in light of the specific social terrain of Latin America, he notes that, "Generally speaking, the task ahead is the construction of collective imaginaries capable of orienting social and political action" and that, "... politically, the task is to foster the democratizing potential of the new social subject" (Escobar, 1992, pp. 41–42). In fostering this potential, power struggles are fundamental. But, what is the process whereby social movement activists politically engage?

Escobar refers to economic, political, and cultural transformation through articulation as a central activity of social movements. We explore and expand on this idea further in the realm of discourse and power. To the extent that social movements are power struggles, and in the case of anti-fracking movements we believe they are, we transfer the key elements of Foucault's ideas about power and struggle to enrich our understanding of social movements as the potentially transformative articulation of political space through discourse. Ultimately, the function of any successful social movement is to seize discourse in a generative fashion. Foucault writes, "... discourse is not simply that which translates struggles or systems of domination, but is the thing for which and by which there is struggle, discourse is the power which is to be seized" (Foucault, 1970, pp. 52–53). In this chapter, we explore the anti-fracking movement in this context as the articulation of a counter-discourse.

The generative aspect, as Escobar notes, is the ultimate seizure in which "categorically subalterned" people can imagine their own existence on their own terms, a re-imagining of the world that both reflects and builds on the everyday and where scalar spheres (i.e., local and global forces of a cultural, political, and economic nature) gain their most resonant meaning (1992). Foucault asserts that:

> ... to speak ... to force the institutionalised networks of information to listen, to produce names, to point the finger of accusation, to find targets, is the first step in the reversal of power and the initiation of new struggles against existing forms of power. (Foucault, in Bouchard, 1972)

This is the "first step." This first step is operationalized as an exercise in framing, particularly in what is known as diagnostic framing (subsequently defined).

Escobar adds in the context of social movements in Latin America, however, that "... social movements are not ruled by the logic of all or nothing; they must consider the contradictory and multiple voices present in such experiences without reducing them to an unitary logic" (Escobar, 1992, p. 48). Hence, social movements may not only call into question the dominant categories constructing their activist subjectivities, but engage in a generative process of forging "new ways of seeing" and ultimately themselves and their communities. What then are the "new ways of seeing" that are forged by anti-fracking movements? Again, we look to framing as a way to explore empirically this aspect of the generative process of the anti-fracking movement, and in forging "new ways of seeing," we engage the concept of prognostic framing (also subsequently defined).

We find in this everyday life a new manifestation of the global market-centered mentality and its technological constructions and exploitation. Hydraulic fracturing drills installed in communities, next to people's homes, eek out the last drops of gas and oil hiding mostly in the fissures of shale rock. The question is can anti-fracking movements overcome the dominant politics and logic that justifies and affirms the exploitation of shale gas and oil by fracking? Does the counter-discourse of the anti-fracking movement make visible the current power relationships structured by global capitalism as well as reconceptualize a more meaningful self-proclaimed existence that simultaneously embraces antagonistic plurality? If so, how? Before turning to the particular framing engaged by the anti-fracking movement, we clarify the terms, "discourse" and "counter-discourse."

Discourse and Counter-Discourse

Discourse, rather than being simply linguistic, constitutes the social spheres of knowledge. Social knowledge is thus constituted by discourse(s) and both creates and limits what one can say, write, and even think within a historical evolution of disciplines. McHoul and Grace summarize Foucault's conception of discourse this way,

> Thus both "the world" and our consciousness of it are effects of the kinds of representations we can make of it. But, at the same time, discourse is not just a form of representation; it is a material condition (or set of conditions) which enables and constrains the socially productive "imagination" (McHoul & Grace, 1995, p. 34).

Discourse(s), they add, are bodies of knowledge revealed by corollary "statements" and their conditions. Statements bounded by disciplinary and discursive knowledge function as the parameters of possibility and impossibility. The fracking debate is an antagonistic zone in which knowledge on the implications of fracking is contested, but also the rules for what constitutes knowledge are asserted. And, as we discuss later, while fracking discourse and counter-discourse debate on the former level, contesting the "truth" about fracking; the discourse and counter-discourse are both stuck in a contradictory methodological quagmire in which the rules for asserting "truths" are both mutually affirmed as well as challenged.

Discourse is characterized by the forces of formation, transformation, and correlation. These forces make it possible for certain objects and concepts to emerge and define the delimiting factors of their evolution and relationship to other discourses and other "nondiscursive contexts" (McHoul & Grace, 1995, p. 44). Thus, discourse is not only what can and can't be said, written, or thought but also has within it the rules for such events and their latency. Thus inherent in a discourse is the nature of its own preservation. The will to truth enunciated by institutions such as science, for example, increasingly is reflected in the social structure of discourse and counter-discourse in which politics is an arena for the struggle to define what is true and what is false based on the common enterprise of reason. For example, "Radical ideas that advocate resistance to the scientific management of everyday life are not restricted but remain unheeded, because they seem illogical, irrational, nonsensical, disorderly, and uncivilized" (Powers, 2007, p. 31). While individual narratives about the experience of fracking are asserted by both dominant and counter-discourses, they are also challenged by the dominant discourse on the grounds that they do not constitute truth about fracking because they are not scientific. Claims made by homeowners who assert that their water is contaminated by fracking, for example, are "tested" and "contested" by authorities such as the United States Environmental Protection Agency and industry, respectively.

We link this conceptualization of discourse to social movements through Foucault's idea of counter-discourse. A counter-discourse is the act of people asserting their own voice in contradistinction to a dominant or authoritarian discourse which "speaks for them." Counter-discourse is the constitution of political action against oppression and the "indignity of speaking for others"[1] (Moussa & Scapp, 1996). In this way, it can be said that, a counter-discourse breaks the transformative nature of a dominant discourse.

Framing

Collective action frames are a fundamental aspect of the discourse of social movements. In the context of fracking, we uncover a "frame war" or what we refer to as an agonistic zone. This zone is characterized by the anti-fracking movement and others (e.g., real-estate interests, food enterprises) in a "counter-discourse" and a dominant discourse perpetrated by particularly governments and the gas and oil industry. Before investigating these discourses in more detail, we look at some of the research on framing and fracking to date[2] to demonstrate our contention that there is a dominant discourse and a counter-discourse on fracking that reflect power in the Foucauldian sense, and locate the anti-fracking discourse within what Escobar refers to as an "articulation of alternative imaginaries."

Framing is widely considered to be the thrust of social movements as movement actors articulate a basis for collective action. Frames, "schemata of interpretation," allow people to interpret and assign meaning (Goffman, 1974, p. 21) while generating support and "demobilize antagonists" (Snow, 2004, p. 384). Furthermore, inherent in frames are the articulation of a problem and its perpetrators and also can include the realm of potential solutions.

The particular success of a social movement is said to be contingent, in fact, on the actors' ability to produce frames that resonate and are aligned with public concerns in what is known as "frame alignment" (Snow & Benford, 1988). Frames can be considered the building blocks of a discourse which further delineates what constitutes "truth" as well as what doesn't and thus inherently delineates the boundaries of understanding and what is and isn't possible.

Framing has been discussed in different ways, but central to framing are diagnostic and prognostic frames. Diagnostic framing refers to the articulation of a problem, whose fault it is, and its resolution. Prognostic framing emphasizes a solution to the problem.

ANTI-FRACKING (COUNTER) DISCOURSE: PROBLEMS AND SOLUTIONS

Without denying the fundamental role that environmental concerns have played in mobilizing various local communities in their protests against fracking (and other methods of unconventional oil and gas production), we posit that it is through the local and global constitutions of the anti-fracking movement that it transcends mainstream environmental movements and finds itself to be more closely aligned with environmental justice and global social justice movements in which "diverse, egalitarian, and non-hierarchical" forums engage "places and sets of relationships that sustain a local community's way of life" (Di Chiro, 1998, pp. 300–306) whereby people could imagine their own existence on their own terms. We particularly focus on systems of "truth" and of knowledge production, especially in the movement framing of problems and solutions, structures of blame, and scenarios of hope.

The central frame of the anti-fracking counter-discourse is that fracking is unsafe; negatively impacts the environment (water, soil, air, and noise pollution) and consequently human health; triggers earthquakes; and that it does not help us move toward a clean(er) and (more) sustainable energy future. These arguments do not solely rest on anecdotal (or personal experience) evidence but also often tap into scientific studies which are then both challenged, especially by industry, on various fronts, but particularly their methodological vulnerabilities. Still, research has shown that while frames across anti-fracking initiatives are similar, their ability to influence decision makers and degree of resonance varies, for example, in a study comparing the United States and Europe (Bomberg, 2013). In Europe, economic benefits at the state level were emphasized over local benefits (which were consistent with the different revenue structures in Europe vs. the United States); support from politicians and the low-carbon industry is more forthcoming in Europe; at the same time, pro-fracking interests are more strongly aligned in the United States; the risks of fracking and general trust in states to secure the public interest are more questioned in Europe than in the United States; and renewables to avoid "fossil fuel lock-in" have stronger traction in Europe (Bomberg, 2013). Other research also affirms the importance of a number of factors in public perceptions regarding fracking which is consistent especially with the movement's framing of the problems with fracking. Research at the Durham Energy Institute, for example, concludes that while risks are important, "other factors like perceptions of the industry structure, judgments about the likely efficacy of regulation, the perceived trustworthiness of government decision-making processes, or feelings of alienation and dependency in relation to seemingly remote expert discussions, are also likely to play a role in structuring public responses to hydraulic fracturing and unconventional fossil fuel exploitation" (Durham Energy Institute, 2014).

Further, under a strong neoliberal-based regime, characterized by deregulation, privatization, commercialization, and consumerism such as the United States, government and industry work from a pre-existing dominant discourse of market benefit in which environmental resources are framed as "sources of profit" which "ensures capital's ability to redefine humans' relationship to land, air, water, each other, and self …" (Hudgins & Poole, 2014, p. 317). Further, the dominant discourse, in which "the indignity of speaking for others" is manifested, portrays people who oppose fracking as "uneducated" and not getting "the facts" about fracking. The assumption is that if people could really understand fracking, they wouldn't be against it. This is further backed up by challenges to the validity of community-based risk narratives. Interestingly enough, however, statistical research on the public perceptions of fracking in the United States concludes that those who are more familiar with hydraulic fracturing and associate it with environmental impacts versus economic benefits (among other factors such as sex, age, and sources of information) are more likely to be against fracking (Boudet et al., 2013). Hudgins and Poole conclude hence that the assertion that if people just knew the facts about fracking, they would not oppose it "may serve as a form of silencing, as the discursive terrain is claimed by those with an ability to speak a specialized language" (Hudgins & Poole, 2014, p. 311). Still, the "discursive terrain" of the anti-fracking movement is not silenced, and has a "specialized language" of its own. In this next section, we engage some of the anti-fracking framing to shed light on the anti-fracking discourse emerging at local and global scales. By engaging an analytical framework using the concept of framing, the discursive terrain of the movement is mapped and the strategy of the movement becomes more evident.

Diagnostic Framing: What is the Problem?

Diagnostic framing is a key part of the local anti-fracking initiatives. It is here where people articulate the problems of fracking that they experience in their everyday lives or their fears and concerns associated with fracking. The specific issues of concern articulated by the anti-fracking movement include health and safety issues impacts on climate change; a distrust of government and industry; and questions about the economic benefits. However, an emerging dominant frame amongst local anti-fracking initiatives is that fracking destroys communities.

Health and Safety Issues

Health and safety issues asserted by anti-fracking initiatives are dominated by stories that depict the experiences of those living next to drilling sites, especially experiences with water. Many of the slogans of local initiatives have a water theme: "No Water. No Future." "What part of turning water into toxic sludge makes sense?," "You can't drink money," "Love your lake ... Don't frack it!," "Water & Fracking Don't Mix," "Let Them Drink Gas," "Fracking Creates A River of Doubt," and "Keep The Frack Out Of My Water."

One of the most impressive visuals used by the movement shows people lighting their tap water on fire. This visual became especially prominent through the movie "Gasland," written and directed by Josh Fox, that won several awards including the 2010 Sundance Film Festival Award (Fig. 1).

There have been several key scientific efforts regarding the issue of water (as well as air and soil) contamination and fracking, including studies by the United States Environmental Protection Agency (2004 and 2014) and Duke University (Osborn, Vengosh, Warner, & Jackson, 2011; and others), but due to the highly controversial nature of the issue, scientists and their work emerge as political actors more than anything else. While the formal science on fracking and contamination is inherently contested (see, e.g., Everley, 2013), experiential knowledge generated by community members is conveyed through stories about flammable, bubbly water, illnesses associated with the introduction and proximity to drilling sites, sick livestock and other

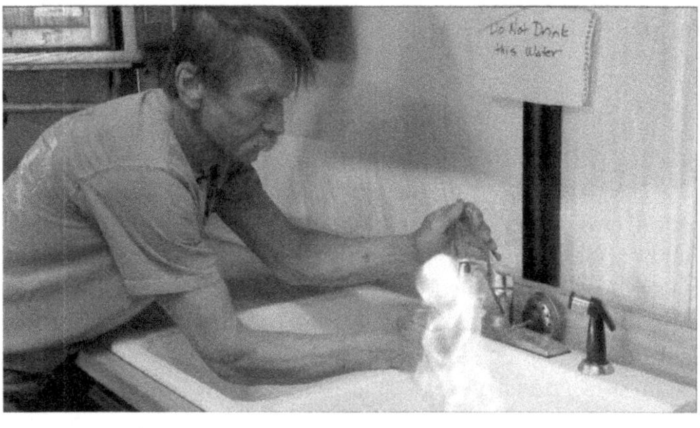

Fig. 1. Photo: "Gasland." *Source*: Josh Fox.

animals, soil pollution with fracking chemicals or through oil spills, terrible odors, increased seismic activity, and noise pollution from increased truck traffic.

While in many ways, fracking could gain framing resonance as a NIMBY issue (Not in My Backyard), focused at the collective community level with industrial counterparts. For example, while individual photos of individual homes and drilling apparati can be found, the more meaningful visual frames are projected in collective protests, landscapes, and industry imagery. A prominent visual of the fracked landscape is seen and shared in the landscape swirls caused by fracking in Wyoming (Fig. 2) and the massive numbers of trucks hauling toxic chemicals used in the process of fracking.

Climate Change and Clean Energy

While the dominant pro-fracking discourse asserts that fracking reduces the impacts on climate change because gas is cleaner than coal, and gas is a viable transition fuel, the scientific debates on this assert otherwise. A controversial study was published from Cornell University and concluded

Fig. 2. Photo: Fracking Drill Sites in Wyoming, Pinedale Anticline, and Jonah. *Source*: EcoFlights.

that from a life-cycle standpoint, fracked gas does not comparatively reduce the impacts on climate change when it comes to emissions (Howarth, Santoro, & Ingraffea, 2011). True to the "frame war," the study was contested by opposing interests (see, e.g., Entine, 2012). The anti-fracking movement, however, often approaches the climate change issue differently and rather than "quantifying" the emissions of CO_2 and other greenhouse gases, asserts that fracking exacerbates climate change by taking us away from developing the renewable energy sector.

Wittmeyer asserts that, "The [anti-fracking] movement seems to have reconnected culture with climate" (2013). Still, it is difficult to articulate climate change as a pressing local issue, and these arguments especially find resonance in the global anti-fracking campaigns. For example, Camp Frack asserts that, "Shale gas directly competes with investment in renewable technologies vital for combating climate change." Global Frackdown writes on their web page, "On October 11, 2014, the Global Frackdown will unite concerned citizens around the globe to tell elected officials that we want a future lit by clean, renewable energy; not dirty, polluting fossil fuels. The journey to a renewable energy future will not be fueled by shale gas. Take a stand against fracking – stay tuned to find a Global Frackdown in your area!" Camp Frack asserts the problem: "Here the maths is unambiguous – we have insufficient budget for the carbon we are already emitting. By the time shale gas is produced in any quantity (5 to 10 years) there will be no emissions space left for it" (Camp Frack, 2014).

Communities and Renewable Energy versus Jobs and Cheaper Energy

The economic benefits are core to the framing debate between pro and anti-frackers. An entity known as EnergyFromShale, an American conglomerate of pro-fracking interests, sums up the benefits of fracking this way, "Fracking has unlocked massive new supplies of oil and clean-burning natural gas from dense deposits of shale – supplies that increase our country's energy security and improve our ability to generate electricity, heat homes, and power vehicles for generations to come. Hydraulic fracturing has also boosted local economies – generating royalty payments to property owners, providing tax revenues to the government and creating much-needed high-paying American jobs" (EnergyFromShale, 2014). The academic research on economic benefits at the local community level, particularly regarding employment, however, raises questions on the topic (see, e.g., Araya et al., 2014; Mauro et al., 2013).[3] The industry is accused by the movement and some scientists alike of over-estimating benefits to local communities.

As social inequality expands across the globe (with an especially significant gap between the rich and the poor in the United States) coinciding with widespread social unrest (e.g., Occupy Movement, Arab Spring, Indignados, etc.), the case of fracking provides an interesting comparison in the mainstream media: "The chief executive of Frac Tech International LLC [Marcus Rowland], a relatively small oilfield services company, earned $24.4 million in 2010, putting him among the highest paid U.S. executives" (Gold, 2011) while a global information company known as IHS asserted that the average household income increased by US$1200 in the year 2012 (Efstathiou Jr., 2013). Social inequality is further engaged as struggling landowners and farmers see no other way out but to accept the much-needed upfront economic benefits of leasing land for drilling. According to the New York Times, one landowner put it this way, "a lot of people who, when we signed and got the upfront money, were going to be losing their land and couldn't pay their taxes" (Hakim, 2012). During the Occupy movement protests in the United States, "no fracking" was prominent among protest themes that generally targeted government corruption and the failing democratic hold of the people on the country.

Anti-Fracking Is Pro-Democracy

What is often questioned locally is not only the "environmental sustainability" of fracking technologies, but also the social and economic impacts of these technologies and the very democratic arrangements of our societies and the countries in which we live. Movement antagonists, while targeted for their pro-fracking stance, are especially meaningfully targeted for obstructing or impeding democracy and justice. This is most clearly visible in eastern European "new democracies" such as Romania and Bulgaria where alongside "environmental demands" for clean/unpolluted water, soil, and air, questions about the distribution of economic benefits are often raised and demands for transparency, corruption-free government, and the protection of citizens over corporate interests are just as prominent (Besliu, 2013; Devey, Goussev, Schwarzenburg, & Althaus, 2014; Frack Off Romania, 2014; O'Brien, 2014; Oprea, 2013; Summerchild, 2013). Romania proves to be a particularly interesting case, where three "separate" movements and protests that emerged in 2012 and 2013 (the anti-government protests in Bucharest, the protests against fracking in Pungesti, Barlad, and other towns, and the protests against gold mining in Rosia Montana) all merged together based on their shared concerns with democracy and corruption (Devey et al., 2014). Furthermore, Vişan (2013) writes that the Romanian Orthodox Church has also, for the first time since the fall of communism, sided with civil society organizations on these issues. Vişan also interestingly summarizes the Romanian anti-fracking movement by calling it "a heterogeneous movement whose environmental core is strengthened by opinion-leaders of leftwing, nationalistic, or anti-capitalist beliefs [...], disgruntled citizens who feel disempowered and [...] a young generation of Romanians who are fed up with Romania's endemic corruption and lack of accountability of the current political elite" (*ibid.*, p. 3). The specificity of the Romanian anti-fracking struggle and the importance of democracy and corruption concerns/frames within it, are also well captured in one of the main slogans used by anti-frackers in Romania, which says, "Stop hydraulic fracking, stop fracking our democracy!" (Frack Off Romania, 2014).

Devey et al. (2014) also discuss the case of Bulgaria, showing how political questions are equally important, if not even outweigh environmental ones in the Bulgarian fracking debate. They assert that, "The shale gas issue did not stand on its own but was enmeshed in much larger controversies about the country's politics and future." (*ibid.*, p. 51). Like in Romania, anti-fracking protests in Bulgaria (e)merged with anti-government protests, and while the protests remained relatively small in size they were "perfectly organized, coordinated, and publicized in the media" (Daborowski, 2012). Culminating in January 2012, the protests resulted in an indefinite ban that was placed on fracking by the Borisov government (Devey et al., 2014).

To a somewhat lesser degree, similar frames are also observable in the "older democracies," particularly the United Kingdom, where fracking is also often contested on political grounds and government representatives are accused of being too closely tied to and having financial interests in the fracking industry. One of the largest corruption scandals, uncovered in 2013, is that of pro-fracking Conservative Party's elections advisor Lynton Crosby whose lobbying firm, as it turns out, represents the Australian Petroleum Production and Exploration Association, an oil and gas lobby group campaigning aggressively for fracking (Dugan & Morris, 2013). The news came out at the same time when conservative MP George Osborne, the Chancellor of the Exchequer and Second Lord of the Treasury of the United Kingdom, announced the "most generous tax breaks in world" for fracking (Macalister & Harvey, 2013). Similar concerns with corruption were more recently raised in South Africa (Prinsloo, 2014).

Corruption scandals have not only enraged anti-fracking protesters/movement but also many citizens who are more broadly concerned with government corruption and questions of democracy, and who now see and support the anti-fracking movement in raising these concerns. It has thus also had the effect of strengthening and further bringing to the fore the democracy and corruption frames within the anti-fracking movement. "Frack Off," the leading national anti-fracking network in the United Kingdom, identifies 20 major impacts of shale oil and gas, and among the many impacts relating to the physical environment they also point to the multiple social and political issues, not least of which include, "corporate profits vs. community costs" (Frack Off, 2013). In effect, they articulate a large base of issues associated with fracking, which extends beyond what might be considered "standard environmental concerns," thus creating a wide platform for involvement of diverse social actors. Citizens are invited to join the protests regardless of what their main concern may be ("environmental" or otherwise) and "solidarity" indeed became a major theme in these articulations that operates not only within the national borders but also across them. "Solidarity protests" have been held in the United Kingdom for both the anti-fracking struggles in Romania and Bulgaria (Occupy London, 2013), and this has echoed back recently with Romanian citizens protesting in solidarity with the people of Barton Moss in the United Kingdom (Frack Off Romania, 2014). The Barton Moss solidarity protest saw people from across several cities and towns within the United Kingdom who joined the protest on that same day, January 26, 2014, the so-called "Solidarity Sunday" (Frack Off, 2014; Greenpeace, 2014).

France, which is considered to hold amongst the largest source of shale gas deposits in Europe next to Poland, now has a ban on fracking despite legal challenges from the Schuepbach Energy Corporation. It is widely hailed amongst the anti-fracking community as a success story and source of inspiration. The British Broadcasting Corporation (BBC) reported that the French Green Member of the European Parliament, José Bové, attended an anti-fracking gathering in the United Kingdom at Balcombe, where anti-fracking protests were launched against Cuadrilla (an independent UK energy company)[4] for test drilling (BBC, 2014). Whereas, the United States, where fracking is most widespread, is used repeatedly by movement activists as an example of the negative impacts of fracking and something to be avoided. Interestingly enough, however, in France where a successful ban was introduced despite their fracking potential, the government (which provides the bulk of the country's energy supply with nuclear energy) was considered supportive and aligned with the movement's general anti-fracking goal. Without such support, however, it is a question whether the movement could have achieved similar results which indicates the key role of governmental interests in promoting or preventing fracking.

The case of Bulgaria, Romania, as well as the United Kingdom and France, signal a shift in how the anti-fracking movements in these countries, but potentially across the globe, constitute themselves and operate. Rather than engaging in reactionary protests that emerge ex-post fracking operations and centered largely on "environmental concerns," the movements in these countries are articulating a wide range of concerns highlighted by "fracking" and thus attempting to mobilize a larger and more diverse set of actors to join or support these movements. Through these diverse articulations, fracking emerges not only as a cause of environmental and social degradation but also as an effect, a symptom as it were, of more durable and structural societal problems that have to do with a crisis of democracy and its manifestations: corruption, violations of human rights and unequal distribution of goods and ills across different parts of society. Such framing of the "fracking issue" has not only enabled wider engagement of national and international citizens but has also resulted in wider institutional support for local anti-fracking protests. The Pungesti anti-fracking protests in Romania, for example, have received support from not only the European Green Party (EGP, 2013), but also the European Peasant Movement (ECVC, 2013) as well as Human Rights organizations such as the Helsinki Committee (APPADOR-CH, 2013a, 2013b).

What Then Constitutes the "Truth" About Fracking?

Diagnostic framing brings to the fore the "social spheres of knowledge" on water contamination (and other health risks) and fracking as they are contested across the antagonistic spheres of industry, government, academia, and communities, whereby the latter two are less cohesive. The debates reflected in knowledge claims on fracking have very specific dynamics that indicate what Foucault would assert are the boundaries of what constitutes "truth" and what doesn't, and what can and can't be said and by whom. The conflict over fracking unveils these boundaries, which put communities in an especially vulnerable position with regards to asserting the "truth." For, the realms of what constitute knowledge marginalize the very methodological resources that community "truths" rely on most: select individual or localized cases; and evidence based on anecdotes and personal experience, what is sometimes also more generally referred to as endogenous knowledge. The global anti-fracking virtual forums, such as Global

Frackdown, are often a platform ultimately for sharing multiple "truths" about fracking and its dangers.

However, as the anti-fracking movement is thereby critiqued on methodological grounds, the movement itself is challenged to overcome the more dominant discourse on what constitutes "truth" by simultaneously exploiting the scientific method for all its weaknesses and strengths. For example, the anti-fracking initiative in South Africa known as Treasure Karoo Action Group, writes on its web page, "South Africans may be well advised to request scientific proof that shale gas is the best option to create sustainable jobs, energy, and allow our children to live out the promise of Section 24 of the Constitution of the Republic of South Africa" (Treasure Karoo Action Group, 2014). As noted in the earlier discussion on diagnostic framing regarding the problems associated with fracking, however, scientific works (i.e., involving researchers with academic affiliations relying on the scientific method and involving peer review processes) are also contested based on methodological grounds. Still, despite the highly contested knowledge base of fracking, pro and anti-frackers still mutually prioritize the notion of scientific knowledge when it comes to the risks of fracking. Mainstream environmental movements have historically depended on arguments that rely on the findings of environmental science and its corresponding paradigm of knowledge (Di Chiro, 1998) – accepting the possibility of direct access to "reality" and assuming that this (scientific) knowledge is neutral, therefore ignoring the connections between "knowledge" and power. They thus operate within a positivist paradigm that sets hard science as a standard against which other forms of knowledge are evaluated. Such an approach produces several outcomes. On the one hand, the privileging of "hard science" also means a privileging of quantitative "knowledges" and an accompanying devaluing of other forms of knowledge, research, and argument-making (Bacchi, 2010).

This hierarchization of knowledge further results in the enshrining of a top-down managerialist form of governance that simultaneously displaces "lay knowledges" (Popay et al., 2003) and puts emphasis on professionalization with an accompanying decrease of commitment to democratization (Parsons, 2002). And, if we have learned anything from the debate on climate change, knowledge-based conflicts throw and take their biggest punches not only strictly in peer-reviewed and reputable academic journals but also in the quantity and prominence of mass media generated across the internet.

Both the role of ICTs in anti-fracking movements and the various and multiple texts on fracking reveal multiple sources of information and knowledge. Knowledge production, widely considered the domain a scientists in the ivory towers of objective academic institutions, is now increasingly pluralistic in its origins. Even as we assert our ongoing faith in modern science, where we get our knowledge is increasingly diverse and ambiguous in terms of its scientific origins. Tim Boersma, a prominent author of several journal articles on fracking, claims in his own profile that he is also affiliated with the Dutch national gas company, Gasterra, and that, "Before becoming a full time academic at the Brookings Institute, Boersma spent five years in the private sector, working as a corporate counsel to the electricity production sector in the Netherlands" (Boersma, 2014). Two of the main sources on fracking on Google are Chevron and Exxon Mobil, the latter being at first obfuscated by its title, Europeunconventional gas.org. Much of the anti-fracking fervor is tied to a documentary film called, "Gasland" by Josh Fox who is identified as a film director and environmental activist. Another source on fracking is a briefing article entitled "Old Story, New Threat: Fracking and the Global Land Grab" from the Transnational Institute Agrarian Justice Programme written by "activist-scholars" who claim among their resources: Frackaction and Friends of the Earth Europe. The media frequently picks up reports from various consulting firms to assert scientific study results that make various claims about fracking.

A common thread across these realms is the assumed bifurcation of objectivity and subjectivity. In this way, especially academics and community members can be equally challenged from pro and anti-fracking standpoints. Hence, the antagonistic spheres of fracking reflect a frame war that is mutually challenged along the same grounds in the ways that knowledge is produced and shared. Hence, the "truth" on fracking only emerges as a consensus in its mutual use and distrust of the same epistemological and methodological approaches, especially across the dominant discourse and the global anti-fracking movements. In this next section, we explore this diagnostic framing aspect of the movement in which the targets of the anti-fracking movement are articulated.

Diagnostic Framing: Anti-Fracking Movement Antagonists

Diagnostic framing also delineates the "sacred and the profane," and defines who is to blame for the asserted problem(s). As a process and a strategy, diagnostic frames "delineate the boundaries between 'good' and 'evil' and construct movement protagonists and antagonists" allowing activists to target their grievances toward a culpable entity (Benford & Snow, 2000, p. 616). When it comes to blaming someone or something for fracking, the targets are the same across local and global efforts: government and industry. However, while movement frames are consistent in their identification of the antagonists, local actors within the movement are confronted with antagonists in place, while actors organizing the forum for the global movement, largely located in Washington, DC, in the United States and London in the United Kingdom, engage their antagonists from a virtual platform. Local anti-fracking struggles do not solely seek national/supranational/global responses in the form of informed and consensual policy solutions, but they articulate and uphold the inviolability of places in which they emerge, of local histories and communities.

Government and industry have long been revolving doors whereby government representatives have industrial links before and/or after their public service terms and vice versa and that they tend to come from the upper income brackets overall. Government officials can move across the boards of gas and oil companies and others related to energy extraction technology (e.g., former United States Vice-President Dick Cheney and Haliburton). As personal financial disclosure data revealed, most members of the US Congress in 2014 are millionaires (see an analysis, e.g., by the Center for Responsive Politics). In the case of government, the question of a sound regulatory framework arises.

In this context, diagnostic anti-fracking frames particularly articulate a crisis of democracy through various slogans such as, "Listen To The Many, Not The Money!" "They Get Rich. You Get Cancer." "I Couldn't Afford My Own Politician, So I Made This Sign Instead," and referring to the water and other environmental health issues, "Health Over Wealth," "You Can't Drink Money," or "Let Them Drink Benzene!." And, as one Camp Frack activist Tina Rothery, among 500 protesters at an anti-fracking camp in Lancashire, United Kingdom, asserted, "... we don't imagine our Government to be squeaky clean with regard to fracking" (Lancashire Evening Post, 2013). Another anti-fracking activist from Ireland put it this way, "A lot of things happen under the radar."

Global Frackdown draws further attention to the issue, "The oil and gas industry has spent millions of dollars on slick PR campaigns and high-profile lobbying efforts to buy the ability to extract fossil fuels from our communities with as little government oversight as possible, all while destroying our water resources and our climate" (Global Frackdown, 2014). Global Frackdown ultimately pits the oil and gas industry against communities and the environment, "It's time to expose the oil and gas industry's desire to profit at the expense of our communities and our environment" (Global Frackdown, 2014).

Global Frackdown and Food and Water Watch point out the collision of government and industry interests which is most widely evident in what is considered a weak and insufficient regulatory structure. Food and Water Watch write, "The industry enjoys exemptions from key federal legislation protecting our air and water, thanks to aggressive lobbying and cozy relationships with our federal decision makers." Through the introduction of the 2005 Energy Policy Act of the United States, in what has become known in the United States as the "Halliburton Loophole," fracking has been exempted at the federal level from certain water protection policy frameworks. Halliburton refers to former United States Vice-President Dick Cheney's affiliated corporation, Halliburton, which is strongly associated with the development and implementation of fracking technology and related services.

Ultimately, anti-fracking initiatives assert a dichotomy in which the struggle is one of communities versus corporations (Fig. 3). The Lock the Gate Alliance, a coalition of anti-fracking initiatives across Australia describe it this way on their web page, "In a David-and-Goliath struggle of farmers against mining giants, everyday citizens against global corporations, our communities are choosing grace under fire and displaying incredible courage, integrity, and imagination" (Lock the Gate Alliance, 2014).

Fig. 3. Photo: Fracking or Community? *Source*: Wendy Lynne Lee, Raging Chicken Press.

At the global level, Food and Water Watch asserts, for example, that "the industry is aggressively clamping down on local and state efforts to regulate fracking by buying influence and even bringing lawsuits to stop them from being implemented. That's why fracking can't be made safe through government oversight or regulations. An all out ban on fracking is the only way to protect our communities" (Food and Water Watch, 2014a, 2014b).

And while diagnostic framing serves to identify and articulate the problems and targets, prognostic framing is where movement members point toward the solutions (Benford & Snow, 2000). In order for these frames to be effective, they require internal consistency. That is, the solutions must address previously identified problems. But, herein, also lays the possibilities, the anticipation and vision, the "alternative social imaginaries," that are to guide the way toward a different and brighter future and away from the nightmarish arrangements of the contested present.

Prognostic Framing: What Is the Solution?

Already by advancing a particular way of seeing "the problem," the possibilities for delineating change emerge, because inherent within the problem is also the solution. For example, if the problem is a question of "dirty fracking technology," then the solution may simply be to make fracking "cleaner" while still maintaining a momentum in favor of fracking. This technocratic framing of the "fracking problem" was for the longest time most prominent within the [anti]fracking discourse and even actively engaged and reproduced by the industry who is willing to accept that fracking is "dirty," but also quick to assert that it is still "safe if done right."

Prognostic framing on fracking from an antagonistic perspective affirms that discourse, as Foucault asserted in his conceptualization of power, is indeed the very thing to be seized. And, consistent with Foucault's conceptualization of power and discourse, the delimitations and connections to other discourses are inherent in the articulation of the problem and its related "objects and concepts." Hence, to assert or agree on the problem is to define the parameters of the discourse on the solutions. For this reason, amongst others, the anti-fracking movement is increasingly making use of a multitude of frames, many of which do not lend themselves to easy technological or isolated policy fixes. As we discussed earlier, the anti-fracking movement(s) simultaneously "diagnoses" fracking as an energy, climate change, environmental health problem, and beyond that, a problem of democracy and social and economic relations within global neoliberal capitalism.

Different solutions correspond to each of these diagnostic frames. The energy and climate change "prognosis," for example, is particularly engaged at the global level, with calls to sustainable and renewable energy resources such as wind and solar power. Global Frackdown summarizes it this way, "The only way for countries to ensure energy independence and security for future generations is to get off fossil fuels and pursue truly sustainable sources of power like wind and solar energy" (Global Frackdown, 2014).

Yet while the delineation of the fracking controversy as an energy/climate change issue or even strictly an economic or an environmental one is a core aspect of the attempt to "seize the discourse," the most important power struggle at the core of the movement lay in its central implicit prognostic frame in which communities, not the elite interests of corporations and government, actively decide on their own fates within an institutional context of nurtured capacities and opportunities. This terrain is most resonantly contested through the local anti-fracking initiatives where citizens, as community members, find their democratic role (and livelihood opportunities) most challenged and most at stake.

As we've discussed earlier, various local anti-fracking struggles are increasingly uncovering this most important theme in which fracking is not just about energy, economic development, and the environment (health), but that at the root of the problem is a fundamental crisis of democracy in which people experience a lack of self-governing power and constrained opportunity or choice, especially through a policy framework that benefits elite, upper-class interests (namely, the 1–10 percent). From this perspective, the problem of fracking is immediate and situated in the effective socio-political alignments and relations sustained by contemporary neoliberal capitalism, which not only promote and allow for the practice of fracking, but also further exacerbate social, economic, and environmental inequalities. Through such articulations, local initiatives are progressively taking on the special role of "foster[ing] the democratizing potential of the new social subjects" (Escobar, 1992, p. 41). Their "repertoires of contention"[5] include gathering and sharing information, raising awareness, getting together in rallies, vigils, demonstrations and sit-ins, as well as engaging legal remedies where possible.

Such framing of the "fracking problem" does not lend itself to easy technocratic policy fixes but brings proper politics back into the arena of contention. It further bridges the global-local and "disciplinary" divides by articulating a shared "social imaginary" that enables coalitions between the anti-fracking and other new social movements such as the alter-globalization or the Occupy movement, for example. Such coalitions indeed proved both possible and successful in newer democracies such as Romania and Bulgaria, but are realized in older democracies and wealthier nations too, such as the United Kingdom and United States. Further, as stories from these countries are being shared within global anti-fracking platforms and networks, they are inspiring other initiatives to expand their framing efforts and make similar coalitions possible. In this way, diagnostic and prognostic framing come together especially when it comes to designating the responsible target and the proposed solution. For, one of the key solutions to fracking lays within the scope of democracy and an engagement in proper political (antagonistic) discussions.

Still the solution to fracking also lies in the details. For, the boldest force to counter fracking and promote democracy and justice is on the ground in particular places where people need to be able to drink, eat, breathe, and sleep in peace. And, such basic everyday life activities that constitute the foundation of our survival link fracking to the democratic subject and the ability to contest violations against the capacities to meet those needs. This is why it is not so shocking perhaps that the homeowner (and CEO of ExxonMobil), Rex Tillerson and his wife, Renda, joined a lawsuit against the local water company (Bartonville Water Supply Corporation, Texas) contesting a water tower construction that will reportedly be used to supply fracking operations near his "luxury property" noting in the legal documentation that, "Furthermore, upon information and belief, BWSC will sell water to oil and gas explorers for fracking shell formations leading to traffic with heavy trucks on FM 407, creating a noise nuisance and traffic hazards" (p. 17).[6] It is "in place" where people bring action to words that is not strictly at what might be considered a policy professional or corporate scale, but where advocacy is concretized in everyday life. It is also additionally here, in place, where fracking-related corporations such as Halliburton, Schlumberger, Baker Hughes, and Frac Tech are alternatively also realized as embodied, human, personified beings doing the work of fracking as local police protect and enforce the implementation and progress of their work.

Industry (and state) efforts to revert to a technocratic framing of the issue, setting it as a specific policy problem and, hence, only requiring if anything a policy solution, deny the political character of various interpretations and needs (Fraser, 1989).

Yet, such prognostic framing seems to be less resonant for communities than in the past (such as with climate change issues or even environmental justice ones). Where states or local governments have responded, the response was that of a (temporary) ban or a moratorium on fracking (i.e. countrywide bans in France and Bulgaria, and many local bans in the United States and Europe). Still, regulatory prognostic framing manifested concretely in the form of proposed legislation that sets standards for where, how or under which circumstances fracking may be utilized and conversely when it may not, is still not part of the policy agenda. "Sustainable fracking" policies are not currently evident in the prognostic frames of policymakers. On the contrary, governments seem to prefer to exempt fracking (e.g., United States) and leave the doors open.

The fashioning of specific policy responses is yet another discursive/framing operation, and as such a power mechanism, in which frames used or constructed in policymaking can have both positive and negative impacts in terms of power relations (Foucault, 1970). This once again reminds us that discourse/framing, through its association to the production of truths, can have significant material and immaterial impacts.

Inasmuch as global or even national debates about fracking are dominated by, or perhaps kept at the level of, narrowly defined environmental (ecological) questions such as the impacts of fracking on "Nature"; local struggles often defy the very society-nature dichotomy on which these debates rest and point to multiple issues that transcend this dichotomy (Di Chiro, 1998). They oppose the "science-politics short-circuiting" (Swyngedouw, 2010, p. 220) which structures most of today's environmental debates (climate change being the prime example) and instead put forward a different set of knowledges and cultural meanings, contesting top-down managerial decision-making processes and engaging in a proper political struggle in which they not only oppose fracking but demand to be heard as "equal citizens in functioning democracies."

In that sense, local anti-fracking struggles also emerge as cultural distribution conflicts that raise the question: "whose norms and meaning-making practices define the terms and values that regulate social life concerning economy, ecology, personhood, body, knowledge, property, and so forth[?]" (Escobar, 2008, p. 13). This is why in the beginning of this chapter we challenge the characterization of the anti-fracking movement as a mainstream environmental movement. For as our analysis has shown, in contrast to other mainstream environmental movements, the anti-fracking movement counters the isolation of nature as separate from society and culture and its corresponding delimitations of the problems. As the anti-fracking movement implicitly notes, fracking is not strictly an environmental problem, or a problem of our surroundings. It is rather both "cause and effect" and symptomatic in terms of our relationships to each other and with(in) nature.

Local anti-fracking initiatives are "place-based struggles" inasmuch as they appear as defenses of place – understood as a particular relation between local people and communities and their physical environments. Environmental, as well as sociocultural and economic changes unfold through these multiple and complex relations as processes that cannot easily (or without consequences) be reduced and reified as thing-like object-causes or singular socio-chemical components the likes of CO_2. As Escobar noted "economic crises are ecological crises are cultural crises … [and] … it is important not to separate these three domains but to let them interpenetrate each other" (2008, p. 14). This means that fracking technologies do not only threaten environments at the ecological level, but also at the cultural level through the "subalternization or even elimination of local grammars and knowledge of the environment" (*ibid.*, p. 9).

In other words, fracking technologies are never implemented in ways that they co-exist with places where they are employed and without any consequences and implications to those places, but they are rather entirely transformative of places: ecologically as well as culturally, economically and throughout these – politically. This is a point that is indeed increasingly recognized and articulated within many local anti-fracking protests, and increasingly being reflected in global anti-fracking platforms and networks.

The unitary frames, as often presented in official political debates, industry responses and even mainstream media texts, sharply contrast with the dynamic reality of multiple (anti)fracking frames at both local and global levels. It thus seems of crucial importance for the study of the anti-fracking movement(s) and the state and industry responses to the movement(s), to further capture the diversity of (anti)fracking frames. In doing so, we will be better able to recognize the role that such diversity plays in either distortions or, more importantly, successes in the anti-fracking struggle.

CONCLUSION

The anti-fracking movement, in both its local and global articulations, reveals what other global movements of our time have told us loud and clear, democracy is in crisis and a prominent symptom of this crisis is the ever widening gap in social equality. This chapter engages a way of understanding social movements that sheds light on their democratizing potential to articulate alternative imaginaries, to engage in a power struggle where power is manifested, namely through discourse, and to draw attention to the everyday world of people who do not see themselves as separate from nature or that nature is all about the market, despite the dominant discourse that attempts to make it so.

The chapter thus engages the anti-fracking discourse through an overview of some of the framing generated from the many local voices and the global virtual forums for social solidarity and exchange. Across the framing of the movement, we identify and reflect on some core themes: (1) what constitutes knowledge; (2) communities struggling in and across places; and (3) the explicit and powerful articulation of a resonant, alternative political and economic imaginary based on democracy and social equality.

In conclusion, one global movement and many local voices reveal various frames of fracking including, a continuation of fossil-fuel-dependence, a technology that destroys local environments by polluting water, soil, and air, a new materialization of global capitalism's unsustainability, an issue of state corruption and governments functioning not on the basis of popular but elitist/private representation, an issue of social, economic, and environmental equity. However, the movement's capacity to resist efforts to narrow or structure the discourse on fracking to technical or strictly environmental terms, let alone what might be considered scientific conclusions, is critical to its democratic and political potential in general.

NOTES

1. The "indignity of speaking for others" was asserted by Gilles Deleuze in a conversation with Michel Foucault in 1972. The transcript is reproduced here http://libcom.org/library/intellectuals-power-a-conversation-between-michel-foucault-and-gilles-deleuze (last visited April 4, 2014) and was published in Foucault (1980).
2. Note that the published research on this topic to date is largely focused on the United States and European context.
3. Scientific studies that challenge the gas and oil industry's position tend to illicit a response: see http://naturalgasnow.org/multi-state-shale-research-collaborative-phony-political-outfit/ (last visited April 12, 2014).
4. Cuadrilla has several gas exploration sites in Lancashire and West Sussex in the UK according to their web page http://www.cuadrillaresources.com/our-sites/ (last visited April 28, 2014).
5. For an overview of "repertoires of contention," see Snow, Soule, and Kriesi (2008).

6. No. 2012. 30982.211. Armey et al. vs. Bartonville Water Supply Corporation et al. Plaintiffs' Second Amended Petition. http://online.wsj.com/public/resources/documents/water20140220.pdf (last visited April 28, 2014).

REFERENCES

APADOR-CH (Asociaţia Pentru Apărarea Drepturilor Omului în România-Comitetul Helsinki). (2013a, December 10). *APADOR-CH and ActiveWatch require the dismissal of the Interior Minister for the abuses committed at Pungeşti*. Retrieved from http://www.apador.org/en/apador-ch-i-activewatch-cer-demiterea-conducerii-ministerului-afacerilor-interne-pentru-abuzurile-de-la-pungeti-2/. Accessed on April 25, 2014.

APADOR-CH (Asociaţia Pentru Apărarea Drepturilor Omului în România-Comitetul Helsinki). (2013b, December 20). *Raport privind evenimentele recente din comuna Pungeşti* (in Romanian). Retrieved from http://www.apador.org/show_report_nf.php?id=338. Accessed on April 25, 2014.

Bacchi, C. (2010). *Foucault, policy and rule: Challenging the problem-solving paradigm*. Feminist Research Centre in Aalborg (FREIA), Institute for History, International and Social Studies, Aalborg University, Denmark. Retrieved from http://vbn.aau.dk/files/33190050/FREIA_wp_74.pdf. Accessed on May 1, 2014.

Benford, R., & Snow, D. (2000). Framing processes and social movements: An overview and assessment. *Annual Review of Sociology*, *26*, 611–639.

Besliu, R. (2013, December 13). Pungesti: Romania's battlefield against Chevron. *Al Jazeera*. Retrieved from http://www.aljazeera.com/indepth/opinion/2013/12/pungesti-romania-battlefield-against-chevron-201312117349323363.html. Accessed on April 24, 2014.

Boersma, T. (2014). Retrieved from http://www.brookings.edu/experts/boersmat. Accessed on February 7.

Bomberg, E. (2013). The comparative politics of fracking: Agenda-setting, networks and framing in the US and Europe. Paper presented at 2013 annual meeting of the UK Political Science Association, March 25–27, 2013, Cardiff, UK.

Boudet, H., Clarke, C., Bugden, D., Maibach, E., Roser-Renouf, C., & Leiserowitz, A. (2013). "Fracking" controversy and communication: Using national survey data to understand public perceptions of hydraulic fracturing. *Energy Policy*. Retrieved from http://dx.doi.org/10.1016/j.enpol.2013.10.017i

British Broadcasting Corporation (BBC). (2014, January 20). *French MEP José Bové leads Balcombe anti-fracking rally*. Retrieved from http://www.bbc.com/news/uk-england-sussex-25805823. Accessed on April 28.

Casey, E. (1996). How to get from space to place in a fairly short stretch of time. In S. Feld & K. Baso (Eds.), *Senses of place* (pp. 14–51). Santa Fe, NM: School of American Research.

Daborowski, T. (2012, January 25). *Bulgaria is no longer interested in shale gas*. OSW Ośrodek Studiów Wschodnich – Centre for Eastern Studies. Retrieved from http://www.osw.waw.pl/en/publikacje/analyses/2012-01-25/bulgaria-no-longer-interested-shale-gas. Accessed on April 30, 2014.

Devey, S., Goussev, V., Schwarzenburg, B., & Althaus, M. (2014). Shale gas U-turns in Bulgaria and Romania: The turbulent politics of energy and protest. *Journal of European Management & Public Affairs Studies*, *1*(2), 47–60.

Di Chiro, G. D. (1998). Nature as community: The convergence of environment and social justice. In D. Faber (Ed.), *The struggle for environmental democracy: Environmental justice movements in the United States*. New York, NY: Guilford.

Dugan, E., & Morris, N. (2013, July 14). Now Cameron aide Lynton Crosby's links to fracking industry are explored. *The Independent*. Retrieved from http://www.independent.co.uk/news/uk/politics/now-cameron-aide-lynton-crosbys-links-to-fracking-industry-are-explored-8708331.html. Accessed on April 23, 2014.

Durham Energy Institute (Durham University). (2014). *Framing "Fracking": Public concerns surrounding hydraulic fracturing and unconventional fossil fuels in the UK*. Retrieved from https://www.dur.ac.uk/dei/projects/framingfracking/. Accessed on March 19.

ECVC (European Coordination Via Campesina). (2013). *The European peasant movement supports the struggle against Chevron's abusive behaviour in Romania*. Open Letter (2013, December 11). Retrieved from http://www.eurovia.org/IMG/article_PDF_article_a815.pdf. Accessed on April 25, 2014.

Efstathiou, J., Jr. (2013, September 4). Fracking boom seen raising household incomes by $1,200. *Bloomberg News*.

EGP (European Green Party). (2013, December 5). *European Greens support anti-fracking protests in Romania*. Retrieved from http://europeangreens.eu/news/european-greens-support-anti-fracking-protests-romania. Accessed on April 25, 2014.

EnergyFromShale. (2014). Retrieved from http://www.energyfromshale.org/fracking-benefits. Accessed on April 12.

Entine, J. (2012). New York Times Reversal: Cornell University Research Undermines Hysteria Contention that Shale Gas is "Dirty". *Forbes Magazine*. Retrieved from http://www.forbes.com/sites/jonentine/2012/03/02/new-york-times-reversal-cornell-university-research-undermines-hysteria-contention-that-shale-gas-is-dirty/. Accessed on April 12, 2014.

Escobar, A. (1992). Imagining a post-development era? Critical thought, development and social movements. Social Text (Third World and Post-Colonial Issues), *31/32*, 20–56.

Escobar, A. (2001). Culture sits in places: Reflections on globalism and subaltern strategies of localization. *Political Geography*, *20*, 139–174.

Escobar, A. (2008). *Territories of difference: Place, movements, life, redes*. Durham: Duke University Press.

Everley, S. (2013). *How Anti-Fracking Activists Deny Science: Water contamination. Energy in Depth*. Retrieved from http://energyindepth.org/national/how-anti-fracking-activists-deny-science-water-contamination/. Accessed on April 10, 2014.

Food and Water Watch. (2014a). Retrieved from http://www.foodandwaterwatch.org/water/fracking/fracking-action-center/map/. Accessed on April 4.

Food and Water Watch. (2014b). Retrieved from http://www.foodandwaterwatch.org/factsheet/ban-fracking-now/. Accessed on April 18.

Foucault, M. (1970). The order of discourse. Inaugural Lecture at the College de France, December 2 in Young, Robert. 1981. Untying the Text: A Post-Structuralist Reader. Boston, MA: Routledge and Kegan Paul Ltd.

Foucault, M. (1980). *Language, counter-memory, practice: Selected essays and interviews*. Cornell University Press.

Frack Off. (2013, July 11). *20 impacts of shale oil & gas*. Retrieved from http://frack-off.org.uk/resource/20-impacts-of-shale-oil-gas/. Accessed on April 23, 2014.

Frack Off. (2014, January 8). *Solidarity Sunday*. Retrieved from http://frack-off.org.uk/event/solidarity-sunday-2/. Accessed on April 23.

Frack Off Romania. (2014). *The live story of the fight against fracking in Romania*. Retrieved from http://frackoffromania.wordpress.com/. Accessed on April 24.

Fraser, N. (1989). *Unruly practices: Power, discourse and gender in contemporary social theory*. Minneapolis, MN: University of Minnesota Press.

Gillens, M., & Page, B. (2014). Testing theories of American politics: Elites, interest groups, and average citizens. *Perspectives on Politics, Fall*.

Global Frackdown. (2014). Retrieved from http://www.globalfrackdown.org/about/. Accessed on April 12.

Gold, R. (2011, September 22). Energy Firm CEO Hits Pay Gusher. *Wall Street Journal*. Retrieved from http://online.wsj.com/news/articles/SB10001424053111904563904576585132870994002. Accessed on April 24, 2014.

Greenpeace. (2014, January 28). *East Bristol supports Barton Moss Solidarity Sunday anti-fracking rally*. Retrieved from http://www.greenpeace.org.uk/groups/bristol-east/blog/east-bristol-supports-barton-mass-solidarity-sunday-anti-fracking-rally. Accessed on April 25.

Hakim, D. (2012, June 13). Cuomo proposal would restrict gas drilling to a struggling area. *New York Times*.

Howarth, R. W., Santoro, R., & Ingraffea, A. (2011). *Methane and the greenhouse gas footprint of natural gas from shale formations*. Climatic Change Letters.

Hudgins, A., & Poole, A. (2014). Framing fracking: Private property, common resources, and regimes of governance. In A. J. Willow & S. Wylie (Eds.), *Energy, environment, engagement: Encounters with hydraulic fracking*. Special section of the Journal of Political Ecology, *21*, 222–348.

Lancashire Evening Post. (2013, May 13). *Hundreds gather at anti-frack camp*. Retrieved from http://www.lep.co.uk/news/hundreds-gather-at-anti-frack-camp-1-5663169. Accessed on April 12, 2014.

Lock the Gate Alliance. (2014). Retrieved from http://www.lockthegate.org.au/. Accessed on April 25.

Macalister, T., & Harvey, F. (2013, July 19). George Osborne unveils 'the most generous tax breaks in world' for fracking. *The Guardian*. Retrieved from http://www.theguardian.com/politics/2013/jul/19/george-osborne-tax-break-fracking-shale-environment. Accessed on April 23, 2014.

Martinez-Alier, J., Anguelovski, I., Bond, P., Del Bene, D., Demaria, F., Gerber, J.-F., ... Yánez, I. (2014). Between activism and science: Grassroots concepts for sustainability coined by environmental justice organizations. *Journal of Political Ecology*, *21*, 19–60.

Mauro, F., Wood, M., Mattingly, M., Price, M., Herzenberg, S., & Ward, S. (2013). Exaggerating the employment impacts of Shale Drilling: How and why.

McHoul, A., & Grace, W. (1995). *A foucault primer: Discourse, power, and the subject*. Melbourne University Press (Taylor & Francis Group).

Moussa, M., & Scapp, R. (1996). *The practical theorizing of Michel Foucault: Politics and counter-discourse*. Cultural Critique. University of Minnesota Press.

O'Brien, P. (2014, February 18). Peasants' revolt – Romania at war with itself over fracking. *Channel 4*. Retrieved from http://www.channel4.com/news/fracking-romania-protests-pungesti-chevron-victor-ponta. Accessed on April 24.

Occupy London. (2013, December 8). *Save Rosia Montana & stop illegal fracking in Romania – Protests highlighting the Romanian gendarmerie as an instrument of oppression*. Retrieved from http://occupylondon.org.uk/events/save-rosia-montana-stop-illegal-fracking-in-romania-protest-highlighting-the-romanian-gendarmerie-as-an-instrument-of-oppression/. Accessed on April 24, 2014.

Oprea, L. (2013, December 13). Fracking, Pungesti, and the destruction of European democracy. *Dissident Voice*. Retrieved from http://dissidentvoice.org/2013/12/fracking-pungesti-and-the-destruction-of-european-democracy/. Accessed on April 24, 2014.

Osborn, S., Vengosh, A., Warner, N., & Jackson, R. (2011). Methane contamination of drinking water accompanying gas drilling and hydro-fracking. *Proceedings of the National Academy of Sciences of United States of America, 108*, 8172–8176. Retrieved from http://sites.nicholas.duke.edu/avnervengosh/duke-study-on-shale-gas-and-fracking/. Accessed on April 10, 2014.

Parsons, W. (2002). From muddling through to muddling up – Evidence based policy making and the modernisation of British government. *Public Policy and Administration, 17*(3), 43–81.

Popay, J., Bennett, S., Thomas, C., Williams, G., Gatrell, A., & Bostock, L. (2003). Beyond "beer, fags, eggs and chips"? Exploring lay understandings of social inequalities in health. *Sociology of Health and Illness, 25*(1), 1–23.

Powers, P. (2007). The Philosphical Foundations of Foucaultian Discourse Analysis. *Critical Approaches to Discourse Analysis across Disciplines, 1*(2), 18–34.

Prinsloo, E. (2014, April 4). Has corruption found its way into fracking applications? *AfriForum*. Retrieved from https://www.afriforum.co.za/corruption-found-way-fracking-applications/. Accessed on April 23.

Snow, D. (2004). Framing process, ideology, and discursive fields. In D. Snow, S. Soule, & H. Kriesi (Eds.), *The Blackwell companion to social movements* (pp. 380–412). Oxford, UK: Blackwell Publishing.

Snow, D., & Benford, R. (1988). Ideology, frame resonance, and participant mobilization. *International Social Movement Research, 1*, 197–217.

Snow, D., Soule, S., & Kriesi, H. (2008). *The blackwell companion to social movements*. John Wiley & Sons, p. 271.

Summerchild, A. (2013, December 12). Pungesti, Romania: People versus Chevron and riot police. *CNN iReport*. Retrieved from http://ireport.cnn.com/docs/DOC-1067976. Accessed on April 24, 2014.

Swyngedouw, E. (2010). Apocalypse forever? Post-political populism and the spectre of climate change. *Theory, Culture, Society, 27*(2–3), 213–232. Los Angeles, London, New Delhi and Singapore: SAGE.

Treasure Karoo Action Group. (2014). Retrieved from http://www.treasurethekaroo.co.za/. Accessed on April 23.

Vişan, G. (2013). Public perception of shale gas in Romania. Policy Brief 3, July 2014, Romania Energy Center. Retrieved from http://www.roec.ro/wp-content/uploads/2013/09/Visan_SG-Public-Perception-July-2013_EN.pdf. Accessed on April 30, 2014.

Weile, R. (2014). Beyond the Fracking ban in France. *Journal of European Management and Public Affairs Studies*, *1*(2), 11–16.

Wittmeyer, H. (2013). Anti-fracking. *Frackwire*. Retrieved from http://frackwire.com/anti-fracking/. Accessed on April 12, 2014.

JAPAN'S EVOLVING CIVIC ENVIRONMENTALISM

Robert J. Mason

ABSTRACT

Japan's civic environmentalism combines a tradition of local protest and activism with a national environmental movement that is limited in size and policy influence. A strong legislative and administrative response to the country's severe pollution crisis of the 1960s and 1970s helped tamp down that era's wave of protests and keep the Liberal Democratic Party (LDP) in power. While the state has generally supported local organizations engaged in environmental improvement activities, it has erected barriers that limit the scope of non-governmental organization (NGO) activities and inhibit the development of an influential national environmental movement. The 1990s reforms, inspired in part by the citizen response to the 1995 Great Hanshin Earthquake, made it easier for NGOs to attain legal status and raise funds. Yet Japan's civic environmentalism – by most measures – still lags well behind that of peer industrialized countries. The 2011 tsunami and nuclear crisis brought another opportunity for major reforms to the nation's civic environmental culture – but the evidence to date indicates that the much anticipated

Occupy the Earth: Global Environmental Movements
Advances in Sustainability and Environmental Justice, Volume 15, 37–61

ISSN: 2051-5030/doi:10.1108/S2051-503020140000015001

transformation is turning out to be of a lesser magnitude than many had initially expected.

Keywords: Japan; environmentalism; environmental movement; NGO; civic; citizen

Japan's environmental movement defies simple characterization; as Broadbent (1998, p. 4) puts it, Japan is a place where a tradition of protest coexists with a tendency to deference. Indeed, describing and predicting the future for Japan's civic environmentalism is something of a mini-industry for researchers, with much scholarly attention having been given not only to the more pedestrian aspects of the country's widespread local environmentalism, but also to the various watershed moments in the evolution of Japan's civic environmentalism. These major "moments" centered around the anti-pollution protests of the late 1960s and early 1970s, the striving for global environmental citizenship in the 1990s, the dramatic impacts of the 1995 Kobe earthquake, and – most recently – the shocks of the 2011 Tōhoku earthquake, tsunami, and Fukushima Daiichi nuclear disaster. The principal questions addressed in this chapter are: in what ways has Japan's environmental movement evolved over the past four to five decades, why is there so much citizen engagement in environmental matters at the local level yet so little voice at the national level, what are the relative roles of "culture" and systemic political forces in shaping Japan's civic environmentalism, and does the 2011 Tōhoku tsunami and nuclear crisis portend some sort of new era for civic environmentalism?

As a broad working definition for "civil society," Pekkanen's (2006, pp. 2–3) representation of it as the "nonstate, nonmarket sector" – largely excluding "labor unions, trade associations, professional associations, … parastatal organizations, and political parties" is quite fitting for this chapter. In no way, however, is this definition meant to diminish the role of the state in defining and shaping civil society. Indeed, NGOs[1] working with the state (Garon, 2003) – however critically or uncritically – are very much within this framework. In the U.S. context, John (1994) develops a notion of civic environmentalism that is collaborative, multi-stakeholder, multi-scaled (but generally bottom-up and local/local-state based), and often ecosystem-oriented. Japan's civil society is at a different place and stage of development, situated somewhere between an idealized "OECD model" of highly evolved, effective NGOs operating in a system characterized by openness, transparency, and legal legitimacy and safeguards (OECD, 2010; Reimann, 2010) versus a system that is more authoritarian and that limits civic space and regards NGOs as extensions of government rather than policy actors and watchdogs. Japan's experience of the past several decades is very much one of evolution and struggle over where the country belongs in this spectrum.

In the post-World War II era, Japan has achieved remarkable economic success, built competent government institutions, and overcome formidable natural and technological hazards – prompting wide amazement and admiration for the nation's self-discipline, resilience, and adaptive capacities. Through all of this, the role of citizen engagement has evolved a good deal, with thousands of environmental and many other types of organizations having emerged, mainly at the local level. High levels of modernization and globalization generally are associated with robust civic cultures, with NGOs playing important parts as national-level agenda setters, policy advocates, and watchdogs (Reimann, 2010). Yet in Japan – very much a wealthy, modernized, democratic, globally integrated country – the extent and influence of NGOs, especially with respect to policymaking, has been and remains rather modest (Broadbent, 1998; Johnson, Saito, & Nishikido, 2009; Kawato, Pekkanen, & Yamamoto, 2011; Mason, 1999; McKean, 1981; Pekkanen, 2006; Reimann, 2010; Schreurs, 2002). Reimann (2010, p. 19) describes Japan's political system as "statist" and relatively closed to NGOs. Pekkanen (2006, p. 27) states that most of Japan's civil society organizations may be described by the "four smalls": "small membership, small number of professional staff, small budget, and a small area of operation." Aldrich (2013, pp. 258–259), drawing on the Johns Hopkins Global Civil Society Index Ratings from 2004, finds that:

> the capacity, impact, and amount of voluntarism in Japan at the beginning of the twenty-first century fell below the levels in economically similar countries (such as the United States, the Netherlands, Ireland) and also below levels found in some developing countries (such as Tanzania and Uganda).

Pekkanen (2006, p. 2) describes Japan's civil society as a "dual civil society," consisting of many small, yet typically very active local groups, but few large, professionalized organizations; a civil society of "members without advocates." Indeed, Japan has a long history of active local organizations, principal among them neighborhood associations, which number in the neighborhood of 300,000 (Pekkanen, 2006); parent–teacher associations, women's groups, and consumer cooperatives. Environmental activities with which local organizations commonly are engaged include cleanup projects, recycling, park management, management of local water resources, and – especially in the case of *Seikatsu* and other consumer cooperative organizations – natural foods, GMOs, and alternative energy (Avenell, 2009, 2012b; Garon, 2003; Haddad, 2007; Lam, 1999; Pekkanen, 2006).

Drawing on a database compiled in 2001 by the Japan Environment Association, Johnson et al. (2009) conducted a comprehensive survey of "environmental movement organizations." They found that over 80 percent of the organizations operated at either the local or prefectural level (p. 492). Further, they report that

> the issues to which Japanese EMOs are most likely to attend are, in order: resource conservation, water pollution, waste and recycling, over-consumption, local issues, and forests. Less than 10 percent of the total sample attends to issues of desertification, global warming, or air pollution. (Johnson et al., 2009, p. 494)

National-level organizations, which comprise a small portion of the sample, are more involved with broader national and global issues than are their counterparts at the local and prefectural levels. Twenty-three percent of all respondents did report being engaged in lobbying activities (p. 493).

These many local NGOs do have close relationships and influence with local governments. Yet they are widely viewed as "do good" organizations; they can expand governmental capacity – or, conversely, government support can enlarge organizational capacity (Mason, 2008). These mutually reinforcing government–NGO relationships are seen at least as acceptable to, or even quite useful for the state (and the local state), particularly when NGOs cooperate and take on duties that are too unimportant for government agencies to be bothered with (Schwartz, 2003, p. 5). They act principally as "service" rather than "advocacy" organizations (Reimann, 2010, pp. 46–47). As Kawato et al. (2011, p. 125) put it:

> the state has promoted the growth and activities of some civil society organizations to facilitate governance, provide social services, manage society, and communicate its stories to the public. The LDP also worked closely with some organizations to remain in power. Civil society organizations have also benefited from this cooperative relationship, with more access to policymaking, information, and state funding.

Avenell (2010, p. 9) argues that Japan's protest movements were not simply co-opted, but that movement intellectuals responded creatively in the face of government barriers by:

> modifying the *shimin* idea and reconstituting or redirecting civic activism, which, in turn, encouraged new forms of mobilization, stimulated new tactics, and promoted new relationships.

Nor does Haddad (2007) view Japan's civil society as weak; while acknowledging Japan's "restrictive institutional environment" (p. 418), she sees strength in Japan's localist movements, which have maintained and even enhanced social capital, in contrast, for example, with the reduced rates of participation in local organizations in the United States (Haddad, 2007; Putnam, 2000; Skocpol, 2003). This dense network of voluntary associations is one factor that has led Curtis (1997, p. 141) to argue that "Japan has always had a stronger civil society than neighboring countries (and a much stronger one than is presumed to be the case, by both Japanese and foreigners)." Future investigation of Japan's civic environmentalism might focus on how reforms could enhance these social capital strengths without further eroding the movement's national-level status (which is taken up below).

The trends in numbers, memberships, activities, and policy effectiveness of Japan's local organizations are unclear. Pekkanen (2006) believes that in contrast with the United States (Putnam, 2000; Skocpol, 2003), numbers may be on the increase. Kawato et al. (2011, p. 122) argue that urban sprawl and increased "individualism" in Japan may be contributing to a decline in citizen engagement and possible reduced effectiveness of local organizations. Neither of these claims is supported by hard evidence.

Clearly, there are and have been exceptions to the characterization of Japan's local NGOs as capacity enhancers and virtual subcontractors to governments, or as very stealth agents of change. As discussed below, a great many local anti-pollution advocacy groups formed in the 1960s and 1970s. More recently, local organizations often have been strong advocates, active in opposing dams, land reclamation activities, airport siting and expansion plans, and nuclear power plants. Japan's "construction state" (*doken kokka*) system has been responsible for immense environmental damage in the post-war era, particularly from construction of dams, channelization of streams, reclamation of wetlands, and consumption of tropical timber (McCormack, 1996; Woodall, 1996). Comprised of the "iron works" triangle of business interests, politicians, and government bureaucrats, this construction empire has long dominated politics under LDP rule. Lam (2011, p. 242) describes the LDP traditional policy framework as:

> technologically-driven, nuclear power-reliant, bureaucratically-led (especially by METI), pro-big business, market-sensitive with financial incentives and subsidies from the state, societal voluntarism rather than mandatory cuts, some initiatives by enterprising local governments, but with very limited inputs from opposition parties, the scientific community, think-tanks, NGOs and public opinion.

Rural regions have benefitted immensely from public works subsidies and disproportionate political power; in turn, rural support for the LDP traditionally has been strong. The construction state was diminished in power and scale when the LDP lost elections in 1993, when reform-minded LDP Prime Minister Koizumi led the country between 2001 and 2006, and when the Democratic Party of Japan – advocating transparency, fairness, and a more democratic, participatory society – was in power between 2009 and 2012 (Kingston, 2013; Waley, 2005). While the construction state has not been vanquished, a return to the iron triangle days of years past is quite inconceivable. Environmental organizations lost many individual battles along the way – reclamation of Isahaya Bay and construction of the Nagara River dam in the 1990s among the more outstanding losses (Mason, 1999) – but can claim at least some credit for this qualified larger victory.

WHY IS JAPAN'S NATIONAL ENVIRONMENTAL MOVEMENT SO WEAKLY DEVELOPED?

Scholars attribute a wide variety of meanings to "culture," ranging from the folkloric to the highly political (McVeigh, 1998). In explaining Japan and other Asian countries' power dynamics and political systems, paternalistic, authority-based Confucian cultural characteristics often are invoked (Imura, 2005; Pye & Pye, 1985). As Schwartz (2003, pp. 4–5) points out, state-centric thinking does indeed have deep societal roots; the notions of self-sacrifice, deep respect for authority, and subordination of individual interests have long traditions and are persistent in Japan's modern era. Yet it is fitting, certainly in Japan's case, to consider how these "cultural" attributes may be "played upon" by the state to circumscribe civic space – rather than simply turning to them to explain civil society's differences or perceived weaknesses in comparison with other highly industrialized countries (Mason, 1999; Pekkanen, 2006).

Japan, it is often argued, lacks both a well-developed charitable ethic and participatory culture – at least with respect to political affairs. Yet Japan does not lack for a history of protest. Contrary to some thinking – buoyed most recently by widespread admiration for the patience, respect, and generally good behavior exhibited by citizens at the height of the Tōhoku crisis – Japan does have a history of riots, strikes, and other forms of protest, over a variety of issues, reaching at least as far back as the Tokugawa

Era (1603–1868) (Bix, 1986; Kuroda, 1972; Najita & Koschmann, 1982; Passin, 1962; Steinhoff, 1989; Vlastos, 1986).

Intense anti-pollution protests took place in the late 1800s and early 1900s in response to widespread air and water contamination from Ashio Dōzan, a sprawling copper mine complex in Tochigi Prefecture, north of Tokyo (Shoji & Sugai, 1992). Diet member Shōzō Tanaka was a leader and supporter of the protests, which inspired later activism in response to mining-related pollution elsewhere in Japan (Nimura, 1997; Strong, 1977). Mercury pollution in Kyushu's Minamata Bay brought local fishermen's protests in the late 1950s – and, as environmental awareness increased in the early to mid-1970s – expanded protests against the Chisso chemical company (George, 2001). Though not explicitly environmental or pollution related, we also have the 15 months of protest over the 1960 renewal of the Japan–U.S. Treaty of Mutual Cooperation and Security (*ANPO* treaty) (Avenell, 2010; Dobson, 2002; Haddad, 2007). Construction of Tokyo's Narita Airport during the late 1960s and 1970s brought intense, often violent protests from local farmers, students, and members of left-wing political parties, whose opposition was not only about the seizure of land, but also embedded the Narita case within a larger narrative regarding the future of the Japanese state (Apter & Sawa, 1984). But beyond these individual landmark protest events, the 1960s and 1970s witnessed thousands of smaller environmental campaigns across the archipelago.

By the mid-1960s, Japan was one of the most polluted places on earth (Ui, 1992) – the culmination of what Broadbent (2002) describes as the "Polluters' Paradise" period that commenced in the 1800s. This led to widespread protests in the late 1960s and early 1970s (Avenell, 2012b; Broadbent, 1998; Hasegawa, 2004; Kawato et al., 2011; Ui, 1989), reaching a peak of over 3,000 in the early 1970s (Lam, 2011, p. 39). Anti-war, human rights, and immigrant assistance groups also were numerous in this era, but "ecology groups" predominated among all protest groups, accounting for 61 percent of groups identified in a 1971 survey reported by Kuroda (1972, p. 948). The main focus for this very decentralized movement actually was pollution and public health; nature conservation and biodiversity issues were taken up – but, as we see below, clearly were secondary concerns.

In addition to the proliferation of protest groups, four major lawsuits were resolved, largely in favor of pollution victims: Itai-itai cadmium poisoning (Toyama), Minamata disease from organic mercury poisoning in Minamata and Niigata, and Yokkaichi asthma, in Mie Prefecture (Broadbent, 2002; Gresser, Fujikura, & Morishima, 1981; McKean, 1981;

Ui, 1992). Local referenda also have been used, from the 1990s onward, typically to express opposition to U.S. military facilities in Okinawa and elsewhere, nuclear power plants, industrial waste facility siting, municipal incinerators, and dams and other structural alterations to rivers (Kawato et al., 2011; Lam, 2011).

While recognizing that most of these groups were local groups, Kuroda (1972) foresaw the potential for major systemic change, arguing – in something of a turn on the traditional culturalist arguments – that "the Japanese people are said to be the type of people who once having set their minds to something, will go all the way to accomplish what they perceive as their goals" (p. 951). If the goal was to dramatically reduce pollution, then the goal was accomplished. The LDP, while not in any conventional sense an "environmentalist party," reacted swiftly and strongly to the pollution crisis (Avenell, 2012b; Broadbent, 2002; Fowler, 2003) – comparably, in its way, to U.S. President Nixon (also not generally regarded as an environmentalist) and the Congress at approximately the same time (Kraft & Vig, 2013). Broadbent (2002) characterizes the mid-1960s to mid-1970s period as "Polluters' Hell." Perhaps the most dramatic "moment" came with the "Pollution Diet" of 1970, which passed 14 far-reaching anti-pollution laws. Beyond these and other regulatory reforms, government officials responded to the protests through "the establishment of nonjudicial, bureaucratic-led mechanisms for dispute resolution; and policies promoting public involvement in *local* governance" (Avenell, 2009, p. 261, emphasis added). These reforms were meant to increase opportunities for citizen engagement, while allowing government to maintain a good measure of control over the process.

The 1970s was a time for civic environmentalism to blossom in Japan. Indeed, the pressure created by local protest groups, lawsuits, and local referenda played their part in creating Broadbent's "polluters' hell." But many of the local groups fizzled after new laws were put into place and as national attention turned toward the oil shocks of the 1970s (which prompted environmental benefits apart from pollution reduction, especially in the domain of industrial energy efficiency). Whereas lawsuits have been a key feature of American environmentalism, this has not been the case in Japan. Despite the success of the "Big 4" cases noted earlier, lawsuits tend to be very laborious and drawn out in Japan (the Minamata lawsuits carried on for years after the early favorable settlements, with some individual cases still being settled in 2014!). Citizen standing to bring lawsuits never was enlarged to include parties not directly affected, as was the case in the United States in the 1970s (Schreurs, 2002). *Nichibenren* – the Japan

Federation of Bar Associations – continues to bring environmentally related lawsuits, but its resources are limited and it relies heavily on volunteer efforts. Local referenda remain important in Japan – but often their results have little or no effect when local citizens seek to confront national policies. Moreover, state authorities have learned to minimize the risk of such confrontations, by seeking to site unwanted facilities and land uses in areas where civil society is comparatively weak and by directing generous subsidies to host localities (Aldrich, 2008). Thus, after the early 1970s, Japan's environmentalism rarely was able to pose serious challenges to the state.

The dramatic government response to the pollution crisis was motivated, at least in part if not in large part, by the desire to maintain power, rather than newfound convictions about the perils of pollution and the importance of a cleaner, safer environment. And, it seems, this is a major reason that the local movements did not transform into a larger, persistent national environmental movement. But these actions alone do not entirely explain this phenomenon; governments in Germany and the United States likewise reacted forcefully to the environmental crisis – and yet those countries did develop strong national environmental movements.

The bureaucratic infrastructure that has run day-to-day Japan through most of the post-war era has little tolerance for a strong, advocacy-oriented civil society (see Johnson, 1995; Wolferen, 1993). As a consequence, Japan's political opportunity structure, particularly for national-level NGOs, has severely constricted their development (Kawato et al., 2011; Mason, 1999; Pekkanen, 2006; Reimann, 2010; Schwartz, 2003). While less restrictive laws covered such organizations as private school corporations, social welfare corporations, religious corporations, and medical corporations, public-interest corporations were not afforded the same treatment under Japan's Civil Code (Schwartz, 2003).

Before passage of the 1998 NPO law, it was exceedingly difficult for NGOs to incorporate; to be recognized officially as a "public interest corporation," or *koeki hojin*. Typically, an organization would have had to possess an endowment equal to or greater than ¥300 million (approximately US $3 million), an annual budget of at least ¥30 million (approximately US $300,000) (Schwartz, 2003, p. 11), and/or a sufficiently large membership base, in the judgment of the government ministry with which the NGO is affiliated. Though specific membership numbers were not generally made explicit, the numbers would have to be quite large. And ministry oversight was generally extensive, reaching to the organization's internal affairs and board membership. Moreover, tax-exempt status was unavailable to all but a very few NGOs. And tax exemptions for contributions made to NGOs were not available. Direct state financial support to NGOs was very limited and foundation funding likewise was minimal. Moreover, bulk postal rates were and still are not an option for Japanese NGOs, while credit card donation options, which had become widespread in the United States and other countries well before the advent of the internet – were slow in coming for Japanese organizations.

Given the daunting requirements, few NGOs had gained legal status before the 1998 law. By the mid-1990s, Japan had only 26,089 groups which had legal status as non-profit public-interest legal persons, while in the United States, the Internal Revenue Service had granted non-profit status to 1,140,000 organizations (Pekkanen, 2000, p. 113). Even some organizations that might have had the resources to gain legal status chose not to, in order to avoid the onerous oversight. The lack of legal status meant that individuals associated with the NGOs would have to sign leases, procure equipment, and rent vehicles in their own names. A few large, old environmental organizations were well established, with generally good relationships with government ministries. These included the Nature Conservation Society of Japan (NACS-J), Wild Bird Society, and World Wide Fund for Nature Japan (WWF-J). But the price for such status and recognition was high. Moreover, even these few flagship organizations were (and still are) characterized by modest membership numbers and funding (Mason, 1999).

THE 1990S TRANSFORMATION

Although Japan had made extraordinary progress in controlling domestic pollution in the 1970s, during the economic bubble years of the 1980s, it developed something of a reputation as global laggard, at best, and pariah, at worst. Japan cast an enormous "environmental shadow"; as the top importer of tropical timber, much of which was used as one-time forms for pouring concrete, it was responsible for massive rainforest losses in Southeast Asia (Dauvergne, 1997). Its fishing and whaling practices came under heavy criticism; as did destructive overseas development assistance (ODA) programs (Hall, 2010; Ohta, 2000). Japan was criticized by the OECD for its failure to comply with international agreements and norms – and to include NGOs in its deliberations (Broadbent, 2002). Japan was behind its industrialized-country peers in signing onto and implementing such international agreements as the Convention on International Trade in Endangered Species of Wild Flora and Fauna (CITES), which it signed in 1980; lagged in designating wetlands sites under terms of the 1971 Ramsar Convention; was a holdout on implementing the 1987 Montreal Protocol on Substances that Deplete the Ozone Layer; did not sign onto the World Heritage Convention until 1992, one to two decades later than most of its peer countries; and lacked a statutory basis for conducting environmental impact assessments until 1997 (Barrett & Therivel, 1991; Mason, 1999). Japan was also a laggard with respect to numbers of NGOs engaged in international activities, as well as their memberships, finances, and consultative status with the United Nations Economic and Social Council (Reimann, 2010).

Japan's leaders, being especially sensitive (selectively at least) to the country's international image and to *gaiatsu* (foreign pressure), seized upon the appropriate moment and sought to reposition the country as a global environmental player (Broadbent, 2002; Dobson, 2002; Hall, 2010; Kawato et al., 2011; Reimann, 2010; Schreurs, 2002). By the early 1990s, Japan had entered Broadbent's (2002) Phase 4 in the evolution of its environmentalism: "New Global and Local Demands." Not only did the government begin to catch up on signing and implementing such treaties as CITES and the World Heritage Convention (Mason, 1999), but also took a very active role at the 1992 Conference on Environment and Development in Rio de Janeiro, signing onto the Agenda 21 sustainable development action plan and the Conventions on Biological Diversity and Combating Desertification. Gradually, and initially very reluctantly, the government began to grant official status to NGOs at major meetings and conventions, such as those of the World Bank and Asian Development Bank (Reimann, 2010). In 1993, the LDP lost the powerful Lower House of the Diet to opposition parties for the first time during the post-war era. As a consequence, the opportunity structure for domestic NGOs expanded and more international NGOs entered Japan (Broadbent, 2002). Later in the decade, Japan hosted the 1997 Kyoto Climate Change Conference and brokered a binding agreement committing participants to mandatory greenhouse gas emission reductions. Japan also announced a massive increase to its ODA at the Rio meeting, with substantial amounts to be earmarked for environmental projects (Broadbent, 2002).

An early victory for environmentalists came in 1990, when the Ministry of Foreign Affairs (MOFA) ended financing for India's controversial Sardar Sarovar dam project; a large share of credit for this success was claimed by the NGO community, particularly Friends of the Earth-Japan

(Hirata, 2002; Mason, 1999). Starting in the early 1990s, MOFA increased its funding of and cooperation with NGOs engaged in international development activities – though this increased support has come with its share of contention (Hirata, 2002; Reimann, 2010). Substantial support also has come from the Environmental Agency (Ministry of the Environment since 2001), via the Japan Fund for the Global Environment, which funds international projects as well as domestic works in the areas of education, recycling, pollution prevention, and afforestation. Additional funding for overseas NGO activities has come through the Ministries of Posts and Telecommunications (now part of the Ministry of Internal Affairs and Communications); Foreign Affairs; Agriculture, Forestry, and Fisheries; and Construction (now part of the Ministry of Land, Infrastructure, Transport, and Tourism). Substantial increases in disbursements from postal savings accounts took place in the 1990s. Citizens could choose to participate in the International Volunteer Savings Account program – but they were given no say in how the monies were spent and the only NGO involvement in those decisions came in the form of a limited oversight role for the NGO Center for International Cooperation (JANIC). Also during the 1990s, *Keidanren* – the Japanese Federation of Economic Organizations – began to make sizeable grants to The Nature Conservancy for conservation projects in Micronesia and Melanesia ... but not so much to domestic Japanese organizations (Mason, 1999; Reimann, 2010). Substantial funding has come from the Toyota and Aeon foundations, as well as such international organizations as Friends of the Earth, World Wildlife Fund, Patagonia, and American foundations, including the Charles Stewart Mott, Alton Jones, MacArthur, Rockefeller, and Asia foundations (Mason, 1999; Reimann, 2010). Among the organizations that relied heavily on American funding during the 1990s and early 2000s are Friends of the Earth-Japan, the Japan Center for Sustainable Environment and Society (JACSES), Mekong Watch, and Japan Tropical Forest Action Network (JATAN) (Reimann, 2010, pp. 138–139). By far, though, the leading funding source for most NGOs has been the modest membership fees they collect.

Japan's civic environmentalism, as we have seen, has been primarily about local pollution and health issues. Although Japan's first national park was created in 1934, nature conservation, wildlife, and biodiversity concerns did not claim as large a place as was the case in Europe, parts of Australasia, and especially the United States, where environmentalists would routinely engage in struggles over protecting places far from their local area (Avenell, 2012a; Knight, 2010; Oyadomari, 1989). Japan's Wild

Bird Society was founded in 1934 and the Nature Conservation Society of Japan (NACS-J) was established only in 1951. U.S. organizations that are in some respects comparable to the Wild Bird Society and NACS-J include the National Audubon Society, founded in 1905, and the Sierra Club, in 1892. Both American organizations have much larger memberships and budgets than their Japanese counterparts (Hasegawa, 2010, p. 86; Imura, 2005, p. 81; Knight, 2010, p. 361; Mason, 1999, p. 195; Pekkanen, 2006, p. 42).

Gradually, however, nature conservation has become more integrated into Japan's broader environmentalism. Tsuru (1999, pp. 124–127) describes how the previously dominant terminology of *kōgai* (pollution) was replaced by *kankyō* (environment) during the late 1970s and early 1980s, to reflect the growing diversity of the movement. In addition to the "site fights" noted above over dams and other projects, local organizations – and to a lesser extent national NGOs – were becoming more proactively engaged with issues of land protection and ecosystem management (Mason, forthcoming). Beginning around the mid-1990s, there has been a proliferation of groups working on river restoration and watershed management (Waley, 2005). Founded in 1992, the Association of National Trusts in Japan[2] is a rather decentralized assemblage of NGOs, inspired by Britain's National Trust, that seeks to protect historic buildings, neighborhoods, and natural landscapes. It is not a national advocacy organization, but does sometimes take forceful local actions. Individuals and corporations are able to deduct charitable contributions to national trust organizations (Kihara, 1989; Takao, 2001).

But the singular key event of the 1990s for domestic civic environmentalism was the 1995 Great Hanshin earthquake, whose epicenter was on northern Awaji Island, and which did immense damage to Kobe and the surrounding region. The importance of civil society was widely displayed in the rescue and recovery efforts following the earthquake. A massive outpouring of citizen volunteers contrasted visibly and dramatically with governmental jurisdictional disputes, delays, and red tape. The world watched, for example, as people remained trapped in the rubble while Swiss sniffer dogs were initially refused entry due to routine quarantine requirements. The central government and other authorities' bumbling and ineffective responses, in combination with their poor preparation, created a policy window for rethinking the role of civic engagement in Japanese society (Ogawa, 2009; Pekkanen, 2006). Still, the emphasis was principally upon building a greater national "volunteer infrastructure"; of NGOs as service providers and capacity enhancers for government agencies, rather than policy advocates and watchdogs. At the local and regional levels, however,

NGOs did become more actively involved in urban planning during the reconstruction period after the earthquake (Shaw & Goda, 2004).

Three major changes followed (Fowler, 2003), though not all of them can entirely or necessarily be linked to the earthquake. The 1998 passage of the Special Nonprofit Activities Law (commonly referred to as the NPO law) *is* widely viewed as a direct outcome of the earthquake and its aftermath; Pekkanen (2006), however, argues that even without the Kobe disaster, an NPO law would likely have emerged within 10 years, given the growth in NGOs and the political processes already underway. The law was developed between 1995 and 1998, with various versions being written by political parties, rather than by bureaucrats, as is the convention with most legislation. Moreover, NPOs had a significant voice in shaping the law (Pekkanen, 2000). What the law provided was a straightforward avenue to legal status for NPOs, something that had been limited to a relatively small number of organizations before 1998. With legal status comes the ability, as an entity, to "open bank accounts, hire staff, own property, sign lease agreements for office space, undertake joint projects with domestic government bodies, or even, on a mundane level, lease a photocopy machine" (Pekkanen, 2000, p. 113). The law was meant to greatly ease the bureaucratic oversight and high capital requirements that formerly had been imposed upon organizations seeking legal status.

After an initially slow start, there was a surge in numbers of NGOs gaining legal status: by July 31, 2004, there had been 19,025 applications and 17,424 approvals (Pekkanen, 2006, p. 55). By late March 2009, approximately 37,000 legally recognized organizations existed, with 29 percent of these working on environmental issues (Hasegawa, 2010, p. 85). In 2008, the NPO law's public-interest certification process was modified, so that it would be handled by an "independent and neutral institution" rather than the ministerial bureaucracy (Kawato et al., 2011, p. 118). Tax-exempt status was not immediately granted for newly formed NGOs under the 1998 law, but rather was delayed for two years. Although the Fiscal Year 2001 Tax Reform did include provisions for "Specified Non-Profit Activities Corporations" (SNCs), created under the 1998 law to achieve tax-exempt status, very few were able to meet the rigorous set of requirements (Pekkanen, 2006). In 2008, SNCs were finally granted the same income tax exemptions and exemptions or reductions on taxes for certain business activities that had been available to Public Interest Corporations that had been legally recognized before the 1998 NPO law (Reimann, 2010). Tax deductibility for charitable gifts to NGOs was granted earlier, in 2001 – but the conditions for eligibility were stringent and rather few organizations were able to achieve this status. After 2008, to achieve these tax benefits, organizations had to become a "Public Interest Incorporated Association" – a new category created for this purpose in 2006. Kawato, Pekkanen, and Yutaka (2012, p. 80) report that only 223 out of more than 42,000 eligible organizations (0.5 percent) have achieved this status. By May 2013, that number had risen to 447 (Japan Association of Charitable Organizations, 2013, p. 2), and may be expected to continue rising in the wake of a loosening of requirements that took effect in April 2012 (Editorial, 2012). Still, Japan's numbers remain far below those for peer industrialized countries (Reimann, 2010).

Even for registered NGOs, including those established before the 1998 NPO law, membership numbers, budgets, and staff are small in comparison with NGOs in peer countries (Kawato et al., 2012). South Korea and Taiwan each have much larger and more influential environmental NGOs than does Japan (Hasegawa, 2010). Japan's largest environmental NGOs at the turn of the century, according to a comparison compiled by Pekkanen (2006, p. 42), were the Wild Bird Society of Japan, with 53,798 members; WWF-J, with 50,000 members, and NACS-J, with 50,275 members. The largest United States groups were the National Wildlife Federation, with 4 million members; and World Wildlife Federation and The Nature Conservancy, with about a million members each. Among the more recently established groups in Japan that have gained their share of attention and are seeking to influence national policy are the Kiko (Climate) Forum and CASA (Citizens' Alliance for Saving the Atmosphere) (Lam, 2011, p. 243). As of May 2013, the total number of Specified Non-Profit Corporations – mostly local groups created under the terms of the 1998 NPO law – was 47,771; about 8 percent of these specify "conservation of the environment" as an organizational activity (Japan Association of Charitable Organizations, 2013, pp. 2, 9).

While the NPO law was meant to reduce – by some early reckonings to eliminate – ministerial oversight of NPOs, this has been accomplished to only a rather limited extent in the law's implementation. Indeed, in the earliest days after passage, bureaucrats were conferring with and making suggestions for changes to applications for legal status (Pekkanen, 2000; Takao, 2001). Subsequent revisions to the law have had the effect of reducing – though not eliminating – ministerial oversight (Reimann, 2010).

Avenell (2009) keenly contests claims that civil society in Japan was transformed in the 1990s (Kingston, 2004). He acknowledges the increase in numbers and activities of citizen groups, but sees these developments as part of a long-term trend, starting in the early 1970s, when the state began to support selected forms of volunteerism and civic activism. Corporate support for "do-good" groups – in the environmental realm, this includes those engaged with river and park cleanup and beautification activities – expanded during the 1980s. Avenell (2009) refers to the activities of these organizations as "proposal-style" activism (*teian-gata*), which by-and-large tend to complement and support governmental policies. As he describes it:

> Civic groups now face a powerful public rhetoric linking their social legitimacy to professionalization, financial sustainability, adaptation of market principles, and service-providing capacity – all encapsulated in the potent symbolism of the NPO. (Avenell, 2012a, pp. 282–283)

As a consequence, environmental NGOs may be cautious about challenging government policies, as seems to be the case with protecting the rights of the *matagi*, traditional users of the ancient beech forests in Shirakami Sanchi World Heritage Area in the Tōhoku region (Mason, forthcoming).

A second major 1990s development was passage of an environmental impact assessment (EIA) law in 1997. Before this, EIA, at the national level at least, was a matter of administrative discretion and guidance – though many localities did enact their own EIA ordinances. The 1997 Environmental Impact Assessment Law expanded the range of projects covered; added screening, scoping, and monitoring provisions; and allowed for a wider inclusion of public opinions, beyond those of immediate local residents. Yet there is no real oversight of the EIA process; the Environmental Agency (Ministry of the Environment since 2001) can express opinions, but has no power beyond this. The law's potential to be transformational is quite limited, given that EIAs are conducted only for large-scale projects and only at the project execution stage, rather than during formulation and planning (Tsuru, 1999; Yoshikazu, 2005). The OECD (2010), in its environmental performance review for Japan, calls for better communication with the public, particularly during the scoping stage of the EIA. Despite its significant shortcomings, the EIA law does provide an important channel for NGOs and individuals to have input on major environmental decisions.

Following earlier passage by every prefecture and major municipality, the national government passed an information disclosure law in 1997, which was implemented in 2001. Although government officials have been reluctant about compliance, the courts have been quite rigorous in enforcing the law (Kingston, 2013). Beyond the three major laws described here, there are the Administrative Procedures Law (1993), Products Liability Law (1994), and revised Code of Civil Procedure (1996) – followed in 2002 by a series of judicial reforms (Kingston, 2013; Schwartz, 2003). Japan's evolving "Quiet Revolution" (Kingston, 2004) was thus underway. Hazards resilience, energy planning, and climate change would be the foci of the next major moment for Japan's evolving civic environmentalism.

THE 3.11 CRISIS

On March 11, 2011, Japan's Tōhoku region experienced a devastating tsunami and nuclear disaster. Just about a year and a half earlier, the Democratic Party of Japan (DPJ) had come into power in a landslide, heralding yet another important era of change for Japan, featuring greater transparency, openness, citizen engagement, and environmental concern. DPJ civil society initiatives included:

> (1) considering the expansion of tax incentives for NPO donors; (2) reducing *amakudari* (a practice where retired bureaucrats obtain positions in the private sector that they used to supervise) in public interest legal persons; and (3) holding special hearings to identify inefficient and unnecessary projects managed by public interest legal persons, and cut state funding to these projects. (Kawato et al., 2011, p. 125)

Post-3.11, Japan found itself at another crossroads (Aldrich, 2013, p. 265; Chandler, Chhor, & Salsberg, 2011; Matanle, 2011), this one with particularly important energy and environmental implications. Yet in 2012, the LDP was restored to power by an overwhelming vote. The case has been made that this was not a vote for the LDP so much as a vote against the DPJ and other parties (Pekkanen, 2012). While it would be simplistic to interpret the vote as an affirmation of historical LDP nuclear, environmental, and civil society policies, it does not appear as if the 3.11 disaster has brought about the sweeping societal transformation that many initially had envisioned. In its post-tsunami coastal reconstruction efforts, the central government is largely favoring structural works, in the old construction-state model, rather than incorporating significant green infrastructure and conservation zones. Although Miyagi and Iwate Prefectures are reviewing the impacts of these projects, the national EIA law does not apply to seawalls, forest buffers, and small coastal roads (Bird, 2013). Moreover, Prime Minister Shinzō Abe is committed to restoring nuclear power, albeit not to the 50 percent level that was anticipated under the 2010 Basic Energy Plan, for 2030 (Matanle, 2011, p. 838). Another potential indication of the government's reluctance to embrace civil society initiatives came in late 2013, with enactment of a strict state secrets law, which has prompted protests and raised concerns that the law might be used to hide details about environmental hazards.

Post-war public hostility toward nuclear weapons did not really carry over to deep opposition to the peaceful development of nuclear energy (Aldrich, 2013; Avenell, 2012a). Nonetheless, Japan's anti-nuclear power movement – long identified with the late nuclear chemist and professor Jinzaburō Takagi, who headed the Citizens' Nuclear Information Center for many years (Sugai, 1992) – has been firmly in place since the 1970s. The Three Mile Island and Chernobyl nuclear accidents, as well as domestic nuclear incidents that include the 1995 sodium leak at the Monju experimental fast-breeder reactor and the 1997 fire and explosion at the Tōkaimura fuel reprocessing plant, did periodically raise serious public concerns about Japan's nuclear energy program (Mason, 1999). But as with Japan's larger environmental movement, the anti-nuclear movement has been relatively ineffectual over the years in shaping national policy (Hasegawa, 2004). The 3.11 disaster seemed to promise significant change – a thorough rethinking of energy policy and sustainability and resilience planning, with local responses possibly coalescing into something more systemic (Aldrich, 2013; Chandler et al., 2011; Matanle, 2011). It might also be seen to be prompting national government responses aimed at calming public concern and quelling mass protests, as had happened more than four decades earlier, in response to growing public outrage about pollution.

Aldrich (2013, p 249) pointed to the disaster's role in stimulating citizen science, large protests, and challenges to governmental authority. Indeed, there was a shift in public opinion, which still favored nuclear power immediately after 3.11, against nuclear energy. In a comparison of public opinion surveys, Aldrich (2013, p. 256) found that on April 4, 2011, 56 percent of respondents favored an increase or at least maintenance of the status quo for nuclear power, whereas by October 28, 2011, 66 percent called for decreasing or abolishing nuclear power. While volunteers poured into the region – though less than half as many as in the wake of the Great Hanshin Earthquake – there was also a tremendous loss of trust in the state. Massive protests, on a near-weekly basis, took place in 2012; local groups filed for referenda on nuclear power questions; and a petition to the government, bearing 5 million signatures, called for an end to nuclear power (Aldrich, 2013).

Following the Fukushima disaster, Prime Minister Angela Merkel called for a phasing out of nuclear energy in Germany. Why Germany and not Japan? Germany arguably has more options and a much higher "reserve margin," or capacity for generating electricity by alternative means, than does Japan (Scalise, 2012). Germany also has a more robust anti-nuclear movement, as well as a green party. Japan's post-Fukushima protests were extraordinarily large – topping out around 20,000 early on and later 80,000 (Aldrich, 2011, 2013, p. 260). But this still pales in comparison with the 1,960 protests over renewal of the Japan–United States security treaty, which had as many as 400,000 protesters, and with Germany's post-Fukushima anti-nuclear protests, which drew upwards of 200,000 participants (Aldrich, 2011).

Avenell (2012a) sees nuclear power, similarly to whaling and biodiversity protection, as a "blind spot" in Japan's environmental activism. The still embryonic movement emerging after the 3.11 disaster, he argues, has the "potential to spearhead a campaign to reform institutions, regulations, and popular attitudes ... but for all its promise, the post-Fukushima antinuclear movement conforms to an established model of reactive, victim-centered environmental activism ... of the 1960s and 1970s" (Avenell, 2012a, p. 372) that does not necessarily lend itself to effectively opposing nuclear power.

Japan's "nuclear village," a formidable force over the years, has been weakened, but not vanquished by the 3.11 disaster. The nuclear village consists of the "LDP ... Ministry of Economy, Trade, and Industry (METI, known as MITI until 2000), the 10 nuclear power companies (the largest of which was Tokyo Electric Power Company: TEPCO), and Keidanren and Doyukai (two leading business institutions representing the interests of major corporations)" (Claremont, 2014, p. 82). The village essentially controls nuclear energy in Japan, is a major supporter of the LDP, and played an important role in the ouster of Naoto Kan, who was the DPJ prime minister during the crisis (Kingston, 2012). The government, according to Aldrich (2013, p. 254), has created a "cycle of addiction" and "culture of dependence" by offering incentives to local communities most likely to be receptive to them – and where citizen opposition to nuclear power is likely to be the weakest (see also Hara, 2013; Juraku, 2013). Nuclear host communities must now decide whether or not to restart nuclear plants shut down after the 3.11 disaster; many of them are inclined to support restarts and resumption of economic benefits – and the nuclear industry will continue to offer attractive incentives to these often struggling communities. But in some cases, companies resorted to more nefarious tactics, in one instance having gone so far as to direct employees to testify in favor of restarts at local hearings (Kingston, 2012).

CONCLUSION

Japan has a strong civic environmental culture, as indicated by its numerous (mainly) local organizations and development of social capital. As a general rule, however, these myriad organizations wield rather little policy influence, particularly at the national level. While it has become easier over the past two decades for organizations to gain official recognition and raise funds, many "soft" barriers still are in place and national-level environmental NGOs today remain relatively small in membership numbers, professional staff, and funding bases. Will Japan's gradual movement toward a stronger civic environmental culture – one that is influential at the national level – continue? Evolution, of course, will carry on – but predicting how this will play out and where it will take Japan is so much a matter of speculation and so likely to be deeply affected by unforeseen events and trends – that it will largely be avoided in this concluding comment. Instead of speculating on the future, a few "advisory" notes are offered here, which might serve as focal points for further scholarly work.

Careful attention should be given to participatory process. What balance of collaborative and confrontational approaches is appropriate? At the national level, especially, is it in the best interests of NGOs to work with government, industry, and private citizens – or should they be working more as watchdogs and challengers to the policy process? Does deep citizen engagement necessarily ensure "better" outcomes than might be achieved by a competent bureaucratic state? Although collaborative processes have found much favor in recent times in the United States, the acclaim is far from universal. For example, former Sierra Club executive director and chairman Michael McCloskey (1999) has called "collaboration a betrayal of democracy, a de facto substitution of minority rule that threatens to undo a century of environmental gains" (Mason, 2008, p. 75). In Japan, these issues may become especially poignant in such areas as nuclear policy, hazards resilience, and protected areas planning.

Given that the "public interest" consists of a multitude of interests, how are these interests to be fairly represented? In what measure should interest groups strive for demographic versus interest representation in their memberships? Likewise, presuming that government agencies are seeking to increase public involvement, how do they structure opportunities for divergent populations and interests? How are foreign interests to be regarded in Japan's civic environmental infrastructure? Are some interests to be discounted altogether? These questions have arisen, for example, with regard to radical property rights organizations in the United States as well as extremist animal rights groups and the Earth Liberation Front. While some of these groups operate well outside conventional civic participation channels, this is not the case for all. Some organizations have been engaged in the participatory process to advance objectives that many environmentalists would find appalling. In Japan, this may be less the case – but questions might be raised about giving voice to potentially violent anti-airport groups, organizations that support unsustainable fishing practices, and groups that oppose immigration.

Takao (2001) calls for greater accountability for NGOs and for diverse assessment criteria. In the end, policy effectiveness – which itself evades simple determinations – would seem to be a laudable goal for civic environmentalism. Yet advocates of collaborative planning processes frequently will consider the enterprise successful simply if a diverse set of stakeholders is actively engaged (Mason, 2008). But this does not ensure equitable and environmentally beneficial outcomes. In comparing Japan, Germany, and the United States, Schreurs (2002) notes that the United States – home to the largest and wealthiest NGOs in the world – has taken the least domestic action on climate change. Among the multiple ways in which we might evaluate outcomes of civic engagement are carbon emissions (which some would argue favor nuclear energy), ecosystem protection and ecosystem services, environmental justice and equity, long-term institutional (NGOs as well as government) capacity building, and hazards resilience (Mason, 2008). Japan's evolving civic environmentalism will continue to provide us with valuable lessons about how we should proceed.

NOTES

1. The term NGO (non-governmental organization) generally is used in this chapter, with the term NPO (non-profit organization) used in reference to specific legislation or sources, as appropriate.
2. The Association was initially affiliated with the Environmental Agency. Soon after its establishment, the Japan National Trust was organized. "This small but well-funded organization, essentially a government interest group that provides amakudari ("descent from heaven") jobs for retired transportation industry executives, is an attempt by the Transportation Ministry to lessen the influence of the Environment Agency" (Mason, 1999, p. 205).

REFERENCES

Aldrich, D. (2008). *Site fights: Divisive facilities and civil society in Japan and the West*. Ithaca, NY: Cornell University Press.

Aldrich, D. (2011). Nuclear power's future in Japan and abroad: The Fukushima accident in social and political perspective. *Paris Tech Review*. Retrieved from www.paristech review.com/2011/08/25/nuclear-fukushima-accident-social-political-perspective/

Aldrich, D. (2013). Rethinking civil society-state relations in Japan after the Fukushima accident. *Polity*, 45(2), 249–264.

Apter, D. E., & Sawa, N. (1984). *Against the state: Politics and social protest in Japan*. Cambridge, MA: Harvard University Press.

Avenell, S. (2009). Civil society and the new civic movements in contemporary Japan: Convergence, collaboration, and transformation. *Journal of Japanese Studies*, *35*(2), 247–283.

Avenell, S. (2010). *Making Japanese citizens: Civil society and the mythology of the Shimin in postwar Japan*. Berkeley, CA: University of California Press.

Avenell, S. (2012a). From fearsome pollution to Fukushima: Environmental activism and the nuclear blind spot in contemporary Japan. *Environmental History*, *17*(2), 244–276.

Avenell, S. (2012b). Japan's long environmental sixties and the birth of a green Leviathan. *Japanese Studies*, *32*(3), 423–444.

Barrett, B. F. D., & Therivel, R. (1991). *Environmental policy and impact assessment in Japan*. London: Routledge.

Bird, W. (2013). In post-tsunami Japan, a push to rebuild coast in concrete. *Yale Environment 360*. Retrieved from http://e360.yale.edu/feature/in_post-tsunami_japan_a_push_to_rebuild_coast_in_concrete/2651/. Accessed on May 16.

Bix, H. (1986). *Peasant protest in Japan*. New Haven, CT: Yale University Press.

Broadbent, J. (1998). *Environmental politics in Japan: Networks of power and protest*. Cambridge: Cambridge University Press.

Broadbent, J. (2002). Japan's environmental regime: The political dynamics of change. In U. Desai (Ed.), *Environmental politics and policy in industrialized countries* (pp. 295–355). Cambridge, MA: MIT Press.

Chandler, C., Chhor, H., & Salsberg, B. (2011). *Reimagining Japan: The quest for a future that works*. San Francisco, CA: VIZ Media, LLC.

Claremont, Y. (2014). Disaster in Japan: A case study. In S. Butt, H. Nasu, & L. Nottage (Eds.), *Asia-pacific disaster management: Comparative and socio-legal perspectives* (pp. 79–99). Berlin: Springer-Verlag.

Curtis, G. (1997). A "recipe" for democratic development. *Journal of Democracy*, *8*(3), 139–145.

Dauvergne, P. (1997). *Shadows in the forest: Japan and the politics of timber in Southeast Asia*. Cambridge, MA: MIT Press.

Dobson, H. (2002). Social movements, civil society and democracy in Japan. In C. Kinnvall & K. Jönsson (Eds.), *Globalization and democratization in Asia: The construction of identity* (pp. 131–148). London: Routledge.

Editorial. (2012). Helping people help NPOs. *Japan Times*. Retrieved from http://www.japan times.co.jp/opinion/2012/05/18/editorials/helping-people-help-npos/. Accessed on May 18.

Fowler, L. (2003). From technical fix to regulatory mix: Japan's new environmental law. *Pacific Affairs*, *12*(2), 441–466.
Garon, S. (2003). From Meiji to Heisei: The state and civil society in Japan. In F. J. Schwartz & S. J. Pharr (Eds.), *The state of civil society in Japan* (pp. 42–62). Cambridge: Cambridge University Press.
George, T. S. (2001). *Minamata: Pollution and the struggle for democracy in postwar Japan*. Cambridge, MA: Harvard University Asia Center.
Gresser, J., Fujikura, K., & Morishima, A. (1981). *Environmental law in Japan*. Cambridge, MA: MIT Press.
Haddad, M. A. (2007). Transformation of Japan's civil society landscape. *Journal of East Asian Studies*, *7*(3), 413–437.
Hall, D. (2010). Japanese lessons and transnational forces: ODA and the environment. In D. Leheney & K. Warren (Eds.), *Japanese aid and the construction of the global environment: Inescapable solutions* (pp. 167–186). London: Routledge.
Hara, T. (2013). Social shaping of nuclear safety: Before and after the disaster. In R. Hindmarsh (Ed.), *Nuclear disaster at Fukushima Daiichi: Social, political and environmental issues* (pp. 22–40). New York, NY: Routledge.
Hasegawa, K. (2004). *Constructing civil society in Japan: Voices of environmental movements*. Melbourne: Trans Pacific Press.
Hasegawa, K. (2010). Collaborative environmentalism in Japan. In H. Vinken, Y. Nishimura, B. L. J. White, & M. Deguchi (Eds.), *Civic engagement in contemporary Japan: Established and emerging repertoires* (pp. 85–99). New York, NY: Springer.
Hirata, K. (2002). *Civil society in Japan: The growing role of NGOs in Tokyo's aid and development policy*. New York, NY: Palgrave Macmillan.
Imura, H. (2005). Japan's environmental policy: Institutions and the interplay of actors. In H. Imura, & M. Schreurs (Eds.), *Environmental policy in Japan* (pp. 49–85). Cheltenham, UK: Edward Elgar.
Japan Association of Charitable Organizations. (2013). *Charitable and non-profit organizations in Japan: Data book*. Tokyo: Japan Association of Charitable Organizations.
John, D. (1994). *Civic environmentalism: Alternatives to regulation in states and communities*. Washington, DC: CQ Press.
Johnson, C. (1995). *Japan: Who governs?: The rise of the developmental state*. New York, NY: W.W. Norton & Company.
Johnson, E. W., Saito, Y., & Nishikido, M. (2009). The organizational demography of Japanese environmentalism. *Sociological Inquiry*, *9*(4), 481–504.
Juraku, K. (2013). Social structure and nuclear power siting problems revealed. In R. Hindmarsh (Ed.), *Nuclear disaster at Fukushima Daiichi: Social, political and environmental issues* (pp. 41–56). New York, NY: Routledge.
Kawato, Y., Pekkanen, R., & Yutaka, T. (2012). Civil society and the triple disasters: Revealed strengths and weaknesses. In J. Kingston (Ed.), *Natural disaster and nuclear crisis in Japan: Response and recovery after Japan's 3/11* (pp. 78–93). London: Routledge.
Kawato, Y., Pekkanen, R. J., & Yamamoto, H. (2011). State and civil society in Japan. In A. Gaunder (Ed.), *Routledge handbook of Japanese politics* (pp. 117–129). London: Routledge.
Kihara, K. (1989). The emerging national trust movement. In S. Tsuru & H. Weidner (Eds.), *Environmental policy in Japan* (pp. 379–386). Berlin: Edition Sigma.

Kingston, J. (2012). The politics of natural disaster, nuclear crisis and recovery. In J. Kingston (Ed.), *Natural disaster and nuclear crisis in Japan: Response and recovery after Japan's 3/11* (pp. 188–206). London: Routledge.

Kingston, J. (2013). *Contemporary Japan: History, politics, and social change since the 1980s* (2nd ed.). Chichester, UK: Wiley-Blackwell.

Knight, C. (2010). The nature conservation movement in post-war Japan. *Environment and History*, *16*(3), 349–370.

Kraft, M. E., & Vig, N. J. (2013). Environmental policy over four decades. In N. J. Vig & M. E. Kraft (Eds.), *Environmental policy: New directions for the 21st century* (8th ed., pp. 2–29). Thousand Oaks, CA: CQ Press.

Kuroda, Y. (1972). Protest movements in Japan: A new politics. *Asian Survey*, *12*(11), 947–952.

Lam, P. E. (1999). *Green politics in Japan*. London: Routledge.

Lam, P. E. (2011). Japan's environmental politics and change: Local, national and global. In A. Gaunder (Ed.), *Routledge handbook of Japanese politics* (pp. 236–245). London: Routledge.

Mason, R. J. (1999). Whither Japan's environmental movement: An assessment of problems and prospects at the national level. *Pacific Affairs*, *72*(2), 187–207.

Mason, R. J. (2008). *Collaborative land use management: The quieter revolution in place-based planning*. Lanham, MD: Rowman Littlefield.

Mason, R. J. (forthcoming). Preservation and preemption in Northern Japan's Shirakami Sanchi world heritage area. *Management of Environmental Quality*. Retrieved from http://nova.campusguides.com/c.php?g=112070&p=726320

Matanle, P. (2011). The great east Japan earthquake, tsunami, and nuclear meltdown: Towards the (re)construction of a safe, sustainable, and compassionate society in Japan's shrinking regions. *Local Environment*, *16*(9), 823–847.

McCloskey, M. (1999). Local communities and the management of public forests. *Ecology Law Quarterly*, *25*(4), 624–629.

McCormack, G. (1996). *The emptiness of Japanese affluence*. Armonk, NY: M.E. Sharpe.

McKean, M. A. (1981). *Environmental protest and citizen politics in Japan*. Berkeley, CA: University of California Press.

McVeigh, B. J. (1998). *The nature of the Japanese state: Rationality and rituality*. London: Routledge.

Najita, T., & Koschmann, J. V. (Eds.). (1982). *Conflict in modern Japanese history*. Princeton, NJ: Princeton University Press.

Nimura, K. (1997). *The Ashio Riot of 1907: A social history of mining in Japan*. Durham, NC: Duke University Press.

OECD. (2010). *OECD environmental performance reviews: Japan*. Paris: OECD.

Ogawa, A. (2009). *The failure of civil society? The third sector and the state in contemporary Japan*. Albany, NY: SUNY Press.

Ohta, H. (2000). Japanese environmental foreign policy. In T. Inoguchi & P. Jain (Eds.), *Japanese foreign policy today* (pp. 96–121). New York, NY: Palgrave Macmillan.

Oyadomari, M. (1989). The rise and fall of the nature conservation movement in Japan. *Environmental Management*, *13*(1), 23–33.

Passin, H. (1962). The sources of protest in Japan. *American Political Science Review*, *56*(2), 391–403.

Pekkanen, R. (2000). Japan's new policies: The case of the NPO law. *Journal of Japanese Studies*, *26*(1), 111–148.

Pekkanen, R. (2006). *Japan's dual civil society: Members without advocates*. Stanford, CA: Stanford University Press.

Pekkanen, R. (2012). *The 2012 Japanese election paradox: How the LDP lost voters and won the election*. The National Bureau of Asian Research. Retrieved from http://www.nbr.org/research/activity.aspx?id=297

Putnam, R. D. (2000). *Bowling alone: The collapse and revival of American community*. New York, NY: Simon & Schuster.

Pye, L. W., & Pye, M. W. (1985). *Asian power and politics: The cultural dimensions of authority*. Cambridge, MA: Belknap Press.

Reimann, K. D. (2010). *The rise of Japanese NGOs: Activism from above*. London: Routledge.

Scalise, P. J. (2012). Hard choices: Japan's post-Fukushima energy policy in the twenty-first century. In J. Kingston (Ed.), *Natural disaster and nuclear crisis in Japan: Response and recovery after Japan's 3/11* (pp. 140–155). London: Routledge.

Schreurs, M. A. (2002). *Environmental politics in Japan, Germany, and the United States*. Cambridge: Cambridge University Press.

Schwartz, F. J. (2003). Introduction: Recognizing civil society in Japan. In F. J. Schwartz & S. J. Pharr (Eds.), *The state of civil society in Japan* (pp. 1–19). Cambridge: Cambridge University Press.

Shaw, R., & Goda, K. (2004). From disaster to sustainable civil society: The Kobe experience. *Disasters*, *28*(1), 16–40.

Shoji, K., & Sugai, M. (1992). The Ashio copper mine pollution case: The origins of environmental destruction. In J. Ui (Ed.), *Industrial pollution in Japan* (pp. 18–63). Tokyo: United Nations University Press.

Skocpol, T. (2003). *Diminished democracy: From membership to management in American civic life*. Norman, OK: University of Oklahoma Press.

Steinhoff, P. G. (1989). Protest and democracy. In T. Ishida & E. S. Krauss (Eds.), *Democracy in Japan* (pp. 171–198). Pittsburgh, PA: University of Pittsburgh Press.

Strong, K. (1977). *Ox against the storm: A biography of Tanaka Shozo, Japan's conservationist pioneer*. Vancouver, BC: University of British Columbia Press.

Sugai, M. (1992). The anti-nuclear power movement in Japan. In H. Krupp, et al. (Eds.), *Energy politics and Schumpeter dynamics: Japan's policy between short-term wealth and long-term global welfare* (pp. 286–306). Tokyo: Springer-Verlag.

Takao, Y. (2001). The rise of the "third sector" in Japan. *Asian Survey*, *41*(2), 290–309.

Tsuru, S. (1999). *The political economy of the environment: The case of Japan*. London: Athlone Press.

Ui, J. (1989). Anti-pollution movements. In S. Tsuru & H. Weidner (Eds.), *Environmental policy in Japan* (pp. 109–120). Berlin: Edition Sigma.

Ui, J. (1992). *Industrial pollution in Japan*. Tokyo: United Nations University Press.

Vlastos, S. (1986). *Peasant protests and uprisings in Tokugawa Japan*. Berkeley, CA: University of California Press.

Waley, P. (2005). Ruining and restoring rivers: The state and civil society in Japan. *Pacific Affairs*, *78*(2), 195–215.

Wolferen, K. V. (1993). *The enigma of Japanese power: People and politics in a stateless nation*. Tokyo: Charles E. Tuttle.

Woodall, B. (1996). *Japan under construction: Corruption, politics, and public works*. Berkeley, CA: University of California Press.

Yoshikazu, F. (2005). Case studies of environmental politics in Japan. In H. Imura & M. Schreurs (Eds.), *Environmental policy in Japan* (pp. 185–214). Cheltenham, UK: Edward Elgar.

ENVIRONMENTAL PROTEST IN IRELAND

Liam Leonard

ABSTRACT

Ireland's economic development not only provided many benefits but also caused difficulties locally and nationally. The economic crash and bailout, which occurred from 2008, have mired the country with huge debts. In addition, development created many environmental challenges in rural, suburban and urban areas. This chapter will examine community responses to those environmental challenges, discussing the mobilisation of the environmental campaigns that shaped the emergence of the Irish environmental movement.

Keywords: Environmental; green; movement; mobilization; community

INTRODUCTION

Ireland has witnessed a series of environmentally based protests which have pitted local communities against multinationals and the state in recent decades. This chapter will examine the significance of local campaigns in

Occupy the Earth: Global Environmental Movements
Advances in Sustainability and Environmental Justice, Volume 15, 63–77

ISSN: 2051-5030/doi:10.1108/S2051-503020140000015003

the communities of contemporary Ireland and argues that these have come to form a modern environmental movement. The chapter articles that in order to understand the basis for these events, an examination of the underlying concept of the 'local sentiment' is required. The chapter addresses these questions through an analysis of the environmental protests which have occurred in the Irish case.

For many, the changes that have occurred in rural Ireland over the last half century have mirrored a disengagement from the traditional patterns of life that had embedded a set of values and practices which allowed rural communities to co-exist with their surrounding environment. With the boom years of the 1990s and 2000s giving way to the austerity years of 2008 onwards, local communities and their environments have suffered episodes of accelerated development and subsequent financial neglect, as investment dwindles in the post-bailout era. With the onset of a technologically driven agri-business sector, mass production and scientisation drove a wedge between rural dwellers and their hinterland. During the 1990s, traditional farming would become synonymous with over production, fish kills derived from slurry spillages and images of European Union subsidies for non-production which stemmed from unsustainable practices. In the era of globalisation, local production for local markets came to be dismissed as small minded thinking. The damage caused to local interactions between communities and hinterlands was significant. The manner in which this trend is reversed due to the current global economic crisis remains to be seen.

One response to this loss of local identity was the concept of 'rural development' which involved community-based initiatives to reinvigorate local discourse in the face of outside challenges, both culturally and environmentally speaking. A combination of grievances relating to depopulation, unemployment and neglect of rural regions provided successive environmental campaigns with a groundswell of dissent to facilitate mobilisation processes against multinationals or the political establishment. When combined with the renewed confidence achieved by new middle class figures bolstered by expertise on rural rejuvenation programmes learned during migrant experiences abroad, a significant form of rural-based resistance emerged. This fundamental response came to be articulated around 'defence of space' territorial campaigns involving the imposition of industrial plants or infrastructural projects.

By addressing the political void which had opened around an increased sense of democratic deficit during the scandal ridden austerity era, rural-based environmental protests reclaimed a dominant sense of community through oppositional politics. This potent mixture of traditional values and local sentiment created a persuasive moral frame for many environmental campaigns to build on. Furthermore, the increased sense of confidence that collective mobilisation in common cause can re-establish a sense of cultural resistance to the globalised hegemony of development and consumerism at a time when uneven economic growth and uneven industrialisation threatened to create unsustainable imbalances between the rural and the urban and between developed sprawl and rural environmental conservation.

COMMUNITY POLITICS IN IRELAND

The backdrop to the emergence of environmental disputes in Ireland has been the development of a form of community politics which has its basis in a rural identity which is embedded in the Irish society. This identity has been born out of a traditionalist discourse which embraces local values over the modernisation projects of colonisers, state officials or European Union (EU) bureaucracy. This in turn has led to an instinctive mistrust of officialdom and technological change in rural communities. One result of this has been a growth of an identity-based community politics which challenges industrial policies and projects in the regions. This type of externalised community expression has taken shape in spite of the dependent nature of the industrialisation policies of the state, and illustrates a rural mindset that holds self sufficiency and local wisdom drawn from interaction with the hinterland in higher regard than the conventional wisdom of the representatives of politics or industry.

This form of 'rural fundamentalism' places an emphasis on localised structures that includes landowners, family-based farms and agrarian small-town life as the unifying component of social and political life in Ireland. While the lineage of this agrarian ideology pre-dates independence, the protectionist policies of post Independent governments and with a vision of a nation built on rural values have bolstered this perspective. The state's policies of modernisation of agriculture through the application of science or through promoting industrialisation served to further strengthen this rural fundamentalism over recent decades. A wider sense of community grievance also developed in response to incidents of environmental and resource contestation.

Extreme emigration was suffered in rural communities between the 1930s and 1980s. Over these decades, a sense of rural decline, due to depopulation and a perceived loss of traditional lifestyles led to a hardening of rural fundamentalism, allowing environmental challengers to construct grievance frames by drawing on the shared sense of injustice that had developed in rural communities. Seen in this light the sense of democratic deficit which existed in Ireland can be better understood, particularly in the context of an historical resistance to authority which had been at the heart of nationalist attempts to subvert the rule of authority, be it based in London, Dublin or even Strasbourg or Brussels.

In the absence of a strong political culture, a parochial form of populism emerged, which created a growing sense of civic-based political competence among citizens who wished to challenge the state. As mistrust of and participation in the formal political sphere went into decline, protest politics emerged as a vehicle for dissent in Ireland. Many issues such as civil rights for nationalists in Northern Ireland, equality for women or better economic conditions for farmers or trades unionists were characterised by public protests and marches across Ireland from the 1960s to the anti-globalisation marches of recent years. The cry of 'another world is possible' is as relevant in rural Ireland as it is in the cities which have witnessed the anti-globalisation and 'Occupy' movements globally.

Within a growing sector of a politicised civil society a reservoir of radicals, pacifists, nationalists and feminists was established, which provided both expertise and prior campaign experience that facilitated the mobilisation of environmental campaigns over the last 40 years in Ireland. The interpersonal contacts between these individuals and groups has created a network which can be mobilised around environmental issues, creating a strata of civil society which assists movements to emerge from their 'NIMBY' (not in my backyard) inception and go on to build more extensive campaigns.

It is important to note, however, that this stratum of politicised civil society is very much in the minority in Ireland. Invariably that society was characterised by a conservative political and social culture where values based on tradition, moralism and a paternalistic authoritarianism held sway over time. While recent years have witnessed more liberalised social norms alongside rapid economic growth, it is important to remember that the 'Celtic Tiger' growth phase of the 1990s has by-passed many rural areas and has remained a largely urban phenomenon. The state's attempts to modernise through industrial or infrastructural projects have been presented as key components of spreading an industrialised wealth base across the country which would stem the tide of emigration. And yet, in spite of historical depopulation and economic stagnation, rural communities have resisted industrial or infrastructural projects which they have deemed to be too much of a risk to the local population or environment.

ENVIRONMENTAL GROUPS AND POLITICS

The basis for their community-based environmental resistance has been non-party or non-political community groups. The social networks which were established across a series of co-operatives and community groups created the platforms for popular discourse which provided fertile ground for the germination of many community challenges in recent years. And while the state had attempted to reach out to communities through the 'partnership' model, the networks of civil society have also provided effective links for the dissemination of information that underscores the mobilisation process. Although the relationship between the state and community partnership groups has been weighted in favour of the authorities, the interaction between civil society and local and national government created a sense of efficiency and confidence amongst community leaders which has bolstered the leadership of environmental campaigns, under the guise of community development. While these individuals may have been involved with community development, they were more likely to be familiar with the marginalisation experienced in their locality.

The position of many community development groups who were directed by the state to develop tourism, agriculture and fisheries was often diametrically opposed to the infrastructural or industrialisation projects which the state wished to impose from above. The emergence of campaigns of rural regeneration, such as the 'Save the West' movement or Irish language campaigns, have further served to create variants of rural fundamental discourse. In the late 1980s and early 1990s, community politics gave rise to the 'single issue candidate' based on the mobilisation of campaigns highlighting the need for or loss of local services, and the political potency of the single issue or independent candidate has been shown to be significantly increased in elections held after the unpopular fiscal bailout and austerity-driven policies became prevalent from 2012 onwards. Conversely, the Irish Green Party entered government as a minority partner in 2007 and was later decimated by voters angered by the economic collapse and austerity which followed in 2008.

In many ways, the Green's emergence into government can be seen as an indicator of the shifts that have occurred in Irish society since the economic boom came to an end. Existing conjunctions between the green movement and its representative party have been stressed by participation in the government, particularly by the coalition with neo-liberal parties. The Greens emerged from the margins at a time when Ireland experienced accelerated growth, transforming the natural and political landscape. The global economic downturn could not have been anticipated by the party leadership when they entered government. (The imbalanced nature of Irish coalitions may also be a factor in the difficulties experienced by some junior government partners.).

The emergence of the Irish Green Party from a coalition of community protesters into a coalition partner in government was one of the more interesting political events in the history of the nation's environmental movement. But the Greens' associations with rural community groups came to be diminished as part of their pursuit of power; the party put its base at risk when it failed to maintain links with those rural campaigns that form the basis of the wider environmental movement, a fact that was borne out in their subsequent electoral demise.

ENVIRONMENTAL MOVEMENT STRATEGY

Strategically, rural community groups attempt to transplant their rural discourses into a legal setting such as the oral hearing or legal challenge, with mixed results. The inability to translate rural sentiment into legal efficacy stymied many environmental campaigns yet the legal route was retained by many campaigners as the primary external strategy. For many campaigns, the ability to create links with influential allies from elite groupings became an essential component of the wider mobilisation process. By obtaining support from a significant political sponsor, community groups could attempt to gain access to the formal political structures of the state and therefore attempt to influence policy makers. By exploiting clientelist political arrangements in this way, community groups hoped to apply pressure on the state.

When local elites are linked with grassroots groups, the combination of 'top-down' and 'bottom up' goes towards creating effective community responses which can build a sense of local populist resistance to both state and corporate entities. This local opposition can draw on rural discourse which can be found in even most non-political community and voluntary groups which form the basis of Irish life in rural areas.

However, a number of factors have traditionally militated against outbreaks of widespread collective action in rural areas. These variables have their basis in a persuasive sense of conservatism that restricted a person's or groups' ability to respond to incidents or disputes. These factors included the intricate set of relationships that existed around property and land ownership as the most significant established institutional arrangement in the regions. Relationships are built around land ownership, so resource disputes impacted upon the very basis of community interaction. This stratified social structure also militated against wholesale class-based collective action and the resultant emphasis on individualism and independence reduced mobilisation potentials, as did practices of clientelism and political brokerage between individuals Accordingly, only the cultivation of grievance frames built on rural fundamentalist discourse facilitated the establishment of a template for community mobilisation against projects or policy initiatives, based on the perception of such projects as an outsider which exploits rural communities, destroys community values and degrades the tranquil beauty of the Irish countryside. The dichotomy between the state's attempt to reduce dependency in rural peripheralities by promoting local self-reliance and the manner in which industrialisation or infrastructural projects were imposed 'from above' also provided campaigners with a basis to construct grievance frames by drawing on resultant societal strain.

COMMUNITY DISSENT

The pool of discontent from which many Irish environmental groups draw support can be understood as part of a wider grievance frame which has emerged from historical circumstance. Culturally, the articulation of rural sentiment has been intertwined with the many versions of nationalism that transpired in opposition to centuries of colonial rule.

Inevitably, the enduring hardship and poverty which had come to characterise life in many rural communities since the imposition of austerity has led to the huge rates of emigration from rural areas and the West of Ireland in particular, added to the sense of societal loss which left many rural communities concerned about the possibility of complete demographic and cultural extinction. Moreover, the deeply rooted sense of connectedness that defined rural communities through their hinterland created a strong sense of what Benedict Anderson has called 'the imagined community', which includes local communities and their diaspora. Within this fulcrum of grievance, folklore and opposition, a sense of place emerged and a set of fundamental beliefs have been cultivated alongside the ditches and hedgerows of the West of Ireland.

This malaise had historical origins. In the years after independence, this sense of parochial grievance was strengthened by what was perceived as the heavy hand of officialdom. The state attempted to impose modernity on rural communities which used traditional methods of production. This model of development emphasises science and technology as a way of maximising the market potential of agricultural produce, with a view towards exports. In order to achieve a modern agri-business sector, the state set out to provide grants and technical assistance which would affect a transition from smaller family-based farms with an emotive attachment to land, towards more 'efficient' larger farms which could service a globalised food-processing industry. One result of these changes was a decline in the numbers involved in agriculture with countless thousands taking the emigrant boat to find work abroad. Viewed in this context, the state's attempts to generate employment through multinational led development were perceived with a degree of scepticism and hostility, even before pollution or health risks were discussed. The economic collapse since 2008 has served to compound these difficulties.

The grievances of rural communities which emerged from this perception of a loss of community have contributed to the growth of populist environmentalism in Ireland, based on the 'rural sentiment' or the collective resistance borne of the community's inherent understanding of their place in the hinterland. Populist environmentalism has been manifested as part of the rural discourse which was a characteristic of anti-toxics and anti-multinational disputes during the pre-boom decades of the 1970's and 1980's. Populist 'rural sentiment' had also been a feature of many of the anti-mining and resource disputes which occurred during the same period. It is quite interesting to see the re-emergence of rural populist discourse during the ongoing dispute about the Corrib gas onshore gas pipeline involving the 'Rossport Five' as calls for local ownership of local resources were combined with concerns about the risk posed by onshore pipelines as part of the framing strategies of that campaign.

The existence of strong parochial rural sentiment has been cited as a factor in the lack of acceptance of official environmental organisations in country-based disputes as communities attempted to mobilise grievance based on local understandings and relationships. However, this localism has also left many populist environmental groups open to the accusation of being mere NIMBYists ('Not in my backyard') as opposition to industrial or infrastructural projects in a communities 'backyard' is identified as the primary rallying point for campaign mobilisation. Further challenges for populist campaigns such as the problems of translating rural discourses as part of normal legal hearings have also been identified.

Essentially, the well of grievance which provides much of the underlying discontent for populist campaigns to exploit is the basis for an understanding of exactly how the various environmental campaigns that have occurred over recent decades can be characterised as components of an overall social movement. While populist environmental campaigns may wax and wane, the significance of each campaign's contribution to an articulation of community grievance has created a movement of sorts, where outcomes can be measured through an understanding of the extent to which populist fundamentalism has come to be seen as the very basis for traditional rural identities in the post-consumption, post-modern era. This is an outcome that can be measured as part of the social capital of all rural and rural-urban communities and which has far greater significance than the outcome measurement models which chart the impact of protest campaigns on policy implementation. The true measurement of the impact of rural populist discourse goes beyond moments of access to political opportunity structures and contributes to the shape and nature of the populist Irish political system itself.

RURAL ENVIRONMENTALISM

There are many definitions of what the term 'rural' means. By and large, rural life is understood through an existence in a hinterland, defined by a relationship with the land, through agricultural production, or traditional cultural values and discourses. In the Irish case, rurality has also been associated with spatial location and community, as the communities west of the Shannon River came to define themselves through their distance from the developed core of society. This sense of local embeddedness and hostility to the ruling elite can be traced back through history in rural areas. It remains a pivotal feature of the ongoing recreation of 'societal grievance or community consciousness', the basic opposition to officialdom due to an emergent sense of both ontological insecurity and structural motivation which stokes the embers of rural discourse, which Irish environmentalism exploits. Invariably, it has been the gradual translation of agrarian discourse into a form of environmentalism shaped by new middle class advocates which has influenced the emergence of a wider eco-consciousness in Irish society in recent years. This motivation can be traced through a succession of Irish environmental campaigns over the past 40 years.

Over time, a pattern of rural resistance to the onset of globalised development becomes discernable in the Irish case. We can see that an overall combination of grievances around perceived threats to traditional processes and identities led to a growing sense of resistance which in many cases surfaced around environmentally based contests and disputes. In many ways, the culture of social and political acquiescence which permeated life in the years after independence gave way to a community-based political articulacy which sought to interrogate the grand narratives of industry or the state in the same manner as had been undertaken around issues of equality and religion across the nation. Moreover, at a time when the cycles of economic growth and recession divided society through emigration and poverty in the pre-boom years or immigration and accelerated growth in the post 'Celtic Tiger' era, the values embedded in traditionalism rural mobilisation and concern for heritage became appealing and achievable for beleaguered communities when faced with the threat from 'outsiders', be they industrial or institutional.

THE IRISH ENVIRONMENTAL MOVEMENT

Of course, the student activist who took part in organising participation in the Carnsore Point anti-nuclear protests in the late 1970s has emerged in adulthood as the concerned citizen protesting against the health risks of incineration for his/her children. In many ways the generation that has their feet planted on either side of the millennium have grown up with and continued to be involved in environmental discourse as a facet of their lives, at a time when emigration and immigration, political scandal and economic success have been encountered. During the previous 40 years, environmentalism became part of the discourse of rights issues that have shaped modern Ireland. Many of the leaders of this generation were women, such as Mary Robinson, who took part in protests against the destruction of Georgian Dublin during her student days and who later became President of Ireland. Others, such as Petra Kelly, would go on to become leading figures in European environmentalism, with Kelly becoming a founder of the German Greens in her native Germany. Many others were teachers or raising families, but all were moved to protest in defence of their environments and communities. A strand of eco-feminism can be traced throughout the successive campaigns that have come to make up the Irish Environmental Movement; this allowed that discourse to be more representative while locating it within the contexts of Celtic mysticism and a deep green connection between the cycles of nature, both of which inform the prevailing 'new age' attitudes now commonplace in alternative culture.

Another facet of the new environmentalism in modern Ireland has been the emergence of risk society and toxicity as a new issue of concern in our daily lives. With the onset of regular air travel to and from the United States, EPA 'Superfund' regulations, and a series of low tax incentives in Ireland, American multinationals began to set up across the country, primarily in rural areas. In the aftermath of the Love Canal and Three Mile Island toxics scandals in the United States, issues surrounding nuclear or toxic pollution came to be a part of modern culture. As multinationals availed of the lax regulatory regime in Ireland, communities grew increasingly concerned about the costs to human health and the environment in the vicinity of established or proposed industrial plants. At the same time, a culture of oppositional politics has become an integral part of civil society in Ireland. Student, feminist and republican protests were routinely featured in the media and on the television news. These new social movements also provided a pool of activists who also found common cause with community and environmental campaigners.

ENVIRONMENTAL PROTESTS IN IRELAND

Having enjoyed the benefits of economic growth associated with the 'Celtic Tiger' boom of the last decade, communities in Ireland are beginning to witness the downside of the property boom. Infrastructural projects came to be perceived as threats to locals and their health and environment, or to the heritage of the nation itself. Offshoots of rapid development such as hyperconsumption, a buoyant property market and increased car ownership led to further demand for roads, waste management sites and water treatment facilities. Such was the link between growth and infrastructure that those communities which voiced concerns about projects were deemed to be backward thinking and against progress. The first of these collective responses emerged in the wake of the attempt to build onshore pipelines on the west coast of County Mayo for the natural gas of the Corrib Field. Local farmers objected to this proposal, and resisted attempts by the multinational involved to gain access to their land. As a result, five local men were imprisoned for 94 days for refusing to agree to an injunction that would allow the agents of the multinational access to their property. What followed was a campaign that gripped the nation's imagination, as 'Shell to Sea' protests sprung up across the island and beyond in response to the community's demand that the gas be processed offshore. The imprisoned men's traditional lifestyle of farming in an Irish-speaking Gaeltacht community represented to many observers an authentic and traditional way of life that had been abandoned in the rush to embrace the consumption-fuelled prosperity which had become prevalent in recent years. The fact that the Mayo farmers put family and community above profit or gain provided many observers with an indication of the wider context that had become the subtext of this dispute; that Ireland had, in some ways, sold its soul in return for multinational investment and growth. The men were released from prison in the summer of 2006, and their campaign for justice continues.

Accelerated growth in Ireland has led to increased waste disposal development. The bourgeoning waste crisis, which has emerged in the aftermath of increased consumption associated with accelerated growth, has led to community action. Ireland had fallen foul of European directives on waste management, due to an over-reliance of up to 93 per cent on landfill. In order to deal with this crisis, waste policies were introduced with plans for regional incinerators across the country. Previous attempts to introduce an all-island incinerator in Northern Ireland were opposed by both nationalists and unionists together. As a result of the networks created from the Northern campaign, communities mobilised in regions such as Galway, Cork and Meath, and more latterly in Dublin. These regional campaigns benefited from links with the global anti-incinerator and 'zero-waste' campaigns, and from the support of experts from the United States and Britain. As a result, the state's regional waste policy was challenged, as communities began to inform themselves of the dangers posed by the dioxins and furans that are released into the air and food chain from incinerator emissions. Nonetheless, there were positive outcomes to the community engagement with the waste issue, and regional recycling was introduced in many areas with great success. Last year, the construction of housing, which had come to provide the second cycle of the 'Celtic Tiger' boom, caused another environmental crisis when the mass construction of new houses in rural areas led to the contamination of water supplies with effluent from archaic treatment plants. Concerns are also being expressed about the possible introduction of fracking in rural areas, and a campaign of opposition to this environmentally destructive process has been mobilised across Ireland.

CONCLUSION

Ultimately, the costs of economic growth have led to greater environmental degradation and an undermining of significant aspects of the nation's heritage. In particular, rural communities have borne the cost of Ireland's transition from a rural agrarian society into an industrialised and globalised economy. While communities have demonstrated a commitment to retaining some aspect of the traditional values and heritage of the 'Emerald Isle', the state's policy of prioritising rapid infrastructural development in order to underpin private consumption and multinational investment over local community services proved to be ill advised as fluctuations on the global markets led to financial collapse and austerity measures being imposed by the International Monetary Fund and the European Central Bank. As the cycle of the country's financial collapse and the 'credit crunch' continues to take its toll, the values and concerns of these local communities may provide a new direction for the future of Ireland's communities.

REFERENCES

Leonard, L. (2005). *Politics inflamed: GSE and the campaign against incineration in Galway* (Vol. 1). Irish Greenhouse Press Ecopolitics Series. Drogheda: Choice Publishing with Greenhouse Press.

Leonard, L. (2006a). *Green nation: The Irish environmental movement in Ireland from Carnsore point to the rossport five* (Vol. 2). Irish Greenhouse Press Ecopolitics Series. Drogheda: Choice Publishing with Greenhouse Press.

Leonard, L. (2006b). The moral framing of a resource dispute. *Studies Irish Quarterly Review*, (Winter).

Leonard, L. (2010a). A challenging transformation: The Irish green's first year in power. *Irish Journal of Sociology*.

Leonard, L. (2010b). Green parties, anti-toxics movements, environmental movements. 3 entries for the SAGE Encyclopedia of the Environment.

Leonard, L. (2010c). Sustainable community development. Entry for the Berkshire Encyclopedia of Sustainability and Business.

Leonard, L. (2010d). The Galway water crisis of 2007. In M. Corcoran & P. Share (Eds.), *Sociological chronicles* (p. 5). Dublin: Liffey Press.

Leonard, L. (2011). *Community campaigns for sustainable living: Health, waste & protest in civil society* (Vol. 8). Advances in Ecopolitics Advances in Ecopolitics Series. Bingley, UK: Emerald Group Publishing Limited.

Leonard, L. (2013a). The political background to and social impact of Ireland's "Economic Bailout". In C. A. Lewis (Ed.), *Ireland: Economic, political and social issues*. New York, NY: NOVA Publishing.

Leonard, L. (2013b). Ecomodern discourse and localized narratives: Waste policy, community mobilization and governmentality in Ireland. In M. J. Zapata & M. Hall (Ed), *Organising waste in the city: International perspectives on narratives and practices.* London: Policy Press.

Leonard, L. (2013c). Cultural criminology, governmentality and the liquidity of the failing state: The view from Ireland. *Critical criminology.*

Leonard, L., & Barry, J. (Eds.). (2009). *The transition to sustainable living* (Vol. 4). Advances in Ecopolitics Series. Bingley, UK: Emerald Group Publishing Limited.

Leonard, L., & Botetgias, I. (2011). Introduction. In *Sustainable politics & the crisis of the peripheries: Ireland & Greece* (p. 8). AiE. Bingley, UK: Emerald Group Publishing Limited.

Leonard, L., & Botetzagias, L. (Eds.). (2012). *Sustainable politics and the crisis of the peripheries: Ireland & Greece* (Vol. 7). Advances in Ecopolitics Advances in Ecopolitics Series. Bingley/Bradford, UK: Emerald Group Publishing Limited.

Leonard, L., Doran, P., & Fagan, H. (2009). The cross-border burning issue. *Environmental movements & waste infrastructure* (*special edition Environmental Politics Journal*), December.

Leonard, L., Doran, P., & Fagan, H. (2010). The cross-border burning issue. *Environmental movements & waste infrastructure.* London: Routledge.

Leonard, L., & Kenny, P. (2011). Power, corruption and lies: Irish economic policy, 1922–2011. In *Sustainable politics & the crisis of the peripheries: Ireland & Greece*, AiE 8. Bingley, UK: Emerald Group Publishing Limited.

Rootes, C., & Leonard, L. (Eds.). (2010). *Environnimental movements & waste infrastructure.* London: Routledge.

FURTHER READING

Andersen, B. (1983). *Imagined communities: Reflections on the origin and spread of nationalism.* New York, NY: Verso.

Leonard, L. (2007a). Ecotopias, identity & place ecopolitics. *Online Journal, 1*(1), (Winter).

Leonard, L. (2007b). Environmentalism in Ireland: Ecomodernism versus ecopopulism. *Environmental Values*, (Winter).

Leonard, L. (2007c). Towards an understanding of a territorial resource dispute: The "shell to sea" campaign in north Mayo. *Irish Journal of Sociology.*

Leonard, L. (2007d). Environmental movements and internet technologies. *Irish Communications Review.*

Leonard, L. (2007e). The Galway water crisis. *Studies Irish Quarterly Review*, (Winter).

Leonard, L. (2008a). Rapid development & community mobilization in the republic of Ireland. *Community Development: Journal of the Community Development Society.* 3(2).

Leonard, L. (2008b). The political economy of growth and green politics in Ireland. International Journal of Green Economics.

Leonard, L. (2008c). Profile: Dancing with the devil or playing for power: The Irish greens in the 2007 general election. *Environmental Politics*, (Winter).

Leonard, L. (2008d). Commentary: The Irish green party and the referendum for the EU reform treaty of Lisbon. Ecopolitics Online Journal. 1(2).

Leonard, L. (2008e). *The environmental movement in Ireland.* Dordrecht: Springer.

Leonard, L. (2009a). Contesting the Irish countryside: Rural sentiment, public space and identity in Ireland. *Nature and Culture Journal*.

Leonard, L. (2009b). Keeping the rural in sight: The future of the environmental movement in Ireland. *New Hibernian Review*, (Summer).

Leonard, L. (2009c). Ireland's environmental policy. *Critical Social Policy*.

Leonard, L. (2013). The water crisis in Ireland: The socio-political contexts of risk in contemporary society. In J. Wagner (Ed.), *The secret life of water*. New York, NY: Berghahn Books.

Leonard, L. (2014). The framing processes and outcomes of environment and conservation campaigns in Ireland. In T. Doyle & S. McGregor (Eds.), *Environmental movements around the world: Shades of green in politics and culture*. Santa Barbara, CA: ABC-Clio/Praeger.

Leonard, L., & Allen, K. (2010). Introduction: Special issue on civil society. *Irish Journal of Sociology, 18*(2).

Rootes, C., & Leonard, L. (2010). Environmental movements and campaigns against waste infrastructure in the United States. In C. Rootes & L. Leonard (Eds.), *Environmental movements & waste infrastructure*. London: Routledge.

LOCATING ENVIRONMENTAL KNOWLEDGE IN ANTIPOLLUTION MOVEMENTS OF NORTHERN INDIA

Sya Buryn Kedzior

ABSTRACT

Recent years have produced significant demand for geographical contributions to the study of social movements in general and of environmental social movement organizations (ESMOs) in particular. Geographical approaches to the study of ESMOs emphasize "the mediation of social movement agency by place" (Miller, 2000; Routledge, 1993) and call attention to the role of place-based environmental knowledge (EK) in the broader "struggle(s) over meaning" that increasingly constitute environmental politics (Buechler, 1997; Escobar, 1992; Rangan, 2000; Watts, 1990). My chapter responds to this call by providing an examination of the reproduction of EK by antipollution organizations in India's central Ganges River Basin (GRB). Through interviews with organization leaders and members, along with analysis of organizational websites and publications, I examine the EK of two key antipollution organizations in the GRB: The Sankat Mochan Foundation (SMF) and Kanpur

Occupy the Earth: Global Environmental Movements
Advances in Sustainability and Environmental Justice, Volume 15, 79–115

ISSN: 2051-5030/doi:10.1108/S2051-503020140000015005

Eco-Friends (KEF). Analysis focuses on methods of knowledge reproduction employed by each organization, their respective framing practices, and the localized natures of the EK they reproduce. I argue that each organization works to reproduce a specific and place-based understanding of pollution in the GRB that informs their framing of the pollution problem, the tactical activities in which their members engage, and the power relations that exist between the two organizations and their leaders. Further, I argue that engaging with EK as both a method of understanding pollution and a tactic for consolidating political power is essential to making sense of the relative success of these movement organizations and the challenges they face in trying to build a broader coalition and mass-mobilization against pollution in the Ganges.

Keywords: Pollution; India; environmental knowledge; geography

The geographies of a social movement organization are reflected in where, when, and how (as well as whether) a movement forms, how they frame the issue(s) they seek to address, how they formulate specific tactics, and whether (or not) their efforts meet with success. Indeed, geographers have been drawing attention to the significance of social movement context for the last few decades, arguing that "social movement processes ... are constituted through space, place, and scale, and that constitution affects how they interact, articulate, and play out" (Miller, 2000, p. 166). Nowhere is this more apparent than with environmental social movement organizations (ESMOs), which often form in response to context-specific, if not localized, environmental degradation or disturbance of society-environment inter-relationships and subsequent patterns of resource use. This chapter examines the geographies of the emergent antipollution movement forming along the banks of India's Ganges (*Ganga*) river, paying particular attention to the contexts in which key ESMOs formed and articulate their agendas.

Through an analysis of the SMF and the Kanpur Eco-Friends (KEF), I consider how two antipollution organizations differently carve out space for widening discursive and material struggles around pollution and pollution abatement. My analysis focuses on the environmental knowledge (EK) (re-)production and contestation activities that seek to both mobilize public opinion in support of particular approaches to pollution abatement and, perhaps more significantly, to inspire change in local water use activities and the notions of social relationships with the river on which they are based. In doing so, I draw primarily from geographical approaches to social movement studies (Miller, 2000; Routledge, 1993) and the recent work on "struggles over meaning" (Buechler, 1997; Escobar, 1992; Rangan, 2000; Watts, 1990, 1997) and "environmental knowledge" (Goodman, Boykoff, & Evered, 2008; Robbins, 2000). The key goals of this discussion are, first, to explore the pollution knowledge being adopted, developed, reproduced, and employed by antipollution ESMOs mobilizing in two very different sites on the river, and, second, to examine how these diverse pollution knowledges shape the agendas and activities of each organization, reflecting both the wider context of antipollution politics in the GRB and the specificities of place in which these politics are grounded.

The following analysis draws principally upon interviews with founders of the SMF and KEF, Dr. Veer Bhadra Mishra and Rakesh Jaiswal, as well as handful of staff, members, volunteers, and local activists. These data are informed by a textual analysis of documents produced by the organizations, including organizational and program websites, brochures, and reports. Other texts include newspaper and magazine articles profiling the organizations. Analysis of these materials focused on the discursive productions of each organization, the tactics or "cultural idioms of protest" they employed (Routledge, 1993), and the stories about historical eco-social practices and relationships or environmental social movements themselves (see Rangan, 2000; Sivamakrishan, 1995). This analysis aimed to develop an understanding of how organizational discourses, tactics, and stories are used to create or reproduce EK and therefore social meaning, to contest existing power relations, to shift the social relations that influence access to natural resources (Eriksson, 2000; Walker, 1998), and "to alter cultural and material practices and relations of power within [a particular] spatial and political context" (Rangan, 2000, p. 41).

GEOGRAPHIES OF KNOWLEDGE AND MEANING IN ENVIRONMENTAL SOCIAL MOVEMENT ORGANIZATIONS

Geographical studies of social movements differ from their sociological counterparts in the emphasis placed on place, space, and movement context (Routledge, 1993, p. xv). In other words, geographers are less concerned than their sociological counterparts with the internal dynamics of social movement organizations (SMOs) and more concerned with understanding why movements occur where (and when) they do, how political resistance is shaped by the particular contexts in which it emerges, and, in turn, how these struggles reshape the spaces in which they occur. It is in this attention to the place-based nature of social movements that geographers find the most ground for contributing to the study of ESMOs. When allied with geographical and geographically informed political ecological approaches to the study of society-environment inter-relationships, geographical social movement studies have much to say about the context-dependent nature of ESMOs (Routledge, 1993). Here, geographers draw attention to the ways in which the ecological dimensions of environmental degradation influence the formation and characteristics of environmental movements, or how society-environment inter-relations grounded in the particularity of place shape the occurrence of struggles at specific times and places. It is this attention to how social movement agency is shaped by the particularity of place-based inter-relationships between society and their environments, and in turn how social movement organizations attempt to affect change in those relationships, that is of particular interest in this chapter.

In examining the efforts of ESMOs attempting to affect and even reconstitute local social relationships with their environments, geographers engage significantly with work on social movements and political contestation as "struggles over meaning" (Escobar, 1992) and from political ecology on "environmental knowledge" (Goodman et al., 2008; Robbins, 2000). The former body of work is grounded in sociological social movement theory, where environmental movements are grouped with other "new" cultural movements that seek to affect change in cultural meaning and social practice (Buechler, 1997). Shared social meaning is here understood to form the basis upon which common responses to events and phenomena (such as environmental degradation) are built, and which form the basis of human societies (Sederberg, 1984). These meanings constitute a "reality that confronts us as external to and independent of ourselves ... turn[ing] back upon us and shap[ing] our responses" (Sederberg, 1984, pp. 4–5). The struggle to control the processes through which social meaning is constructed is ultimately a struggle to influence how people respond to environmental or social events and what actions are deemed appropriate in the face of environmental and social change. ESMOs are viewed as protracted responses to environmental degradation that engage in these "struggles over meaning" through forms of organization and protest in order to influence the terms under which social relationships with particular resources, or with the "environment" more generally, are determined and access to those resources are mediated (Walker, 1998; Watts, 1990, 1997).

This emphasis on the centrality of social meaning in larger struggles over political power coincided with the development of the latter body of work on "environmental knowledge" in geography and political ecology.[1] EK includes not only ideas and information about the natural environment but also the complex and often contested social and material processes through which these understandings are constructed and reproduced (Goodman et al., 2008). EK therefore shapes how individuals, social groups, institutions, governments, and other organizations explain the natural world around them, how they identify and assign value to various environmental attributes, how they judge appropriate relationships with the natural world, how they account for evidence of environmental change, and how they act upon that thinking (Robbins, 2000). Emerging out of the constructionist turn in geography, work on EK draws attention to the complex socio-political contexts that produce multiple contemporaneous (and often conflicting) methods of understanding and interpreting the various inter-relationships between human societies and their natural environments and resources. Advocates argue that it is this consideration of "multiple socially constructed rationalities" that lends the approach strength and allows for more "detailed and embedded accounts of environmental problems and public concern" by recognizing the diverse ways of knowing, structuring, and acting upon ideas (Walker, 2005, p. 49, 62).

While many EK studies focus on explaining "how certain accounts of environmental process became dominant and to what effect" (Robbins, 2000, p. 120), others interrogate the broader *politics of knowledge*. The emphasis here is less on the content of knowledge and more on the contexts in which knowledge is produced, through which it gains social purchase and is deemed valid, and (perhaps most importantly) through which knowledge is contested and reproduced. We see here the natural overlap between work on the politics of EK, which examine how ESMOs work to reproduce particular ways of understanding society-environment relations and acting upon that knowledge, and the study of "struggles over meaning," which examine ESMOs as one of many actors engaged in a larger struggle over how societies attribute meaning to their environments and formulate common social practices based on those meanings. Both of these bodies of literature draw inspiration from "Foucauldian notions of the relationship between power and knowledge," which call for attention to how knowledge is created and disseminated through an "economy of discourse" and how it functions as a form of power (Foucault, 1980; McEwan, 1998, p. 372). I draw upon both of these approaches in the analysis presented in the following pages, where I examine how two antipollution organizations in India's central Ganges River Basin (GRB) seek to reproduce particular environmental knowledges or way of knowing pollution in the river that reflect the relatively localized context of the city in which each organization works. These contexts produce distinct ways of explaining and framing the problem of pollution, of valuing water and human social relationships with the river, of engaging with other organizations and members of their communities, and of proscribing various methods of engaging with pollution. Understanding these organizational EKs helps make sense of the shape that environmental issues take in various social settings and why responses to environmental problems change over time, space, and place. Fundamentally, they also elucidate the broader dynamics of the struggles over meaning associated with pollution in the GRB, where antipollution ESMOs contest and partner with state, religious, civic, and other interests to influence how people make sense of pollution in order to exercise power over the broader politics of pollution abatement.

ANTIPOLLUTION ORGANIZATION IN THE GRB

Flowing across the north Indian plain, the Ganges is the largest and arguably most significant river in India. Its river basin supports a population of nearly 800 million people and has served as the cultural, economic, agricultural, and industrial center of the nation for much of recorded history (Chapman, 1995). Rising rates of pollution in the Ganges River received much attention from domestic and international media since the mid-1980s, when the nation's first large-scale river water pollution abatement program was launched. This Ganga Action Plan (GAP) sought to address rising levels of industrial and municipal pollution in the river and was launched in a manner sensitive to the religious significance of the river and the devotional respect afforded to Ganga in Hindu religious dogma. Ganga is the most sacred of Indian Rivers and the goddess of purity and purification in the Hindu religious pantheon. To wade on her banks, swim in her streams, or even to take a drink of containerized *ganga jal* (Ganges water), is to be cleansed by the waters of the Gods, sent to earth to rid humankind of evil and impurity. Images of Ganga are therefore intimately intertwined with ideas of cleanliness, filth, and pollution. Many religious practices reflect this understanding, from the ritual bathing and drinking of river water performed by pilgrims, to the practice of placing (often partially) cremated bodies in the river to insure liberation of the soul.

For many believers, the river cannot be polluted, as the goddess bears the power to purify any person or material that enters her stream. Yet, there is a clear and noticeable disjuncture between this belief and the tangible evidence of river water pollution in the central basin, where water quality falls far short of that recommended for either drinking or bathing at most testing sites (Ministry of Environment and Forests, 2005). Major sources of pollution in the river are untreated sewage and effluents from leather tanneries, soap factories, and other industries, which cause and exacerbate rashes and yellowed patches of skin, eyesight problems, cholera, dysentery, typhoid, intestinal worms, trachoma, and diarrhea[2] (Sengupta, 2006; WHO, 2004). Despite the investment of millions of dollars the initial GAP abatement efforts and their successors, GAP-II, and the National Ganga River Basin Authority (NGRBA), water quality in the central GRB is decreasing at many sites and government-sponsored policies and programs are frequently critiqued as ineffectual, unsuitable or even corrupt (*The Hindu,* 2004). In addition to instances of protests at major religious festivals, a small number of social movement organizations have mobilized in response to ongoing grievances with poor water quality and ineffective river water management. Key among these are the SMF of Varanasi and the Kanpur Eco-Friends (KEF) of Kanpur (see Fig. 1).

Varanasi is one of the holiest cities in India and one of the most auspicious sites for devotees to die, be cremated, or bathe in the Ganges.

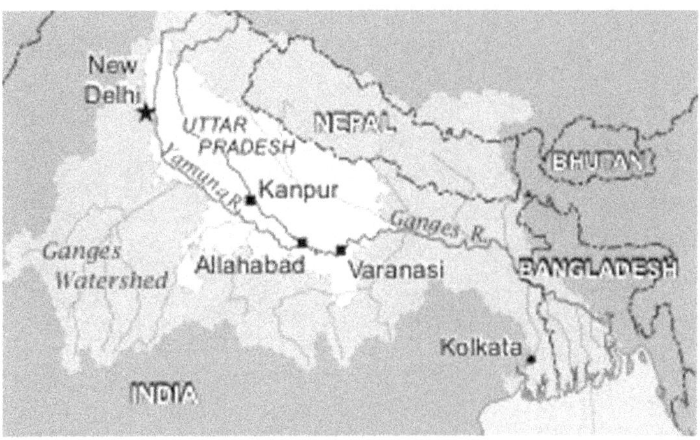

Fig. 1. Ganges River Watershed. *Source*: Modified from original map by Jeff Levy, University of Kentucky.

Varanasi is widely known as a city of worship, learning, literature, and light. But, more than anything else, the city is associated with the Ganges River. Images of the city highlight the *ghats* or stairs that line nearly the entire waterfront of the city, the boats that line up to carry passengers into and across the river, and the numerous temples that make up the city skyline (see Fig. 2). Tourism is the single most important industry in Varanasi, bringing more than 1.4 domestic and 125,000 foreign overnight stay tourists into the city each year (Ministry of Tourism, 2006). Most of these come to see the Ganges, one of the city's many temples or to participate in arts and cultural activities. The SMF takes advantage of this affiliation to promote the city as the primary site for antipollution abatement efforts and investment. The SMF is a quasi-religious organization that seeks "to restore the [Ganges River] to its pristine purity and glory by preventing [the] discharge of pollutants into it" (SMF, 2006). The SMF is the best known and most publicized organization currently fighting against pollution of the Ganges river. The SMF attempts to combine scientific and Hindu religious interpretations of river water pollution under the instruction of Dr. Veer Bhadra Mishra, former Professor of Hydrology and current Mahant of the Sankat Mochan Temple.

Around 300 km upstream from Varanasi is the industrial city of Kanpur. The city is the largest in the state of Uttar Pradesh and second largest on the Ganges, with a population just over 6.3 million (Census of India, 2011).

Fig. 2. The Ghats of Varanasi.

The city's textile mills, sugar refineries, and tanneries dominate the region economically, earning the city the moniker "Manchester of India" (Cumming, 1994, p. 30). Jajmau, the industrial heart of the city, houses around 300 tanneries that produce around 1,500 tons of chromium sulfate discharge and 146,000 tons of solid waste (Schjolden, 2000). This adds to 339 MLD of domestic wastewater from the city, agricultural runoff, and much of the uncollected an unburned trash that flows down to the river through the *nalas* or drains of the city (see Fig. 3). The city is subsequently one of the most polluted sites along the Ganges, where the river is infamous for its poor quality and color, often described as "black," "tar-like," or dead. The KEF seeks to address water pollution in the GRB through awareness programs, media advocacy, public protests, and civil litigation (EcoFriends, 2008). But, the tension between religious and secular interpretations of water use, the state's dual commitments to promote industrial growth and pollution abatement, and the political clout enjoyed by factory owners, all complicate efforts to improve local river water quality. The city's reputation as a major polluter also complicates relationships with cities, water users, NGOs, and others downstream who point to Kanpur as a source of pollution and a site of contamination.

Fig. 3. Kanpur Riverfront, with Debris and Small Drain or *Nala*.

In the following pages, I examine each of these organizations and provide an analysis of the EK they seek to reproduce.

THE SANKAT MOCHAN FOUNDATION (VARANASI/BENARAS)

The SMK[3] is the best known, best funded, and by some measures most successful officially registered NGO campaigning against river water pollution in the GRB. Founded in 1982, the SMF's stated goal is "to restore the Ganga to its pristine purity and glory by preventing discharge of pollutants into it" (SMF, 2006). The SMF's success comes from its ability to capitalize on the dynamism of its leader and from his portrayal as a "savior" of the Ganges. Founder Dr. Veer Bhadra Mishra promotes a technological approach to solving pollution problems in the Varanasi region combined with educational and awareness-raising programs that attempt to change how Hindu devotees understand and act upon their relationship with the holy river. Dr. Mishra explains that this approach, like his life as former Professor of Civil Engineering at Benaras Hindu University and current Mahant[4] of the Sankat Mochan Temple, rests on a combination of faith, in terms of religion, tradition, and reverence for Ganga, with a scientific and technological approach to treating and eliminating sources of pollution (Mishra, 2005).

The SMF is critical of the shortcomings of the Ganga Action Plan (GAP) and its failure to improve river water quality. However, the SMF does not oppose the GAP per se, but rather envisions an extension of GAP projects, along with increased government funding for the program and adoption of more cost-efficient technologies. The SMF views itself as a watchdog agency for the Uttar Pradesh state Pollution Control Board (UP-PCB), Varanasi Jal Nigam, and local GAP officials. By collecting and analyzing water samples[5] from the ghats where people bathe and the banks of the river where people gather water, SMF volunteers provide a cross-check against government testing practices and published findings, and challenging state claims of river cleanliness and GAP success (see Fig. 4).

Tactically, the organization is skillful at targeting its message to an international audience of both individual environmentalists and large environmental organizations. Brochures, flyers, reports, and a website produced by the Foundation are largely available in English and designed in a manner that appeals to foreign tourists[6]: website links include links to Uttar

Fig. 4. Swatcha Ganga Lab with Water Quality Notice Board. *Source*: Photo by author.

Pradesh's state tourism board, as well as mountaineering and rafting service providers. Appealing to a foreign audience enables the SMF to build relationships with international collaborators and foreign donors and volunteers.[7] While foreign volunteerism is welcomed by the SMF, there is no mechanism through which local people are able to become involved in the organization. The handful of permanent staff consists of Dr. Mishra's colleagues from BHU and a few associates hired for their technical or financial expertise. The water users targeted by SMF awareness-raising programs are not incorporated into the organizational structure as members but rather are encouraged to participate through passive attendance at organizational events (Ahmed, 1994). Awareness-raising efforts are meant to influence the water use practices of a target audience primarily composed of religious worshippers and pilgrims, and to encourage their general awareness of pollution and environmental issues not to promote their further involvement in organizational or other politics.

The Swatcha Ganga Abhiyan (SGA or Clean Ganges Campaign) is perhaps the key public project undertaken by the SMF. Through the SGA, the SMF seeks to draw attention to the problems of pollution in the Ganges River, to encourage the Government of India to renew and pursue the objectives of the GAP, to generate a database on water quality, and to find "local" solutions to the problem of water pollution (Clean Ganga, n.d). The SGA is fundamentally an educational program that focuses on raising public awareness in Varanasi about the causes of nonpoint pollution, especially "open defecation, laundering activities (*dhobi ghats*) and removal of corpses and carcasses from the waterway" (Clean Ganga, n.d). The main projects of the SGA include riverfront *ghat* cleanup events, a conference where local priests discussed "altering ingrained social habits that contribute to Ganga pollution and agreed to [spread] the word [about pollution]" (*ibid.*), and community workshops where local people received education on their "rights under the Indian [Constitution] to take charge of environmental issues, including Ganga pollution" (SMF, 2006). Other events sponsored through the main branch of the SMF include children's education and painting competitions and poetry and music festivals where "poems, songs and hymns in praise of Mother Ganga are united with a more important environmental message" (Ahmed, 1994, p. 11).

Critics of the SMF argue that these efforts are at best sporadic and at worst, produced as a display for foreign contributors or as an effort to draw attention to the SMF and increase esteem for Mishra. Ahmed (*ibid.*) points out that the SMF office and its "educational center" are rarely open and relatively inaccessible to local people.[8] An anonymous environmental activist suggests that the organization misuses donations, fails to provide transparent reporting on how funds are spent, and consolidates power in Mishra as its sole leader (Good and Evil, n.d.). Some of these concerns were echoed in interviews with local water users:

> "Mohant issues a few statements, but in practicality does nothing." (retired male, personal communication, 16 February, 2009)
>
> "[The SMF] doesn't do anything, [it] just takes money." (Varanasi businessman, personal communication, 26 February, 2009)
>
> "They just take money." (male dhobi, personal communication, 10 March, 2009)
>
> "The government gives them money, but it is not being used to clean the water." (male dhobi, personal communication, 26 February, 2009)

Despite these critiques, the Foundation enjoys praise from domestic and international media, environmental groups, and some government agencies. In most accounts, Dr. Mishra is credited for championing the cause to prevent pollution of the Ganges and improve river water quality. In 1992, Mishra was included on the UNEP's Global 500 Forum Roll of Honour for "outstanding achievements in the protection and improvement of the environment" (UNEP, 2010). In 1999, he was named a "Hero of the Planet" by Time Magazine (Ganguly, 1999), and in 2000, Mishra was praised by former US President Bill Clinton during ceremonies surrounding the signing of a memorandum of understanding between India and the United States (Jain, 2000). Dubbed a "holy man with a secular plan"

Fig. 5. Mishra Speaking on World Water Day. *Source*: SMF new website.

(Ridge, 2008), Mishra remains the focus of media attention and praise bestowed upon the SMF (see Fig. 5).

Balancing Religion and Science: The SMF Campaigns to Save "Ma Ganga"

The SMK reproduces EK that rests on a set of dualisms. They define their approach to combating pollution in the Ganges as a unique combination of science and faith. But, in practice, these principles are the basis of a categorization of pollution into opposing camps that reflect organizationally identified divisions between point and nonpoint pollutants, sewage and community-sourced pollution, state and public responsibility, and their own litigious and educational program responses (see Table 1). These dualisms form the basis of a two-pronged approach to pollution abatement through which the Foundation attempts to change the actions of both the State and Central Pollution Control Boards and local water users. However, these approaches are not integrated into a cohesive response, but rather operate as two separate programs run by a single organization: a "scientific" program in which the SMF collaborates with municipal partners and development agencies to sponsor antistate litigation and formulate proposals for alternative wastewater treatment facilities, and

Table 1. Dualisms Identified in the Sankat Mochan Foundation's Antipollution Discourse.

Classification	*Point*	*Nonpoint*
Identified sources	*Sewage*	*Community use*
Responsibility	*The state*	*The people*
Discursive response	*Science*	*Religion/faith*
Programmatic response	*Litigation/AIWPS proposal*	*Education programs*

a "faith-based" program in which the SMF solicits foreign donors to support "culturally sensitive" education and awareness-raising programs targeted to local waters.

While it is not unusual for a single organization to coordinate multiple campaigns, the Foundation's programs are distinguished by the distinct discourses they employ. Foundation staff "talk science" with municipal and development partners. In their work with local people, and in the publications geared to foreign donors, volunteers, and tourists, they appeal to a particular vision of Hinduism and Hindu devotees that allows them to portray the organization, and Mishra, as saviors of the Ganga. Only in publications and interviews geared to foreign media are these discourses merged to support the claim that the SMF is best equipped with the appropriate response to pollution abatement. Otherwise, these discourses, and their operation in isolation from one another, have significant consequences in terms of the success of the Foundation, its relationships with local water users, and its ability to reproduce its distinct form of pollution knowledge.

The Scientific Campaign Against Point-Source Pollution

In defining pollution, the SMF distinguishes two main categories of pollutants: point-source sewage, specified as the untreated sewage emanating from city drains or *nalas* that empty directly into the river, and nonpoint pollution, which includes other "community uses" (SMF, 2007). The Foundation identifies point-source sewage pollution as the primary source of pollution in the Ganges River:

> "[I]n recent decades the river has become severely polluted. This is primarily because of the flow of untreated sewage into Ganga." (SMF, n.d.[a], para. 4).

> "The main cause is untreated sewage which in Varanasi and 113 other cities is dumped directly in the river." (Clean Ganga, (n.d), para. 5).

Preventing this discharge of this raw sewage into the river is one of the key goals of the organization, which argues that discharge occurs because of the government's failure to provide basic services (SMF New Website, Homepage). Organization staff criticize the government for insufficient investment in sewage treatment infrastructure and for investing in "inappropriate" technologies and treatment systems:

> The government followed Western scientists' advice to build treatment tanks, when that solution isn't appropriate ... They also don't work because of the regular power outages in Varanasi. The treatment tanks will work about 12 hours each day, and when they are not working, they allow sewage to flow directly into the river. (personal communication, 2009)

They argue that the state's failure to invest in sufficient sewage treatment infrastructure is a result of government apathy and lack of commitment to addressing the problem of pollution in the river. According to an SMF staffer:

> The government wants to write a policy, put some money to build treatment plants, so that they can point and say, 'Here. We are taking care of pollution.' Then, they would say that the water is clean and safe for bathing. But, we do our own testing and show that it is not, so they can no longer deny that their plans aren't working. (personal communication, 2009)

The organization's efforts to combat point-source pollution therefore involve litigation against state government agencies over municipal water and sewage treatment decision-making rights, alongside collaborative efforts with foreign scientists and donors to develop and champion their proposals for alternative wastewater treatment facilities.[9] In this legal-scientific branch of the organization's work, the SMF joined with the local municipal government, the Varanasi Nagar Nigam (VNN), to argue for devolved, localized governance of sewage treatment in particular, and water governance in general. They argue that state agencies superseded the municipal authority of the VNN and therefore ignored local needs and rights by funding only conventional sewage treatment programs. They characterize the state government as resistant to power-sharing and governmental devolution.

Despite the organization's criticism of the state government, the SMF is decidedly not antistate, antigovernment, or even anti-GAP. They champion a form of localized governance that would not place decision-making power directly into the hands of local people, or even NGOs, but rather to municipal authorities over which the SMF and similar NGOs would have significantly more influence. Their discourse is generally supportive of centrally conceived plans to address pollution in the Ganga, describing the GAP as a "well-financed plan" whose well-conceived objectives could not be achieved due to political and public indifference (SMF New Website). Ultimately, SMF would like to see an extension of the Plan and increased governmental funding for the program, along with heightened governmental monitoring and management of river water and its use (SMF, 2006; Wallace, 2006). This is in keeping with the Foundation's ideology, which envisions an extension of centralized management and control over India's river water resources.

Using the slogan, "Water for Development," Mishra (2005) argues that water resources fuel national economic development and water pollution problems are best addressed through the combined "processes of development, science and technology" (758). He calls for a neoliberal approach to water management, identifying development projects and associated funding as the appropriate solution for water pollution problems in the GRB. Mishra is quick to point out that the World Bank shares this perspective on neoliberal water management and that it is the "only way" to achieve a balanced approach to development that incorporates environmental protection (*ibid.*). It is the Foundation's faith-based campaign, he maintains, that is necessary to change the mindset of Hindu worshippers who he describes as "ignorant of the processes of development" and unaware of the "good things" they offer (*ibid.*, pp. 755–758).

The Faith-Based Campaign Against "Community" Pollution

The SMF's second campaign is a faith-based and community-centered effort that is subordinate to their legal-scientific crusade against point-source pollution. Nonetheless, it is this aspect of their work that often receives the most praise and international attention. The campaign targets sources of indirect or nonpoint pollution caused by local people or "community use," especially the effluents produced by open defecation, washing (bodies, animals, laundry, and household goods), as well as the direct disposal of "garbage, human and animal corpses, and other pollutants from non-point sources" (SMF, n.d.[a]). Activities coordinated under this program focus on educating and raising the awareness of local people who use river water in their daily activities.

The goal of the SMF's community program is twofold, aiming to protect both water users from the health risks posed by bathing in or ingesting contaminated water, and to protect the river from contaminants introduced by river water users. In Foundation discourse, these water users – the people of the city – are represented as being both those who the river needs to be saved *for* and those who the river needs to be saved *from*:

> Community using the Ganga river directly also pollute the river. This is termed as non-point source of pollution ... and has to be cleaned by the direct users – the people of Varanasi. For cleaning the pollution caused by direct users the Sankat Mochan Foundation works with the community and uses culturally consistent ways to motivate the people to take care of the river. To sensitise the people to take care of the non-point pollution is our "spiritual work with the people" requiring patience and innovation. (SMF, 2007, p. 10)

SMF community awareness and educational activities do not ask water users to reject Hindu faith-based interpretations of the river and its relationship to the country or its people, but rather seek to change how people understand their role in that relationship and act on the basis of that knowledge. In Foundation discourse, the river remains *Ma Ganga,* the mother of Hindus, the nation and the world. In one organizational flyer, the SMF asserts, "The problem is, Ganga Ma (Mother Ganges) isn't feeling well," and then reassures, "She can be cured" in very large text (Clean Ganga, n.d, inside cover). While her devotees are still encouraged to think of themselves as her children, they are asked to shift from thinking of themselves as protected by the Ganga, to thinking of themselves as protectors of the Ganga. In a 2007 interview, Mishra explains that Hindu pilgrims to the Ganges:

> want to touch the water, rub their bodies in the water, sip the water ... If you tell them 'the Ganga is polluted,' they say, 'we don't want to hear that.' But if you take them to the places where open sewers are giving the river the night soil of the whole city, they say, 'this is disrespect done to our mother, and it must be stopped.' (Hammer, 2007)

By appealing to the significance of these religious practices, the Foundation claims that they are able to overcome the resistance that some observers argue is posed by long-sustained religious beliefs about the nature of the river as goddess (see Alley, 2002; Haberman, 2006). The SMF does not try to force devotees to change how they think of the river, but how they think of themselves in relation to the river.

But, the relationship between the SMF and the people it claims to represent, protect, and educate is complicated by negative perceptions of Mishra and by his generally paternalistic attitude toward those who use the water to fill both material and spiritual needs. Accusations abound among local water users that Mishra misuses funds donated to the Foundation and that he caters the Foundation's activities to foreigners' visions of what the city should be, rather than local people's needs. One Foundation staffer explained:

SBK: Why doesn't the SMF have more popular support, among the people here [in Varanasi]?

Sharma: Because Mohantji is not entertaining any local people, that is a big problem.

SBK: Is that why is there no mass movement?

Sharma: Some people want to join movement, but the government has chosen Mahantji … . The central Foundation has five people doing all projects …[lists some roles]. Only five key people. Then people like you [and others who support the cause but are not part of the organization]. (personal communication, 2009)

Even Mishra recognizes that as Mahant, a separation exists between him and the people of Varanasi. While his position in society demands that many local Hindus pay him obeisance, this means that people rarely engage him directly or challenge him openly. Having inherited his position at 15 years of age, Mishra explains that he always felt separate from the people and that it was shocking to many in his family and in the Temple when he chose to enroll at University and pursue a degree in Engineering. While he feels that this unconventional path left him uniquely equipped to address pollution in the Ganges from the dual perspectives of science and religion, his lack of identification with the people of the city and his position of superiority have no doubt contributed to his, and by extension the Foundation's, representation of local city dwellers, worshippers, and water users as naive, ignorant and even backward:

> Practicing Hindus believe that Ganga is mother and goddess. Like fish living in water they can not live without Gangaji. They don't know pollution and can't believe that Ganga is polluted yet. They are suffering on account of Ganga pollution. In today's modern world only few have the sensitivity to understand them and their spiritual life. These practicing Hindus have become endangered specie [sic] of human beings …. Common people in India are not ready to believe that Ganga can be polluted. (SMF, 2007, p. 9)

By portraying local people as ignorant, and as the source of pollution problems, Mishra and the SMF reproduce a rhetoric that disempowers and marginalizes local people and local forms of EK. It is this assumption of local ignorance that at least one staffer argued is a misconception held by Mishra and the Foundation:

SBK: So, do people who bathe and drink *ganga jal*, do they know about pollution?

Tripathy: Yes. Every people know. Because when they take a bath in Ganga, they see dead bodies float in the Ganga and the polluted items on the top [of the water]. (personal communication, 2009)

Yet, the idea that water users and river worshippers are not aware of the problem of pollution informs the guiding rationale on which the SMF's community programs are built. Water users are viewed not only as uninformed or ignorant of the pollution problem, but unwilling to recognize or admit that it exists. By emphasizing aspects of the organization's faith-based campaign to foreign donors and media, the SMF argues that their activities are most "consistent with the cultural beliefs and lifestyle of the people" (SMF, n.d.[a]), and that they are uniquely able to "harmonize the benefits of a Ganga cleanup with traditional Indian culture that regards the Ganga as pure" (Sankat Mochan Foundation, n.d.[b]). But, in "talking faith" to donors, media, and local worshippers, while speaking the language of science and development with government officials and AIWPS collaborators, the SMF actually effectively prevents the devolution of water quality governance and the involvement of local people in the campaign to save the river. Officials, local water users, and pilgrims are enabled and encouraged by the Foundation to continue speaking different languages, with only Mishra and his staff serving as translators. This allows the SMF to maintain a vital position as intermediary agent responsible for education programs that appeal to local people to change their water use practices and that explain the work of government in terms amenable to a (presumably homogenized) community, for channeling people's energy into nonthreatening shows of resistance, and for rallying public support for both NGO and government-sponsored programs. The SMF's positioning for supremacy among antipollution NGOs in the GRB is perhaps challenged only by their counterpart upstream, the KEF.

THE KANPUR ECO-FRIENDS (KANPUR)

Operating from the most populous and industrial city in Uttar Pradesh, the KEF is a relatively small but historically active organization. The group was formed in 1993 by environmental scientist and activist Rakesh Jaiswal after he began investigating the source of problems with the quality of his own tap water. The KEF describes itself as an environmental NGO that "performs all such acts that may assist in or be conducive to fostering environmental education, protection and security" (KEF, 2008). Although the organization seeks to address broader environmental issues, the KEF's main activities concern "protecting the river Ganga from various sources of pollution" and mobilizing local people to guard against the health risks posed by chemical pollution (*ibid.*).

According to Jaiswal, the Ganges issue came to the forefront only because it is the most pressing problem being faced locally in Kanpur. But, Jaiswal is adamant that pollution of the Ganges cannot be understood or responded to as a stand-alone problem. Instead, poor water quality is viewed as just one of the many negative environmental outcomes that results from human misuse of nature, especially rapid and unchecked population growth and industrial economic development. Jaiswal explains that pollution in the Ganges is both part-and-parcel and symptomatic of an imbalanced approach to development and growth:

> It is not just pollution – this is just one aspect of the problems – just one aspect, pollution. It is not only the pollution which is killing the river, but if we see the problems in totality, miss. The problem begins right from its source – the global warming, the melting of glaciers which make it a perennial river, which feed river Ganges throughout the year ....Then various dams or hydroelectric projects are coming across the river Ganges and its tributaries ... and then the diversion of the river ... into canals ... to agricultural fields. So hardly any water is left for the downstream towns. And *then* pollution ... And in case of India ... the exponential growth of the population ... even the river plains are being taken away for urbanization and industrialization. And also in many other ways the river is being exploited. (personal communication, 2009)

Through its work, the KEF draws attention to this "big picture" of environmental health in the GRB, asking local water users, factory owners, and government officials to consider not only how to reduce pollution in the river and mitigate its effects but also how to ensure adequate stream flow, protect endangered species, reforest the riverbank and essentially rebuild a healthy ecosystem. It is this focus on the multiple challenges faced by the Ganges, the multiple threats to local environmental health and their interconnection, which distinguishes KEF's approach.

Jaiswal uses the term "holistic ecology" to describe the KEF's vision and believes that its widespread acceptance will promote the adoption of basin-wide multiuse and multi-institutional watershed governance. Together with a six-member "Think Tank," or an advisory panel consisting of current or retired scientists and faculty from local institutions, Jaiswal is supported by a network of members drawn from friends, colleagues, and local citizens but whom even Jaiswal admits do not constitute an organizational membership. The KEF also attracts volunteers and interns, mostly from abroad, who usually arrange to work on a particular campaign for a few weeks.[10] The Think Tanks informs, but does not directly guide, the actions of the organization. Instead, the day-to-day activities are overseen by Jaiswal, who serves as the group's primary representative and decision-maker. Indeed, during the months when he was recovering from heart surgery, the organization's activities all but ceased.

Despite its centralized organization around Jaiswal, the KEF is able to consistently circulate projects and engagements – last month they might have worked on a scheme to bring clean drinking water to a district affected by poor water quality, today Jaiswal might be sitting in on a meeting to redraft the city development plan, and next month Think Tank members and volunteers may visit local area schools to sponsor an art contest. Some of KEF's ongoing activities include publishing water quality and GAP-monitoring reports, public interest litigation (PIL),[11] filing injunctions against industrial effluent and sand mining in the riverbed, media advocacy, river cleanup and an adopt-a-ghat programs, educational outreach, training *Ganga Praharis* (River Watchdogs) from people who live and work along the riverbank, and various public awareness-raising campaigns, including programs promoting the burial of corpses (rather than cremation) (see Fig. 6). The KEF has also created a Ganga Task Force

Fig. 6. KEF-Sponsored Corpse Removal Cleanup Drive. *Source*: KEF (2008).

(Ganga Vahini) to work in 20 rural villages near the city of Kanpur, where they organize public meetings, hearings, and rallies.

In most of these campaigns, KEF staff and volunteers work with people affected by pollution, referred to as "stakeholders" in organizational discourse (see Fig. 7). These diverse groups, including boatmen, ghat dwellers, tannery owners, and virtually anyone in the city with water access or quality concerns, are embraced as essential to the efforts of the KEF but are not invited to join the organization as members or to participate in intra-organizational decision-making processes. Even those ghat dwellers who are trained by the KEF as "river watchdogs" do not become official members of the group. Instead, KEF organizers see the real value of these events not as opportunities for promoting direct action, but rather in their capacity to educate participants, onlookers, and public officials about river pollution. Proof of event success comes later, when KEF staff hope they will see the public adopting the practices advocated by the organization:

> Ecofriends have been carrying out these kinds of campaign over the years ....Earlier, people used to dump discarded clay items into the river. This year they buried those items and also the worship materials .... And maybe if this campaign is carried out, maybe in a decade or two this can be eliminated completely, this kind of practice. (Rakesh Jaiswal, personal communication, 8 January, 2009)

Fig. 7. Rakesh Jaiswal Pictured with Cleanup Campaign Volunteers. *Source*: KEF (2008).

> Through this campaign, we succeeded in garnering the support of the Ganga communities whom we reckon as the custodians of the river. (KEF, 2004)

According to one KEF member, these efforts to raise awareness are only marginally successful because, in Kanpur, people are already aware of the problem of pollution:

> People are sensitized and know what to do, but need to get into action ....Everybody is aware. They'll all say something should be done, but who will do, what they will [do], no one knows. (personal communication, 2009)

Another local activist agreed, stating that, "Everybody is aware. They'll all say something should be done, but who will do, what they will, no one knows" (personal communication, 2009).

In response to a question about his future outlook, Jaiswal spoke like a man without hope: "I do not see a hope. Because things have only worsened since I started working 15 years ago. The quantity of the water in the river and also the quality, it has gone down. The situation has only worsened. Maybe some miracle. I don't know" (personal communication, 2009). When asked what miracle he would like to see, Jaiswal identified two: that the government would calculate a new (and improved) minimum ecological flow at every station along the river that it would be monitored for compliance daily, and then bringing an end to the practice of allowing untreated sewage and industrial effluent to run directly into the river. While little progress has been made on the latter goal, the former has been one of the top priorities articulated by the newly formed National Ganga River Basin Authority (NGRBA).

Holistic Ecology in the City of Industry: The KEF's Campaign for a New River

The KEF's adoption of a holistic ecology perspective leads them to consider the ecological health of the river basin as a whole, and guides them away from confronting river water pollution as a stand-alone issue. As a guiding principle, the holistic ecology perspective also directs the organization's interpretation of the problem of environmental degradation in the GRB as a direct outcome of the imbalance present between humans and their natural environments. Water pollution is therefore understood as just one manifestation of the imbalance present in the GRB. While this imbalance is not inevitable or universal, it is described as common in urban environments where the concentration of people and industry overtax local resources and the environment's natural ability to absorb waste, and where appropriate measures are not taken to reduce this pressure (KEF, 2008). Pollution, referred to in organizational discourse as waste from factories and human settlements, cannot therefore be addressed on its own, but requires a comprehensive response to the problems associated with the overuse of resources and the overproduction of waste.

The KEF's holistic ecology approach is strikingly similar to scientific systems or ecological systems, in that it encourages examination of the river at a basin- or watershed-scale and draws attention to the complex systems of inputs and outputs that make up the GRB ecosystem. Organizational discourse makes use of this language of inputs and outputs, arguing that environmental degradation occurs when one falls out of balance with the other. In the GRB, this imbalance is caused by both the overproduction of inputs to the river (more domestic and industrial waste than the river can process or dilute) and the overuse of river outputs (too much water is withdrawn from the river for agriculture, wildlife and water quality are disturbed when sand and silt are mined from the riverbed). But the politics associated with water use and management, as well as water quality assurance, are also regarded as ecosystem inputs in their own right:

> River Action Plans can be considered as one of the several inputs that are needed to keep the towns and rivers clean. Other inputs, like management and handling of garbage and slums, regular operation & maintenance of sewerage systems and provision of adequate number of toilets for the masses to minimise the practice of open defecation, minimisation of use of rivers for cattle wallowing etc. are the primary responsibilities of the respective local self-governments. Unless concurrent measures are taken to address all these issues, full benefits of the river ... cannot be realised. (Jaiswal, 2007, p. 23)

A comprehensive response to ecosystem degradation therefore involves balancing inputs and outputs, including management and policy responses.

In promoting a comprehensive response to river water management, the KEF asks local residents, politicians, factory owners, and other "stakeholders" to come together in creating a new vision of the river. This vision is not something that the KEF seeks to dictate, but one that they believe should be drafted by all stakeholders and should account for the many various uses of the Ganges. Jaiswal explains that river water management and pollution abatement efforts in the GRB have failed because there is no comprehensive vision of the river, and because the plans that currently exist prioritize certain type of river water use or values, such as its use as a holy river or its value as a clean river (personal communication, 2009). But, Jaiswal argues that these plans fail because they do not consider the multiple aspects, purposes, and incarnations of the Ganges. Because politicians see only the goal of a clean river, and not an agricultural river or an industrial river, plans like the GAP ignore the other demands that are placed on the system and are subsequently doomed to fail:

> There is not one single reason which can be attributed to the failure of Ganga Action Plan. First, we do not have a vision for the river. Still *we do not know what kind of river do we want* and what's practical, what is achievable ... it is not impossible, we have the resources, we have the talent, we have the competence, we have the science and technology ....So first we should have a clear vision, then we should have a plan, and then a commitment. (Jaiswal, personal communicatyion, 2009)

> Before any action can be initiated, all concerned should start thinking in terms of a new vision for Ganga. How do we want Ganga to be and what can be done to achieve that vision is the question posed to all of us. A new vision for a pristine and pure Ganga has to pour forth and translated on the ground ....A new vision that can inspire the masses to action. A new vision that needs to reconcile the competing demands on the precious waters of the river with sustainability. It needs to think of the river as one organic entity where tinkering in one- part affects the entire body of the river. A new vision which believes that if we as humans wish to survive, Ganga needs to survive ....A whole new approach is required to restore the river. (Jaiswal, 2007, p. 48)

There are a number of similarities between the type of comprehensive vision promoted by KEF and the approach to river management ushered in by the NGRBA. Both approaches advocate a basin-wide scale of planning and the need to address issues of water quantity in addition to water quality by maintaining "minimum ecological flows" in the river. More importantly, they share a desire to replace or subsume the "piecemeal" approach to river management that heretofore characterized the GoI's treatment of the Ganges.[12]

Jaiswal and the KEF argue that the complexity of water management and water quality governance actually leads to inaction or solely symbolic action, rather than the flurry of activity that one might expect. They describe the GoI's approach to river management as being characterized by "frightening ad-hocism" (KEF, 2008) and inaction:

> [I]nstead of a comprehensive, sustained and intensive attack on pollution in the river, action could not proceed beyond taking of symbolic steps. There exists a wide chasm between the promise of Ganga Action Plan and the reality of millions of litres of all kinds of pollution meeting the river every single minute. Inaccurate, partial and self-laudatory reports have become the norm but do precious little to make significant forward movement on reversing the flow of pollution in the river. (KEF, 2008)

Jaiswal argues that official efforts, like the GAP, to prevent and abate pollution in the Ganges are embarked on with genuine intentions, but these are gradually eroded as planners are faced with the challenge of needing to balance the river's many uses:

> Had there been an honest implementation of GAP, the story would have been different, Ganga would have looked different. The enthusiasm waned with the passage of time and GAP became just one of the government projects ....Ganga can not be cleaned like this. We need drastic steps, bold decisions and a visionary and committed leadership. GAP is a dead horse, and there is no point in flogging it any more. (KEF, 2008)

By encouraging GRB stakeholders to reconsider and reformulate their collective vision of the river, Jaiswal and the KEF are really asking people to rethink what kind of relationship they want to have with the river in its many capacities as goddess, power plant, breadbasket, or industrial powerhouse. They also draw critical attention to the question of whether the Ganges is a national resource to be tapped in the pursuit of economic development, or whether the river is a part of the nation's heritage and dominant culture that should be protected. But, in their efforts to promote a holistic view of the Ganges, KEF members must contend with, and ultimately address, her use and prioritization as a goddess.

Pointing to a Paradox: The KEF's Campaign Against Ritual Bathing

When KEF members talk about the practice of ritual bathing, they refer to what they perceive as a paradox – people polluting the very river that they worship:

> When we attempt to reconcile the significance of the sacred river in the past to its present reality, a most tragic paradox is encountered. Ganga today is being worshipped and defiled simultaneously. In fact, at most times, the process of worship itself has a polluting influence since bulk of the worship materials are disposed [of] in the river in ugly non-biodegradable polythene bags and in other unthinking ways ....The coexistence of worship and defilement of the Ganga defies logic and reason and leaves most observers confused. (Jaiswal, 2007, p. 47)

> It can only be an irony of history that with the passage of time this mighty river is today a victim of defilement and pollution by the very human race that finds sustenance by it. (KEF, 2008)

In KEF discourse then, bathing and river worship are described as both causing pollution, by contributing waste and contaminates to the water, and preventing people from taking action against pollution, by convincing them that the river is a great purifier and unable to be polluted. While the organization engages in a variety of activities, it is their struggle to end the practice of bathing and belief in the holiness of the Ganges that is granted discursive priority. This aspect of the organization's work is also the most controversial.

The KEF describes practices of river worship as "ancient traditions" and "ancient superstitions" that, while shaping "Indian civilization," seem nonetheless outdated or backward (KEF, 2008). Worship of Ganga is frequently described in the third person, making clear that organization members neither share these beliefs nor participate in the ritual practices of bathing in the river or drinking *ganga jal*:

> "Many Indians depend on this great river for their physical, psychological and spiritual sustenance." (*ibid.*)

> "[P]eople have a great belief in her powers of healing and regeneration." (*ibid.*)

Jaiswal describes river worship as a "polluting social practice" that started in ancient times and became perverted over the years:

> Its not just a river, but its a deity or goddess for Hindus. They have a lot of faith [in] that, and this belief that river is a deity, Ganga is a goddess, and if you put anything into it, it will clean it. That river Ganges is a cleanser par excellence, this belief has done the damage most. (personal communication, 2009)

The KEF's response to this perceived contradiction of river worship and pollution involves an educational campaign that aims to end bathing practices associated with worship of Ganga. In order to achieve this goal, KEF members believe that Hindu devotees must abandon their religious beliefs and adopt a scientific or ecological view of the river, or at least concede that the river is no longer a goddess:

> The Ganga devotees who consider the river as a cleanser par excellence and treat Ganga as a deity who gives salvation *need to be taught that Ganga has lost its divine role, Ganga has lost its cleansing properties* and Ganga herself needs salvation. (Jaiswal, 2007, p. 48, emphasis added)

The specific programs associated with this aspect of the organization's work include educational programs held in schools that emphasize scientific interpretations of the pollution issue, community workshops that emphasize the negative impacts of bathing and river worship, and a community burial program in which corpses are removed from the river and buried on the sandy banks and people are encouraged to bury (rather than cremate) the recently departed (see Fig. 3).

When KEF members undertake these programs, they are careful to emphasize the negative environmental impacts from ritual practices, rather than the health risk that is posed to people who live along the river and participate in these activities. People are therefore encouraged to make a selfless sacrifice in abandoning, or no longer acting on, their religious beliefs, rather than understanding this action as a self-serving effort to protect one's health. A KEF member explained that, for most Kanpurites, the health risks associated with bathing and water use are ontologically tied to the factories. In other words, if people are asked not to bathe in order to protect their health, they may not comply because of a perception that the health risk is posed by factories and not people's practices. If, however, people are shown that their activities damage a river they love, they may be willing to relinquish or substitute those activities and the beliefs upon which they are based. The member emphasized that "there are two separate issues in Kanpur: the sacredness of the river and the factory issue" and these must be kept separate in order to encourage local engagement in anti-pollution activities.

There exists significant doubt about the practicability of an antibathing and antiworship campaign in a country with 800 million practicing Hindus who come to the river by the tens of millions each year to bathe and worship. But Jaiswal believes that it is possible to change these practices, given enough time:

> It is [a] problem you see, practices take a lot of time in evolving. Maybe it took thousands of years ....So, if it took hundreds or thousands of years getting evolved, it would also take years, maybe some decades, in getting eradicated. (personal communication, 2009)

From a social movement standpoint, the logic of risking alienation of one's potential support base by demonizing their most sacred identity-forming cultural beliefs and practices is questionable, especially when the success of this campaign may, at best, achieve the elimination of only a small percentage of the pollutants in the river.[13] Indeed, the KEF even supports criminalization of people's religious activities in order to further the goal of cleaning the river:

> Those who are found polluting the river should be punished on the spot like those violating the traffic rules are punished. This will be an effective deterrent. (Jaiswal, 2007, p. 46)

But, as one KEF member pointed out, success may not be the main goal or purpose of the antibathing campaign. They explained that through the campaign, people are made aware of their activities' negative effects on the river, then, "when people become sensitized, then they will care ... when they are sensitized and [it is] known what [they need] to do ... [it will be easier] to get them into action" (personal communication, 2009). The anti-bathing campaign, they hope, will get Ganga worshippers involved in the broader efforts to cleanup the river by breaking through a wall of apathy that exists when encountering a seemingly insurmountable task: "The average man has no real power, no real influence and [no real place in NGOs or other organizations that] have their own politics, their own hierarchy" (personal communication, 2009). By teaching people that their activities have an impact, this KEF member hopes that people along the river will adopt an ecological, if not environmental mindset. This could inspire people to start making smaller, incremental changes to their activities that may eventually build to a more meaningful engagement with antipollution politics in the GRB.

ORGANIZATIONS, PROTESTS, AND MOVEMENTS: STRUCTURES OF RESISTANCE IN THE GRB

Both the KEF and Sankat Mochan Foundation SMF work to reproduce a specific and place-based, often localized, understanding of and relationship to the problem of pollution in the Ganges River. Engagement with the issue of ritual bathing is particularly indicative of the differences between these organizations. As the KEF attacks the practice of ritual bathing, SMF activists defend it as a core water use activity in the region and a window through which to change the knowledge of local water users. The act of bathing, in particular, is used by SMF members as a point of inspiration, through which the organization attempts to tie water users' devotion to Ganga with their observational and experiential knowledge of pollution. As a religious leader, Mahantji is careful to point out that bathing or washing in the river during ritual or religious practice does not cause or create pollution:

> Other people view people's bathing as pollution. In this sense, fish in the water, birds flying over and shitting in the river cause "pollution." Any of these things can be viewed as pollution, but they are not. (personal communication, 2009)

The SMF reinterprets and reframes traditional ways of knowing and interacting with the river in their effort to convince local water users and pilgrim-tourists to change their water use practices. They adopt the language and concepts dominant in preexisting Hindu religious dogma, and reinterpret these in order to convince water users to change their interactions with the river, not the underlying beliefs on which these activities are based. While their failure to merge this faith-based aspect of their worship with the scientific discourse they speak with government officials may ultimately marginalize local water users from the decision-making process, they are nonetheless able to take advantage of their strategic location in one of the nation's top tourist destinations to build valuable bridges with foreign donors and collaborators. These efforts have been successful in bringing attention of international press and like-minded foreign environmentalists and environmental NGOs that have helped the SMF to pressure state authorities into considering alternate sewage treatment technologies.

On the other hand, the KEF is able to articulate a unified system of knowing and acting with river water pollution that reflects their framing of pollution as one of many problems caused by lack of cohesive vision for river management, an overtaxation of river systems, and a fundamental imbalance between humanity and the natural environment. But their promotion of holistic ecology attempts to supplant core religious practices, like bathing and cremation, through which local water users relate with and worship the river. In their view, these ways of seeing and knowing the river prevent the recognition of pollution and contribute to the degradation of river ecology. While the KEF maintains that replacing this EK with one based on holistic ecology will directly reduce the pollution load in the river and inspire local people to act as environmental agents in protection of the river ecosystem, they run the risk of alienating local water users and of reducing the wider debate over river water quality to a simplistic conflict over religious beliefs and practices.

While the KEF's antipollution campaign faces clear challenges in Kanpur, their approach would likely be untenable in a tourist-pilgrim city like Varanasi, or at a basin-wide scale. The antibathing campaign is easier to undertake (and more likely to be successful) in Kanpur, where many people have already abandoned the practice of ritual bathing in the river and where few people bathe locally.[14] At least one KEF member echoed this sentiment, saying that an antibathing campaign must be site-specific and would not be successful in other places on the river: "In Varanasi, they are fanatic. In Kanpur, we are not so concerned with religion." (personal communication, 2009). While this member sees the antibathing campaign as pragmatic only at the local level, Jaiswal and official KEF discourse argue that it is a basin-wide vision for a new relationship between people and the river that is based on a scientific, rather than spiritual, worldview: "[E]veryone who goes to the river washes themself or selves contributes towards the pollution – and this is scientific fact and these are scientific findings ....Why can't we be honest and warn people that Ganga waters are not worth bathing and drinking?" (personal communication, 2009).

As long as the KEF continues to fight against the practice of bathing, while the SMF defends it, it is unlikely that these groups will be able to come together in order to create a shared, cohesive vision around which to structure social movement frames or build a broad-base of popular support. The place-based nature of these respective efforts becomes a barrier to inter-organizational coordination, as groups not only emphasize different aspects of the problem or propose different solutions, but also as they struggle against one another for primacy and to bring national political and media attention to their local struggles. If these groups choose to coordinate their activities or attempt to form a broader mass movement, they will ultimately need to set aside or somehow reconcile their different environmental knowledges. Because the knowledges reproduced by these groups is not only dissimilar, but actually contradictory, they may have to adopt a sort of strategic essentialism in order to achieve this goal (see Spivak, 1987). This would involve setting aside their differences and agreeing on some simplified, essentialized knowledge framework that could, even temporarily, form the basis of a social movement ideology.

For both groups, the reproduction of EK remains a key element of the larger political struggle surrounding pollution and pollution abatement in the GRB. This work involves not only identifying particular sources of pollution and promoting abatement activities in their respective cities, but also recasting the values associated with Hindu ritual bathing practices and the notions of society-environment inter-relationships on which they are based. Here, we see the significance of local context in shaping not only how these groups make sense of pollution and attempt to shift both larger social meaning associated with pollution and human use of river water, but also in the use of EK reproduction activities as a key method of consolidating political power and affecting the political discourse surrounding resource use and protection. Highlighting these processes reveals that the political debate surrounding water pollution consists of ideological struggles over the meanings associated with river water and pollution, and that these ideologies and their successful or unsuccessful reproduction are high imbedded in and reflective of the places and spaces where struggles take place.

NOTES

1. Geographers have been concerned with human understanding of social relationships with the environment for most of the past century. Early work in this vein tended to focus on human responses to environmental phenomena (i.e. Barrows,

1923) and the relationship between human perception of the environmental phenomena and decision-making practices (i.e., Larimore, 1969; Whyte, 1985). Later work built on this foundation to consider relationships between perception and social responses to environmental change (i.e., El Ramly, 2003), relationships between perception and willingness to participate in environmental projects (i.e., Hema & Jamal, 2004), and the affects of environmental "awareness" as a factor in the decision-making process (i.e., Brookfield, 1969; see DuPuis, 2004). Many of these approaches viewed "nature" or environmental degradation as part of a measurable "reality" of which humans are at least partially aware (or perceptive), and to which we respond (individually and collectively) through resource use and management activities (see Cloke, Philo, & Sadler, 1991; Johnston, 1991 for discussion of these developments).

2. In the Gangeatic region, an average of one person dies every minute due to diarrhea. Secondary health effects can also be caused by ingesting freshwater fish from the Ganges, which can contain concentrated levels of mercury and pesticides (Sampat, 1996).

3. Sankat Mochan translates in English as "removal of suffering."

4. Hereditary head priest.

5. The organization tests its water samples at their independent lab housed at organizational offices on Tulsi Ghat and often, but not consistently, posts their findings on a board located just outside of the lab's entrance for public viewing.

6. The SMF also receives prominent mention in the Lonely Planet's guide to India, where visitors are encouraged "to make a contribution, financially or through voluntary work efforts" to the organization (Lonely Planet, 2010).

7. Since its inception, the SMF has worked in close partnership with international NGOs, including Oz-Green (Australia), Friends of the Ganges (the United States), the Asia Foundation (the United States), and the Pacific Institute (the United States). Collaborative projects have also been conducted with researchers from Montana State University's microbiology department, and funding for SMF projects and facilities was contributed by Svenska Naturskyddsföreningen (SNF), a Swedish nature conservation society, and the Swedish International Development Cooperation Agency (SIDA). Through these partnerships, the SMF has been able to secure funding for its projects, expand its facilities, and draw international attention to its efforts.

8. During my visits to SMF, I also found the educational center closed during morning and midday, understaffed, and stocked with pamphlets, brochures, and educational material printing primarily in English and clearly intended for foreign visitors or prospective donors. The only brochure seemingly directed to a local audience (and printed in Hindi) was a pamphlet for women on domestic and personal hygiene. Ahmed (1994) points out that the office limited hours, particularly morning and midday closings reduces its accessibility by women and children. The organization has no Indian female staff, though I was told that many women (mostly worshippers at the Sankat Mochan Temple) participate in the ghat cleaning events as volunteers.

9. The SMF's alternate sewage treatment proposal involves development of an Advanced Integrated Wastewater Pond Systems (AIWPS), which would begin with a single facility constructed outside Varanasi city limits. The facility would consist

of a series of ponds that use gravity, sunlight, oxygen, bacteria, and algae to remove sewage and "other pollutants" from water over the course of about 45 days (Mishra, 2005). Unlike traditional systems of sewage treatment, the AIWPS requires no consistent power source and is therefore argued to be more technologically appropriate for a city like Varanasi with an intermittent power supply. After years of championing this program to the UP-PCB and Varanasi Nagar Nigam, the SMF finally received approval and funding for the AIWPS project from the Central Government in 2009, as part of PM Singh's new National Ganga River Basin Authority (NGRBA) plan.

10. One volunteer explained that her contribution to the group would be to shoot footage for a documentary on the KEF, and that the young man who came before her participated in organizing a riverside cleanup event. Few of these intern-volunteers remain active with the organization in the long term. Their accounts of the Ganges or their time spent with the KEF do, however, constitute a segment of the organization's website. Grouped together with environmentalist interviews under the heading, "Eco-Talk," these accounts emphasize foreigner's impressions of the river's plight or draw readers' attention to the universality of water pollution: "I only wish the camera could record the stench that went with the sights-breath-holding instead of breath-taking ... How lucky some of us are to have clean water ... My thanks to Eco-Friends for their Clean the Ganga project" (KEF, 2008). "Michigan, like India, has struggled with pollution and the abuse of its resources ... We have taken our access to clean water for granted, but we are abusing our most precious resource" (*ibid.*). These accounts by volunteers and interns often convey emotions of disgust or shame associated with the river. Many talk of a visceral reaction to common river scenes, of corpses being cremated, children swimming near floating trash, or effluent pouring out of drains into the river: "Ganga looks like a huge open drain in Kanpur ... without any doubt the worst I've seen in Kanpur were the floating dead bodies in the river ... their sight is truly irritating and disgusting" (*ibid.*).

11. Public Interest Litigation (PIL) began in the mid-1990 to late-1990s, after Jaiswal wrote a letter to the Allahabad High Court drawing attention to the lack of implementation of GAP-Phase I in Kanpur and singling out specific factories and drains that violated GAP programs. Over the course of the following year, the Court released a number of important orders that eventually led to closure of a handful of gross polluting factories in Kanpur.

12. While water quality management and pollution abatement are overseen by the CPCB under the Ministry of Environment and Forests, a host of other government bodies and offices also regulate some aspect of the river or the use of its waters: The Ministry of Agriculture is responsible for regulating the use of pesticides and fertilizers that contribute pollutants to the river through agricultural run-off; The Ministry of Power (and to some degree the Ministry of New and Renewable Energy) is able to plan hydropower projects that can impact the flow and the course of the river; The Ministry of Urban Development establishes laws and policies that dictate where settlements can be constructed and what infrastructure must be in place; The Ministry of Textiles regulates the textile industry, including standards related to the treatment and processing of material that can produce effluent; The Ministry of Micro, Small and Medium Enterprises, along with The

Ministry of Heavy Industries and Public Enterprises, oversee various size industries and address issues related to production standards and profitability; The Ministry of Tourism can promote certain destinations and leverage political pressure on behalf of the growing tourist industry; and then The Ministry of Water Resources actually oversees river water management.

13. While it is difficult to determine the environmental impact of bathing and worship activities along the river, they are believed by most to be only a small percentage of the total pollution load in the Ganges.

14. As a local activist explained, "The river is too polluted. The local arrangement has been to canalize water for bathing and cremation, but pollution is common knowledge, so no one bathes in the river" (personal communication, 8 January, 2009). Indeed, I had a difficult time identifying active water users to participate in the study in Kanpur. While river worship still occurs in Kanpur, many water users adjust their activities by reducing direct contact with river water by splashing water on their head/body and avoiding full bathing submersion, or by leaving town and traveling to bathing sites upstream from Kanpur and its many gross polluting industries. People with the financial resources to do so travel upstream, usually to Bithoor, to bathe and worship the river. There, the Ganges can be sacred, whereas in Kanpur, there is no space for a sacred river in a city where "nothing is holy anymore" (personal communication, 2009).

REFERENCES

Ahmed, S. (1994). The rhetoric of participation re-examined: The state, NGOs, and water users at Varanasi, Uttar Pradesh, India. *The Environmentalist*, *14*(1), 3–16.

Alley, K. D. (2002). *On the banks of the Ganga: When wastewater meets a sacred river*. Ann Arbor, MI: University of Michigan Press.

Barrows, H. H. (1923). Geography as human ecology. *Annals of the Association of American Geographers*, *13*(1), 1–14.

Brookfield, H. C. (1969). On the environment as perceived. *Progress in Human Geography*, *1*, 51–80.

Buechler, S. M. (1997). New social movement theories. In S. M. Buechler & F. K. Cylke, Jr. (Eds.), *Social movements: Perspectives and issues* (pp. 295–319). Mountain View, CA: Mayfield Publishing.

Census of India. (2011). *Census data summary*. Retrieved from http://www.censusindia.gov.in/2011-common/CensusDataSummary.html

Chapman, G. P. (1995). Introduction: The Ganges-Brahmaputra basin. In G. P. Chapman & M. Thompson (Eds.), *Water and the quest for sustainable development in the Ganges valley* (pp. 3–24). London: Mansell Publishing.

Clean Ganga. (n.d.). *The holy river Ganga (Ganges) is divine goddess for one billion believers the world over*. Varanasi: Sankat Mochan Foundation.

Cloke, P., Philo, C., & Sadler, D. (1991). *Approaching human geography: An introduction to contemporary theoretical debates*. New York, NY: Guilford Press.

Cumming, D. (1994). *The Ganges delta and its people*. London: Wayland.

DuPuis, E. M. (2004). *Smoke and mirrors: The politics and culture of air pollution.* New York, NY: New York University Press.

EcoFriends. (2008). *ECOFriends website.* Retrieved from http://www.ecofriends.org/

El Ramly, E. (2003). Women's perceptions of environmental change in Egypt. (Cairo paper in Social Science, Vol. 23, No. 4).

Eriksson, P. (2000). Territorial conflicts in the Northern atlantic coast of Nicaragua: Indigenous peoples' struggles over resources and representations. *Fennia, 178*(2), 215–225.

Escobar, A. (1992). Imagining a postdevelopment era? Critical thought, development, and social movements. Social Text 31/32: 20–56. Reprinted in J. Crush (Ed.). (1995). *The power of development* (pp. 211–227). New York, NY: Routledge.

Foucault, M. (1980). Questions on geography. In C. Gordon (Ed.). Power/Knowledge: Selected interviews and other writings, *1972–1977* (pp. 63–77). New York, NY: Pantheon Books.

Ganguly, M. (1999). *Holy war for "My Mother."* Time magazine. Retrieved from http://www.time.com/time/reports/environment/heroes/heroesgallery/0,2967,mishra,00.html

Good and Evil. (n.d.). *A call for comments.* Retrieved from sankatmochanfoundation.org

Goodman, M. K., Boykoff, M. T., & Evered, K. T. (Eds.). (2008). *Contentious geographies: Environmental knowledge, meaning, scale.* Burlington, VT: Ashgate.

Haberman, D. L. (2006). *River of love in an age of pollution: The Yumuna river of Northern India.* Berkeley, CA: University of California Press.

Hammer, J. (2007). A prayer for the Ganges. *Smithsonian, 38*(8), 74–82.

Hema, B., & Jamal, S. (2004). *Environmental perception of slum dwellers.* New Delhi: Mittal.

Jain, S. (2000). *Give green a chance, says Clinton with Taj as his backdrop.* Express India, April 3. Retrieved from http://www.expressindia.com/ie/daily/20000403/icl23079.html

Jaiswal, R. (2007). Ganga action plan: A critical analysis. Kanpur: *Kanpur ECOFriends.* Retrieved from http://www.ecofriends.org/main/eganga/images/Critical%20analysis%20of%20GAP.pdf

Johnston, R. J. (1991). *Geography and geographers: Anglo-American human geography since 1945* (4th ed.). London: Edward Arnold.

Kanpur ECOFriends (KEF). (2004). *Ganga is littered with corpses.* Retrieved from http://www.ecofriends.org/reports/032gangacorpses.htm

Kanpur ECOFriends (KEF). (2008). *Kanpur ECOFriends website.* Retrieved from http://www.ecofriends.org/main/about.htm

Larimore, A. E. (1969). Environment and spatial perception: An approach to research in African rural geography. *Canadian Journal of African Studies, 3*(1), 276–280.

Lonely Planet. (2010). *India: Varanasi.* Retrieved from http://www.lonelyplanet.com/india/uttar-pradesh/varanasi

McEwan, C. (1998). Cutting power lines within the palace? Countering paternity and eurocentrism in the 'Geographical Tradition'. *Transactions of the Institute of British Geographers, 23*(3), 371–384.

Miller, B. A. (2000). *Geography and social movements: Comparing anti-nuclear activism in the Boston area.* Minneapolis, MN: University of Minnesota Press.

Ministry of Environment and Forests. (2005). *Water quality.* Annual report, 2004–2005. Retrieved from http://www.cpcb.nic.in/annualreport04–05/ar2004-ch5.htm. Accessed on January 25, 2006.

Ministry of Tourism. (2006). *Survey on tourism statistics in U.P.* (April 2005 – March 2006). Retrieved from http://tourism.gov.in/CMSPagePicture/file/marketresearch/statisticalsurveys/06%20up.pdf

Mishra, V. B. (2005). The Ganga at Varanasi and a travail to stop her abuse. *Current Science, 89*(5), 755–763.

Rangan, H. (2000). *Of myths and movements: Rewriting chipko into himalayan history*. London: Verso.

Ridge, M. (2008). *Holy man, secular plan: Clean up the river Ganges*. Christian Science Monitor, July 23. Retrieved from http://www.csmonitor.com/World/Asia-South-Central/2008/0723/p01s01-wosc.html

Robbins, P. (2000). The practical politics of knowing: State environmental knowledge and local political economy. *Economic Geography, 76*(2), 126–144.

Routledge, P. (1993). *Terrains of resistance: Nonviolent social movements and the contestation of place in India*. Westport, CN: Praeger Publishers.

Sampat, P. (1996). The river Ganges' long decline. *World Watch, 9*(14), 24–32.

Sankat Mochan Foundation (SMF). *New website*. Retrieved from http://www.sankatmochan foundationonline.org/

Sankat Mochan Foundation (SMF). (2004). The holy river GANGA and her ancient GHATS are befouled. Varanasi, India.

Sankat Mochan Foundation (SMF). (2006). *Sankat Mochan Foundation [online]*. Retrieved from http://members.tripod.com/sankatmochan/index.htm

Sankat Mochan Foundation (SMF). (2007). Sankat Mochan Foundation clean Ganga campaign souvenir. Varanasi, India.

Sankat Mochan Foundation (SMF). (n.d. [a]). Sankat Mochan Foundation. Varanasi, India.

Sankat Mochan Foundation (SMF). (n.d. [b]). Sankat Mochan Foundation and its clean Ganga campaign. Varanasi, India.

Schjolden, A. (2000). *Leather tanning in India: Environmental regulations and firms' compliance*. FIL Working Papers, No. 21. Retrieved from http://www.cicero.uio.no/media/1677.pdf

Sederberg, P. (1984). *The politics of meaning*. Tucson, AZ: University of Arizona Press.

Sengupta, R. (2006). *The tragedy of the Ganges. TerraGreen 121*. Retrieved from http://www.teriin.org/terragreen/index.php?option=com_content&task=view&id=367&Itemid=3

Sivaramakrishnan, K. (1995). Colonialism and forestry in India: Imagining the past in present politics. *Comparative Studies in Society and History, 37*(1), 3–40.

Spivak, G. C. (1987). *In other worlds: Essays in cultural politics*. New York, NY: Routledge.

The Hindu. (2004). *Ganga action plan bears no fruit*. August 28. Retrieved from http://www.hinduonnet.com/thehindu/thscrip/print.pl?file=2004082807430400.htm&date=2004/08/28/&prd=th&. Accessed on June 26, 2010.

UNEP (United Nations Environment Programme). (2010). *Roll of honor/laureates*. Global 500 forum. Retrieved from http://www.global500.org/rollofhonour.html

Walker, P. (1998). Politics of nature: An overview of political ecology. *Capitalism, Nature, Socialism: A Journal of Socialist Ecology, 9*(1), 131–144.

Walker, P. (2005). Political ecology: Where is the ecology? *Progress in Human Geography, 29*(1), 73–82.

Wallace, B. (2006). *Wade in the water: Veer Bhadra Mishra and life's shallow pool*. Namaste: A journey through spiritual India [online], April 22. Retrieved from http://www.covering religion.org/2006/04/22/wade_in_the_water_veer_bhadra.html. Acessed on November 2, 2007.

Watts, M. (1990). Review of land degradation and society by piers blaikie and harold brookfield, 1987. *Capitalism, Nature, Socialism, 4*, 123–131.

Watts, M. (1997). Classics in human geography revisited: P.M. Blaikie's the political economy of soil erosion in developing countries. *Progress in Human Geography*, *21*, 75–80.

Whyte, A. V. T. (1985). *Guidelines for field studies in environmental perception.* Retrieved from http://scholar.lib.vt.edu/theses/available/etd-03072002-181243/unrestricted/04Chapter3LiteratureReview.pdf

World Health Organization (WHO). (2004). *The environment: Where's the risk, and where are children safe?* WHO press release [online], June 22. Retrieved from http://www.who.int/mediacentre/news/releases/2004/pr43/en/

THE MAYANGNA RESOLVE TO SAVE THE RAINFOREST, THEIR HOMELANDS

Nan Marie Greer

I dedicate this chapter to
The Mayangna of the past, present, and future.
Your leadership transforms the world.
May we all come to thank you.

ABSTRACT

For over 40 years in Nicaragua, the Mayangna indigenous group has fought for legal rights to traditional lands with the expressed purpose of protecting their rainforest. On December 21, 2009, the last of nine Mayangna territories were granted rights by Nicaragua to a majority of their historical claims, in addition to rights to have illegal colonists removed by Nicaraguan police and military. Indigenous leaders pursued land rights as a measure for cultural survival and the protection of their broadleaf rainforest, also the site of a UNESCO International Biosphere Reserve, the BOSAWAS. While Indigenous lands are encroached upon by the frontline of imperialistic consumerism, people like the Mayangna ask for international and national respect for their autonomy, self-determination, land ownership, and even sovereignty.

Occupy the Earth: Global Environmental Movements
Advances in Sustainability and Environmental Justice, Volume 15, 117–136

ISSN: 2051-5030/doi:10.1108/S2051-503020140000015004

The Mayangna lead the way to understand necessary steps for protecting the rainforest. Their actions demonstrate the possibility for social justice given respect for true ecologically sustainability. To begin, they fought to obtain ownership of their homelands, thereafter, they battled legally and even with their lives to defend their boundaries and everything within them. The Mayangna insist indigenous land ownership, the protection of their rights, and a respect for their traditional forms of management lead to the continued protection of the rainforest and other areas critical to the survival of the global ecosystem.

Keywords: Conservation; human rights; indigenous land rights; Mayangna; Nicaragua; U.N.D.R.I.P

Human rights are forever evolving in light of economic development and changes in the natural environment. Protecting their sovereignty, indigenous peoples are often forced to go outside the nation state for international persuasion and coercion upholding agreed upon laws/rights (both in national and international law). Hegemony or coercion in this case, must be used to enforce universal adherence to human rights. When outside coercion is no longer working, and ruling governments pay no credence to foreign persuasion, what is the next step? Is force justified? What is the duty of the international community in protecting the human rights of groups such as indigenous peoples? Can special attention be given to the evolutionary nature of human rights of indigenous peoples in light of economic development and environmental change, especially when the indigenous are the surviving guardians of the world ecology?

The Mayangna, a small indigenous group of only 40,000 in the heartland of the rainforest in Nicaragua, Central America, is gravely endangered, as is their home. This area was the frontline of the beginning of the Sandinista-Contra War in 1980, and it was in indigenous dissatisfaction with government assimilation policies (termed *Reincorporación*) geared toward national economic development that spurred the beginning of the Mayangna struggle for land, autonomy, and self-determination.

In November of 1991, the BOSAWAS Biosphere Reserve was created by Executive Decree 44-91, shortly after the end of the Sandinista-Contra war and the installation of President Violeta Chamorro. The purposes of the decree were twofold: (1) to preserve the biodiversity and natural resources of the Nicaraguan Humid Tropical Forests and (2) to secure and protect the lifestyles and cultures of the Miskitú and Mayangna indigenous people living within the Reserve. Additionally, the BOSAWAS Reserve was also established under UNESCO's Man and the Biosphere Program. This program establishes reserve areas where critical biological and human habitats are threatened.

The BOSAWAS Reserve is located in the northeastern region of Nicaragua. Totaling 8,000 km^2, it is one of the largest protected areas in Central America. It is the southernmost portion of a tract of tropical broadleaf forest that stretches to the northern coast of Honduras. This tract is the largest extension of tropical forest north of the Amazon basin and is the subject of several conservation efforts besides BOSAWAS, such as the Platano River Biosphere Reserve, the Tawahka Forest Reserve, and the Patuca River National Park, all in Honduras (Stocks, 1996). In addition to portions of five major rivers, the BOSAWAS' mountainous upland rainforest (ranging from 200 to 750 m in elevation) contains both coniferous and deciduous forest communities. The core area of the Reserve was until recently, little altered by human disturbance and is not very well inventoried, but it is known to harbor a wide diversity of species including jaguar (*Panthera onca*), tapir (*Tapirus bairdii*), harpy eagle (*Harpia harpyja*), and other endangered wildlife (TNC, 1996). The total area of the BOSAWAS Biosphere Reserve, including the buffer zone, is 18,000 km^2 and makes up 15% of the Nicaraguan territory. There are a number of distinct ethnic groups in and around the buffer zones of the BOSAWAS Reserve making up a population of just over 160,000. The most prominent indigenous people within the reserve are the Miskitú and the Mayangna, who make up 93% of the inhabitants.

INESCAPABLE DISASTER WITHOUT PREDICTIVE SOLUTIONS

Human rights have changed dramatically over time while nation states redefine them in accordance to the national economic interest. As national economies progress, human rights are purported to trend toward improvement, especially collective human rights. As national economies plunder, human rights often bear the brunt. Indeed, it is even the measure of human rights which provides validity to the nation state. As one notable human rights leader comments, "How a state treats minority populations has become a key factor in assessing its viability as a modern state (Gelazis, 2012)."

Throughout Nicaragua's history, as with most countries, its government has taken strides to accomplish what it articulates as *sustainable economic development*. When exploring this concept, it becomes evident that it cannot be achieved without the acquisition, commodification, and sale of nature. "Sustainable development draws on two frequently opposed intellectual traditions: 1) limits which nature presents to human beings; and 2) the potential for human material development locked up in nature (Redclift, 1987)." Many claim there can be no such thing as sustainable development, as it is clearly an oxymoron. Economic systems arise from exploiting nature, and it is just that exploitation that both liberates and violates human rights. Most notably, minority groups suffer human rights violations under the guise of sustainable economic development. "Indeed, the argument could be put that containing economic demands for material advance, in a highly unequal world, requires political measures that are so *authoritarian* they would immediately contradict the liberating, humane objectives that would make development sustainable in the first place (Redclift, 1987, pp. 199–200)." In Nicaragua, this has come to mean an invasion of the lands of indigenous peoples located in areas the government has referred to as "the agricultural frontier" or as Mayangna (Sumu/Sumo/Smu)[1] call them, "Home," and as conservationists call them "the Lungs of the World" (The Economist, 2010).

BBC notes the complicated nature of the country's economic history: "Nicaragua has traditionally relied on agricultural exports to sustain its economy. But exports benefited mainly a few elite families of Spanish descent, primarily the Somoza family, which ruled the country with the United States backing between 1937 and the Sandinista revolution in 1979 (BBC, 2012)." Indeed, over the years, Nicaragua suffered from decades of dictatorship rule, revolutions, and even civil war. By 1990, when the Sandinistas were defeated in elections held as part of a peace agreement, Nicaragua's per capita income had plummeted and its infrastructure was in ruins. Notable, at this time, children in grade schools learned to count with bullets and were trained to clean weapons. Peace after the Sandinista-Contra civil war was the progenitor of subsequent economic growth, lower inflation, and lower unemployment. However, in 1998 Hurricane Mitch hit the country and "rendered 20% of the population homeless and caused billions of dollars worth of damage (BBC, 2012)."

The disaster brought on by Hurricane Mitch affected the indigenous communities of the North Atlantic Autonomous Region (R.A.A.N) greatly in addition to other communities in the region. The inhabitant's houses were destroyed, all of their food crops were eliminated, and almost all of their domestic animals were lost. The tremendous ruin brought on by the hurricane was devastating to these communities that relied heavily on subsistence strategies of fishing, hunting, gathering, and swidden agriculture, a highly adaptive form of agriculture for eco-regions of heavy rains causing leeching of nutrients (also known as shifting agriculture). In addition to this, the minimal support they gained from regional and national sources was temporarily cut off, due to destroyed radio communication and transportation lines. Even those bathing unawares in the dirty river waters encountered landmines from the 1980s that had dislodged in the floods of the hurricane, and floated downstream, occasionally blowing up upon contact. Certainly the chaos of dictatorship, revolution, civil war, and natural disasters have caused great poverty among the indigenous and colonists.

GOVERNMENTS TAKE ADVANTAGE TO IMPOSE NEW SYSTEMS

To escape the hardships recorded in its history, the Nicaraguan government has turned to "sustainable" economic development, choosing to emphasize the beef industry, grains, as well as logging and gold-mining, made possible by taking lands from rainforest indigenous people.

According to the US Central Intelligence Agency, Nicaragua is the poorest country in Central America and the second poorest in the hemisphere, with widespread underemployment and poverty (2014). With a population of 5.8 million, and a Gross National Income per capita of only US$1,650 annually, the country's primary exports are coffee, meat, shellfish, sugar, tobacco, cattle, and gold (World DataBank, 2012).

Aside from its limited local economy, Nicaragua relies on international assistance to balance its national budget and spending. On July 20, 2003, *La Prensa* reported that the World Bank provided over 66% of Nicaragua's annual income. This percentage rose in 2010, falling only slightly in 2011. The United Nations reported $963.4 million in Nicaraguan public debt, making up 78.5% of GDP in 2010, falling slightly to $1.259 billion or 70.5% of GDP in 2013.

Even though in early 2004, Nicaragua secured some $4.5 billion in foreign debt reduction under the Heavily Indebted Poor Countries (HIPC) initiative, much of the country continues to remain below the World Bank-accepted poverty level, with 46.2% recorded in poverty in 2005. Though the reported unemployment rate for 2011 was 7.3%, slightly lower than 7.8% in 2010, UNDER-employment was a striking 46.5% in 2008 (UN, 2012).

Certainly much of the country is embroiled in the struggling nature of the national economy, yet its current financial depression is felt no greater than in the indigenous Atlantic Autonomous regions of Nicaragua. "Puerto Cabezas is the poorest corner of Nicaragua. Unemployment stands at around 80% (BBC, 2009)." Indigenous Mayangna, Rama, Kriol (or Creole), and Miskitu suffer massive human rights violations due to the incursion of illegal colonists in search of the environmental riches of their lands, after having suffered a grueling history of such violations throughout colonial and post-colonial times (Williamson, 1997). Indeed, indigenous lands have been viewed as an escape valve for population and economic pressures in urban areas of the country. In this regard, indigenous human rights in Nicaragua have repeatedly been redefined and reinterpreted in light of the economic development of the nation state.

AGREEMENT ON ECONOMIC DEVELOPMENT WITHOUT AGREEMENT ON HUMAN RIGHTS

Though the United Nations and its member states have agreed upon concepts of universal human rights (Universal Declaration of Human Rights of 1948, UN Declaration of the Rights of Indigenous Peoples of 2007 among the examples), their enforcement and interpretation varies widely from country to country. Human rights being based on basic needs for survival, also vary from country to country, and culture to culture, and include the need to satisfy rights for both the individual, and the collective. Human rights ensure access to agreed upon needs for survival, and by nature are socially constructed. Deeply embedded in the socio-cultural relativity, needs vary from place to place, and temporally. Needs certainly are not described universally, as there is great variation, and yet, to survive, humans must mutually agree on some form of political state or laws mandating their adherence. Author Donnelly (2003) suggests that perhaps economic development may require a sacrifice of some human rights. As a controversial matter, this moral dilemma outlines an appeal to the constantly transforming nature of the universal human rights model. Concern must turn toward the issue that, "those unable to enjoy human rights, will be estranged from their moral nature (Donnelly, 2003, p. 14)."

If the world struggles on agreement as to the universality of human rights and needs, can there possibly be an agreement about a specific form of economic development as a *natural right of man*? Such a pursuit appears even more puzzling. As an example, the Declaration of the Natural Rights of Man and of the Citizen was agreed upon in 1789, yet great human injustices have occurred until today where indigenous human peoples' human rights have been sacrificed in the name of colonial or national sustainable economic development. Even colonizing "Americans" viewed their conquests as divine. They relied on *The Doctrine of Discovery* utilized first by Catholic Pope Nicholas V in 1452 to allow Portugal to claim lands in West Africa, and again by Catholic Pope Alexander VI in 1493 to allow Spain to claim lands in the Americas. This doctrine, along with philosophies of *Manifest Destiny* suggested it was the natural superiority of (Christian/civilized) Europeans over others that lead to the logical conclusion of their right to colonize. The colonial pursuit was rather seen as "civilizing native savages/barbarians" and creating economic beachheads of their lands and resources that would tap the outer peripheral areas of the globe and bring their riches home to the motherland. Colonists in Nicaragua, as elsewhere, established a globalizing imperialistic economic force that would never leave the indigenous territories, and would continue violating their rights. Ironically, this world-systems global economic inequality is exactly what the World Bank used in 2003 to force Nicaragua to legally title lands to the Mayangna territory of Awas Tingni (ultimately awarded by the Nicaraguan government in 2009).

Though common elsewhere in Central America, Spanish colonization did not occur on the eastern Atlantic coast due to a perceived lack of gold in the coastal inlet areas, the challenges of the humid/tropical climate, and the geographically inaccessible nature of the highlands (Romero Vargas, 1992, 1995). It wasn't until the late 1500s that the first pirate camps were established by European buccaneer expeditions intent on exploiting the great riches of the Atlantic coast of what is now Nicaragua. In the 1740s, after a period of slow extraction of minerals by foreign buccaneers, British land settlements began to be established for the purpose of trading, forming more buccaneer trading stations, woodcutting, and some plantation development. This established a colonizing beach head, initiating a succession of imperial powers that would dominate the region's politics, economy, natural resource extraction, and even regional identity (Hale, 1994; Romero Vargas, 1992, 1995, 1996). At the heart of this external presence was the desire to maintain an ongoing source of natural resource extraction for the growing foreign and colonial economies. This tradition has continued through today, as the pacific federal government continues to utilize indigenous lands to satisfy the national debt, or fails to control resource extraction to the detriment of the indigenous people such as the Mayangna. Unfortunately, this is no different from the global experience of indigenous groups.

As of 2014, the Nicaraguan Ministry of Environment and Natural Resources (MARENA) has calculated that 35.2% of the rainforest's conservation zone had been cut/destroyed by illegal colonists (loggers) (*La Prensa*, Bermúdez, 2014). Strikingly, notable politicians from the National Assembly have commented: "We are inept to protect the riches of the BOSAWAS (*La Prensa*, Aug., 2012)." This recognition provides a unique challenge to Nicaragua and the protection of the world's precious rainforests. It points to an obvious strategy of relying on indigenous land ownership and control to protect what they (nation states) are unable to protect. What a twist of fate indeed to recognize ineptitude and the need to return that which was taken in "*divine right*," to its original stewards and caretakers, those who have been marginalized and dispossessed of these same lands. The global community must clearly recognize how hard it is to admit fault, yet to heal the rainforest and its people and species, it is perhaps the initial step.

As documented in previous census and territorial studies (BOSAWAS/MARENA/GTZ, 1996; Greer, 2004; Greer, Murrar, & Stocks, 2004; TNC/USAID, 1996–1998) indigenous inhabitants in rainforest areas of Nicaragua have significantly outnumbered colonists (*mestizos*), while colonists have greatly outnumbered indigenous in the buffer zones of the rainforest areas. In 2009, *La Prensa* reported an estimated 18,000 illegal armed colonists within the core conservation zone of Mayangna heartlands. While the Mayangna depend on the rainforest for over 73% of their household income and consumption (Greer, 2004; Greer, Murrar, & Stocks, 2004), MARENA calculates 60% of the buffer zone of rainforest lands have been destroyed by illegal colonists (Bermúdez, 2014). MARENA, military, and police have been unable to completely study populations in the core conservation zone due to armed illegal loggers, *mestizos*, or excombatientes.

Putting political agendas aside, the Mayangna are found to conserve the most, while simultaneously being dependent on the rainforest for their subsistence and cultural survival (Greer, 2004; Stocks, 2007). Using Satellite Imagery from three different time periods, analyzing forest cover and conservation practices, indigenous farmers are found to deforest the area significantly less than colonists. (Stocks, 2007, p. 6). As Stocks states: "Over the 15 years covered by the satellite images in indigenous territories

0.24 ha/capita and 0.15 ha/capita were deforested in 1995 and 2002, respectively. In contrast, in the colonist area, deforestation rose from 1.65 ha/capita in 1995 to 2.50 ha/capita in 2002, over 16 times the indigenous ratio (*ibid.*)." In fact, Stocks reports (2007, p. 6) from 1995–2002, indigenous per capita deforestation declined!

When indigenous leaders protest for government support, the state turns a blind-eye on the regions, allowing illegal natural resource extraction and the destruction of the rainforest ("lungs of the world") to occur despite its clear and violent infringement of collective human rights, and consequentially, global human rights. In the meantime, Mayangna culture and lifestyle are severely threatened by logging, mining, and hydroelectric projects geared to benefit the entire country, and its citizens and beyond. The Nicaraguan federal government has historically insisted on the need to harvest trees, gold, etc. on such lands in the name of national economic development.

UNACHIEVABLE "SUSTAINABLE ECONOMIC DEVELOPMENT" AND THE VIOLATION OF HUMAN RIGHTS

Conversely, if sustainable economic development is not obtainable, such as in the case of severely indebted countries, clearly human rights are violated all around (as Amnesty International has affirmed on many occasions in Nicaragua, most notably in 2010). Nicaragua has stated that illegal colonists will be removed by law if no other resolution can be achieved, (Nicaraguan laws 28 and 445). While these laws, if upheld, would protect the rights of indigenous Mayangna to their lands (and resources), clearly the rights of *mestizo* excombatants (soldiers) would be violated, as members of the nation state guaranteed them such lands (illegally offering indigenous lands) to help them demobilize from the civil war ending in 1990 with new Presidential elections.

Notably, President Arnaldo Aleman (president of Nicaragua from 1997 to 2002) illegally granted rights of rainforest lands to 800 *mestizo* families of ex-combatants in the indigenous territory of the Bawihka-Mayangna (Greer, 2008; Greer, 2009a, 2009b). Since then, armed excombatants have come in droves to claim free "frontier" lands. Various members and administrations of the federal government have guaranteed excombatants these lands over the past decade as a means of ensuring peace with soldiers who had no economic source of income. Fighting against this, indigenous land laws passing the National Assembly (notably 28 and 445) have included eventual procedures for removal of illegal colonists (nonindigenous) from titled lands. With all nine territorial lands titled to the Mayangna as of 2010, all their lands have been eligible for the colonist removal phase. In fact, this last phase is legally structured to occur within six months of receipt of indigenous title, yet this process has not started. On the contrary, colonization is accelerating.

PROTECTING SOVEREIGNTY, INDIGENOUS TURN TO THE INTERNATIONAL COMMUNITY

Stuck in the conundrum of how to support indigenous rights to land (Nicaraguan laws 28 and 445; UN Declaration of the Rights of Indigenous People, International Labor Organization – Convention 169), the Nicaraguan government is embroiled in a new reality and judicial mess: Will the human rights of colonists be violated with a potential forced removal, as described in Nicaraguan law 445? To where will the excombatientes/illegal colonists/loggers be moved? (Some have been there for over nine years, beyond that required for Nicaraguan squatter's rights.) While the federal government delays the last phase of indigenous land titling, indigenous peoples in Nicaragua often are forced to go outside the nation state for international persuasion and coercion on prior agreed upon laws/rights (both in national law and international law) – to protect their sovereignty (Anaya, 2009a, 2009b; Anaya & Grossman, 2002; Greer, 2009a, 2009b). Even the US government supported indigenous militia throughout the Sandinista-Contra War in their resistance against national projects of *reincorporation* or, a giant national land grab. Hegemony/Coercion in some cases is used for enforcing universal adherence to human rights.

Nation state indigenous group interaction has been perpetually and continuously conflictive. Indigenous groups by their nature challenge the overall autonomous authority and even at times, wealth, of the nation state. Indeed, even its proposed existence is a result of colonial history and dominance of one human group over others. Mayangna lands make up 17% of the entire country of Nicaragua; overall indigenous lands claim a total of nearly half of the country, with the R.A.A.N. and R.A.A.S. regions (North Atlantic Autonomous Region and South Atlantic Autonomous Region). Mayangna lands hold the nation's riches, namely, gold, Mahogany (*Swietenia macrophylla*), Mexican Cedar (*Cedrela odorata*), Pine (*Pinnus caribaea, Pinnus oocarpa*), petroleum, and high-gradient rivers capable of producing great hydropower. Lands of such nature surely are priceless to the federal government and national collective riches.

This became clear when two days post Hurricane Mitch in 1998, President Arnoldo Aleman proposed a land law to "give" title to the Mayangna of over 8,000 km^2, but not allow them rights to the resources of those lands (Greer, 1999, Nicaraguan Law: "The Law that Regulates the Demarcation of Communal Property Regime of Indigenous Communities of the Atlantic Coast and BOSAWAS"). On October 13, 1998, only a week before the hurricane, Nicaraguan President Arnoldo Aleman introduced to the National Assembly, his administration's version of the "Law that Regulates the Demarcation of the Communal Property Regime of the Indigenous Communities of the Atlantic Coast and BOSAWAS." This decree, was in part in direct response to the World Bank GEF (Global Environment Facility) Atlantic Biological Corridor program, which insisted the government stabilize the land tenure system in the Atlantic Autonomous Regions to protect the ecosystems which made up the regional corridor's forest lands. While attempting to comply with the World Bank's demands, Aleman proposed a law that had many elements that, "allowed the non-renewable natural resources to continue to be owned by the state and permitted the leasing of these lands by outsiders, respecting the rights of those who are already in possession of those lands to be decreed as indigenous communal properties" (comments made by prominent Nicaraguan Environmental Lawyer, Lilliam Jarquín). This law, though not widely shared and communicated with the indigenous communities of the 16 Atlantic Coast, did not satisfactorily address the issues of direct land tenure in the BOSAWAS of the Atlantic Coast. At the onset of the political proceedings on this law in the National Assembly, militia groups began forming in protest, creating yet another inflammation of violence and a return to the politics of prewar indigenous coalitions.

Because of this egregious law, the Mayangna and others consulted indigenous communities in the process of developing Nicaraguan land law 445 (2003) in support of law 28 (1987), guaranteeing their rights to traditional lands, and establishing a judicial method for eventual removal of illegal colonists. While these laws passed the National Assembly, outsiders looked on with a suspicious eye to the future of such legislation.

Examples such as this bring to light as to how global civil society makes allowances for national human rights violations. Indigenous territories more frequently combat injustices by forming and relying on bilateral, multinational nongovernmental organizations (NGOs) to support the human rights of those under oppressive/hegemonic political systems. In such a case, international assistance has been partly responsible for enforcement, more so than the nation state/politic.

AWAS TINGNI RECEIVES TITLE TO TRADITIONAL LANDS AND MAKES HISTORY

After a 35-year struggle for land rights, in 2001, Awas Tingni won its landmark case in the Inter-American Human Rights Court (IAHRC) against the Nicaraguan state, and began a land titling process with the federal government that would astound the international community. Nicaragua initially refused to act on the regional international court decision. Shockingly, in 2003, the World Bank announced they would remove all financial assistance given to the government of Nicaragua if they were not to title ancestral lands to Awas Tingni, as mandated by the IAHRC of the Organization of American States (O.A.S.). This amounted to approximately 66% of the national budget at that time, and an overwhelming figure that twisted the arm of the federal government, and established Awas Tingni as the first Indigenous Multicommunal Territory in Nicaragua to have their traditional land claim adjudicated successfully in the Inter-American Human Rights Court (Anaya, 2009a, 2009b).

As such, Awas Tingni was challenged to present a methodology of court documentation worthy of sustaining an award of their lands in perpetuity, with inalienable character. In addition to the other Mayangna territories, Awas Tingni utilized an interdisciplinary approach including triangulation methodologies to ensure accuracy and limit bias. In addition, research was conducted utilizing a community-participation approach, incorporating several disciplines and scientific branches in the analysis of multiple streams of data from various epistemologies, verified against one another to validate the results in an interdisciplinary manner. With the help of geography, sociology, ethnography, socioeconomic studies, anthropology, and environmental sciences, the court report allowed for the systematization of an integrative approach toward a description of the ecological and social life in an interactive process. This process established a claim to land, in addition to reporting the adaptive management model of land use and planning utilized by the territory. In analyzing the land use patterns, an assessment was made of values, customs, uses of land, and present and historical occupation of the territory. It included an assessment of land and natural resources relating to the foundation of the culture, spiritual life, and economic survival of the territory. These methods focused on oral testimony, focus groups of men and women, empirical data collection of census and forest resource harvest rates. Ultimately, a map of the land claim, use and management was created through training community forest guards to utilize Global Positioning Systems (GPS). In addition, territorial members received technical assistance managing their own Geographic Information System (GIS) and digital display of their homelands. In a process of "conflict negotiation," Awas Tingni met with other communities who expressed rights/crossing-claims to the same lands – exploring resolutions for illegal colonist removal and allowances for colonist presence.

Empirical data were collected with the full involvement of trained local investigators appointed by the community. Data collection, processing, interpretation, and validation occurred with the direct involvement of the territory. In this regard, the territorial residents participated in the activities of the process, both expressing themselves in general, and participating in association with outside scientists. These residents made a formal claim to their ancestral lands, even training other Mayangna territories in the methodology to further their claim to lands as a sovereign Mayangna Nation.

Awas Tingni's efforts spanned the years of 2003 and 2004, offering a year's worth of data collection on household census and forest harvest rates that would "justify" awarding 90,000 hectares of ancestral lands to this Mayangna territory. Ultimately, the original claim of 150,000 hectares was dramatically reduced, as government officials argued it was simply "Too much land, for too few people (Stocks, 2005)," an argument that clearly had no logic with an indigenous group who had cared for the lands for millennia.

Despite this effort, it was not until 2009 that the lengthy diagnostic study of the territory was reviewed by the regional government. As a result of this review, Awas Tingni finally received title to 90,000 acres, 60,000 acres shy of their original ancestral claim. Additionally, as of 2014 at the time of this writing, the Mayangna, including the Awas Tingni, have sent over 10 letters to Nicaragua President, Daniel Ortega, to protest the lack of resolution of land conflicts with colonists within their legal claim, guaranteed within six months of receiving title (D. N. Coleman, personal communication, April 23, 2014). In addition to other Mayangna territories, Awas Tingni leaders have contacted the office of the UN Special Rapporteur on the Rights of Indigenous Peoples several times in this year alone to report continual violation of the 2001 Inter-American Human Rights Court decision. Sadly, Awas Tingni is one of 9 Mayangna and 21 other indigenous territories experiencing the same response from the government. Mayangna indigenous members recall sadly the times of the Sandinista-Contra War when their territories were torn apart between political groups and forced to fight each other. Brothers and sisters fought cousins, aunties, and uncles on their territory that had become the frontline of civil war. Some have commented as to the development of post-traumatic stress disorder (PTSD) in these indigenous communities, as a reaction to the dire trauma experienced by community members during the 10-year civil war on their lands (Ross, 2005). Indigenous community members report they do not want a return to such times, yet they are bound by their identity to fight for their lands and those of their grandchildren (Greer, 2009a, 2009b).

Surprisingly, while national and indigenous governments remain imbued with the reality of illegal colonization of indigenous lands, international and national NGOs helping Mayangna territories have had their bank accounts frozen and their projects re-routed through the federal government where funds disappear. This was the case of the 20 million dollar Green Heart/*Corazon Verde* project of MARENA (Greer, 2008, 2009a, 2009b; Gutiérrez, La Prensa, June 4, 2012). In an attempt to promote global enforcement of indigenous human rights, it appears many of these NGOs are hegemonic, perhaps even violating the national sovereignty of the country of Nicaragua, as claimed. Yet both the indigenous peoples and the international community recognize that the only method of enforcement is to garner international support with organizations such as the UN and the OAS, perhaps requiring multiple returns to the IAHRC.

WHEN OUTSIDE COERCION NO LONGER FUNCTIONS

External efforts to protect indigenous human rights in Nicaragua have continued for well over 40 years, and yet it appears outside coercion is not functional. President Ortega, often called a dictator, appears to pay no credence to foreign coercion and persuasion – pursuant to asserting national sovereignty perhaps. Admittedly, the imperialistic Yankees to the North have long been an overly hegemonic and aggressive force. However, in 2009, in a dramatic show of protest, foreign aid agencies left due to Ortega's lack of addressing human rights violations. The countries of Denmark, Switzerland, Sweden, and Great Britain, among others, threatened to leave the country they had been so dedicated to in their foreign assistance. The same year, the US Consulate made a surprising and shocking public statement: "Though we have invested millions in Nicaragua over the past 20 years, nothing has changed (Callahan, 2011; Rogers, 2011)." In 2012, even the German aid agency GIZ (previous GTZ), one of the largest foreign state donor to Nicaragua, announced their threat to depart from the country due to problems with dictatorial rule. According to Mikkelsen (2014, p. 96), "Financial support from international cooperation declined by 67% in 2013 ... and was at its lowest for 19 years. Austria, the Netherlands and Finland all withdrew ....The space in which Nicaraguan NGOs are able to operate has been reduced. Not only is financial cooperation disappearing; bodies that are not aligned with official government policy are being persecuted, and some indigenous authorities have received orders not to work with NGOs."

As international agencies leave en-mass, and years have gone by where the Mayangna, among other indigenous groups, have fought incessantly for their lands and basic needs, and while the Nicaraguan state has persisted as the second poorest country in their hemisphere, what is left of indigenous human rights in Nicaragua? How are the Mayangna protected, when most bilateral human rights agencies and their funding have fled the country, and the Nicaraguan president refuses to address human rights concerns, such as those of the Mayangna?

Donnelly argues that the international community has a moral responsibility to combat the violations of human rights even with force, if necessary (2003). Obviously this just-war condoning of violence is anti-Gandhian, against the notable examples of UN leader Ban Ki Moon, and South Africa's Nelson Mandela, among others. Yet, as some social groups are stuck with surviving on much less than the globally agreed upon poverty level of $1.25/day (World Bank, 2014), it is clearly understandable why they may rise up and defend their ability to survive, even in areas first world nations strip with voracious and obsessive consumerism. What should be the international community's responsibility to this dilemma?

WHAT IS THE NEXT STRATEGY?

Foreign Aid agencies fled the country to pressure Nicaragua in a first step toward a country embargo. However, this has only resulted in an increase in human rights violations among indigenous people like the Mayangna, having no support and little help in an extremely volatile case, the same that sparked CONTRA involvement (or outside US-CIA presence) initiating the civil war in 1980. It was indeed in the Mayangna territories that the resistance of the 1980s was first born. Surely, Mayangna human rights cannot be ignored, for to do so would be to repeat the horrific history of war, as indigenous leaders promise, not to mention destroying critical rainforest. Yet, most actors on the scene agree, there is no good plan on how to litigate and ensure indigenous rights guaranteed in both national and international laws. It is clear that removing colonists in these lands will cause violence, as will not removing them. While Nicaragua strives to improve its national economy, indigenous human rights, and those of other minorities, will be violated. Unfortunately, this has been the case for centuries, and yet no new model has formed to protect both the individual right and the rights of the collective, with truly successful results.

A new strategy, however, has emerged in the case of the Mayangna. The last two territories to become titled to the Mayangna conducted their own "Self-Demarcation" project, whereby lands were mapped, demarcated, and a court case presented with virtually no funding and support from the outside. As the last two territories of Mayangna, Sauni Arungka (Matumbak) and Tuahka, stood to receive title in December of 2010, a new sense of pride emanated from their autochthonous/locally derived independence efforts – they had proven the ability to defend themselves legally with close to no outside help, using the legal mechanisms available to them to do so. Their success fueled further flames of eco-brigades, youth protests, and loud cries of protest that the international community could not, and cannot ignore. Concurrently, the Mayangna experience in land titling has provided an international example of viable approaches to recovering indigenous land rights. While the Mayangna land titling process has hit a standstill regarding the removal of illegal colonists, the strides they have made up to this point suggest a solution is in the making. The Mayangna continue to work in their territories to further defend their rights to land and peace without the trauma or threat of violence or war, as politicians continue with their power struggles.

Due to the constant influx of colonists, in February of 2013, "the Sumu-Mayangna people – in alliance with civil society organisations – held public protests and organised road blocks demanding the establishment of the Special Regularisation Commission of CONADETI along with a state budget for its operations. The immediate outcome of this was that five people were wounded in clashes although, over the year, one Mayangna and two settlers died as a consequence of conflict on the territories (Mikkelsen, 2014, p. 93)." Due to these actions, and alongside protests of Mayangna leaders, Decree 15-2013 was enacted, and the Inter-Institutional Commission for the Defense of Mother Earth on Indigenous and Afro-descendant Territories in the Caribbean and Alto Wanki-Bocay was established. "It is notable that, during these processes, the F.S.L.N.'s (Sandinista National Liberation Front) political secretary for the Mining Triangle in the R.A.A.N. (Mayangna territory), promised the *mestizos* that no-one would be evicted from the lands they were in possession of because it was not the policy of President Ortega's government to do so; moreover, at least 17 trucks containing timber from indigenous territories were stopped in the R.A.A.N. and found to have no documentation or extraction permits. Days later, it emerged that the wood belonged to the Alba-Forestal company, linked to the President of the Republic, and the trucks and timber were subsequently released under police and army escort (Mikkelsen, 2014, p. 96)." James Anaya, UN Special Rapportuer on the Rights of Indigenous Peoples, observed the case and noted his concern for the lack of effective measures, and the deforestation on reserves that put the habitat of the indigenous peoples living there at risk (*ibid.*). The Mayangna now fear a reform of the 445 law, ultimately legislating "cohabitation" of indigenous with the illegal colonists in their territories, quite different than an "offer of coexistence" extended by the territory to colonists after the territorial authorities themselves come to a cohabitation conclusion.

In an unusual move, the Atlantic Autonomous regions were flooded by Pacific coast *mestizos* bent on transforming the make-up of the regional indigenous government. Shortly thereafter, a regional election was called, and political power thus transferred from indigenous to colonist. While Nicaraguan law 445 allowed for the transformation of municipalities in the Autonomous Regions, this now means the takeover of indigenous territorial governments and their respective lands (representing multicommunal ancestral land claims) by the outnumbering colonists/*mestizos*/excombatants in a legalized approach, supported by the Pacific national government (Alvaréz, 2014; Artola, 2014; Confidencial, 2014; Picón, 2014; Romero, 2014).

Perhaps not surprisingly then, in April of 2014, after a request by UN Special Rapporteur on the Rights of Indigenous Peoples, Dr. James Anaya, to enter the country of Nicaragua for indigenous rights negotiations, President Daniel Ortega issued a response: The UN human rights negotiator was not welcome and would be turned away at Nicaragua's border. No longer did the nation state entertain the enlightened approach of liberating indigenous communities and their lands. Despite this, the Mayangna continue to claim, "We have never been subject to anyone's rule (D. N. Coleman, personal communication, 2014)." Despite the federal, Pacific approach, Nicaragua's indigenous peoples remain on the forefront of the fight for indigenous land rights.

Clearly the Mayangna experience stands as a global example of indigenous human rights struggles to the approximate 370 million indigenous peoples worldwide. A methodology has worked, and in one of the most poverty-stricken countries in their hemisphere – indigenous people have filed their claim, and acquired recognition of their land rights. While the colonist issue remains, the Mayangna have accomplished what seemed the impossible – they recovered their ancestral land claim and are its rightful inalienable owners now and in perpetuity.

It was written in the most famous book of the world: "Blessed are the meek, for they will inherit the earth (Matthew 5:5)." And now, I stand smiling from ear to ear, truly my prayers and those of countless others, are being answered.

The time seems ripe to tackle the difficult question of colonist occupation and Mayangna rights – to seek solutions in line with the examples set by those previously mentioned – Gandhi, Nelson Mandela, and a host of others whose nonviolent philosophies achieved major victories for all people. This is where efforts and energy may be expended in the near future to encourage the Mayangna people in pursuit of a permanent resolution to the problems with both indigenous rights and destruction of an irreplaceable eco-region – the BOSAWAS and other areas of this Central American endangered rainforest.

NOTE

1. The current name "Mayangna" is a new term meaning "us," which has come to replace prior terms used by colonists and Miskitú to describe the indigenous group, using terms such as "monkey's in the trees" and "barbaric" (previous terms: Sumu, Sumo, Smu).

ACKNOWLEDGMENT

This writing has been sponsored by the University of Hawai "i – Kaua"i Community College's National Science Foundation – Tribal Colleges and Universities Program (T.C.U.P.), grant #HRD – 1265462. Special thanks to the Mayangna people, Noe Coleman, Eng. Agustin Jarquin Anaya, Ignacio Javier Hernandez Policarpo, Claus Kjærby, Dr. James Dire, Dr. James McManis, and my family, dad, mom, Michael and, my daughter Kristina.

REFERENCES

Alvaréz, L. (2014). Mujeres de Yatama impede de lucha. *La Prensa*, 10 de Mayo, 2014.

Anaya, J. (2009a). *International human rights and indigenous peoples*. New York, NY: Aspen Publishers.

Anaya, J. (2009b). Nicaragua's titling of communal lands marks major step for indigenous rights. *Indian Country Today*, January 5, 2009.

Anaya, J., & Grossman, C. (2002). The case of Awas Tingni v. Nicaragua: A new step in the international law of indigenous peoples. *Arizona Journal of International and Comparative Law*, *19*(1), 1–15.

Artola, G. A. (2014). Invasión de tierras indígenas es alarmante. *El Nuevo Diario*, 1ero de Mayo, 2014.

BBC. (2009). Nicaragua's Miskitos seek independence. *BBC*, August 3, 2009.

BBC. (2012). Nicaragua country profile. *BBC*, August 20, 2012.

Bermúdez, J. C. (2014). Le quedan 23 años a Bosawas. *La Prensa*, Mayo 21, 2014.

Callahan, R. J. (2011). Why Nicaragua matters to the United States. *Washington Post*, November 3, 2011.

Confidencial. (2014). Mujeres del partido indígena reclaman por fraude electoral en el Caribe: CSJ desestima recurso de amparo de Yatama. 20 de Mayo, 2014.

Donnelly, J. (2003). *Universal human rights in theory & practice*. Ithaca, NY: Cornell University Press.

Gelazis. (2012). Rights of persons belonging to national minorities, including: National minorities preventing aggressive nationalism, racism and chauvinism. Remarks from Scholar Nida Gelazis at OSCE Session on National Minority Rights, Woodrow Wilson Center Senior Associate Nida Gelazis, Warsaw, Poland, October 3, 2012, Session 12, *OSCE Human Dimension Implementation Meeting*.

Greer, N. (2004). Awas Tingni: Project review and discussion. *Report of the Awas Tingni land titling project*. Managua, Nicaragua: World Bank-Alistar International.

Greer, N. (1999). *The indigenous responses to HurricaneMitch, in Light of their Historical Struggle for Self-Determination and Land Tenure in the BOSAWAS International Biosphere Reserve, Nicaragua*. Seattle, WA. University of Washington.

Greer, N. (2008). *Mayangna indigenous land titling project review and discussion*. Manuscript held by Alistar International, Managua, Nicaragua.

Greer, N. (2009a). *The season of the ants*. Managua, Nicaragua: Alistar International.

Greer, N. (2009b). *Fighting for peace in the season of the ants*. Managua, Nicaragua: Alistar International.

Greer, N., Murrar, G. A., & Stocks, A. (2004). Informe Sobre el Reclamo de Tierra de la Comunidad de Awas Tingni. [Report of the land title claims of the community of Awas Tingni, Nicaragua]. Project completed for PRODEP, a Nicaraguan government inter-institutional organization under funding by the World Bank. Project completed by Alistar Nicaragua, CIDCA and Idaho State University, 2004.

Gutiérrez, F. J. (2012). Día del ambiente, día del antifaz verde (Bye Bosawás). *La Prensa*, June 4, 2012.

Hale, C. (1994). *Resistance and contradiction: Miskitu Indians and the Nicaraguan state, 1894–1987*. Stanford, CA: Stanford University Press.

La Prensa. (2012, August). MARENA incapaz de vigilar BOSAWAS.

Mikkelsen, C. (2014). The indigenous world 2014. International Working Group of Indigenous Affairs, United Nations.

Picón, G. (2014). Mujeres Yatama rompen portones de CSJ. *La Prensa*, 19 de Mayo, 2014.

Redclift, M. (1987). *Sustainable development: exploring the contradictions*. London: Routledge.

Rogers, T. (2011). Callahan: Nicaragua's democracy has 'serious problems'. *Nicaragua Dispatch*, October 18, 2011.

Romero, E. (2014). CENIDH acompañará a mujeres de Yatama a CIDH. *La Prensa*, 20 de Mayo, 2014.

Romero Vargas, G. (1992). *Persistencia Indígena en Nicaragua*. Managua, Nicaragua: CIDCA-UCA.

Romero Vargas, G. (1995). *Las Sociedades del Atlántico de Nicaragua en los SiglosXVII y XVIII Colección Cultural*. Managua, Nicaragua: Banco Nicaragüense.

Romero Vargas, G. (1996). *Historia de la Costa Atlántica*. Managua, Nicaragua: CIDCA-UCA.

Ross. (2005). Nicaragua's crazy sickness. *The Walrus* (pp. 74–79).

Stocks, A. (1996). The Bosawas natural reserve and the Mayangna of Nicaragua. In K. Redford & J. Mansour (Eds.), *Traditional peoples and biodiversity conservation in large tropical landscapes*. Washington, DC: America Verde Publications.

Stocks, A. (2005). Too much for too few: Problems of indigenous land rights in Latin America. *Annual Reviews of Anthropology*, *34*, 85–104.

Stocks, A. (2007). Indigenous, colonists, and government impacts on Nicaragua's Bosawas reserve. In *Conservation biology*, *21*(6), 1495–1505.

The Economist. (2010). Forests and how to save them: The world's lungs. *The Economist*, September 25–October 1.

The Economist. (2011). The tormented Isthmus. *The Economist*, April 14, 2011.

The Nature Conservancy. (1997a). *Mayangna Sauni As: Tradición oral de la historia y estudio socioeconómico de las comunidades Mayangna de la cuenca del Río Wapuk*. Managua: Impresiones Modernas.

The Nature Conservancy. (1997b). *Mayangna Sauni Bu: Documentación del reclamo histórico de las comunidades Mayangna de la cuenca del Río Bocay*. Managua: Impresiones Modernas.

The Nature Conservancy. (1997c). *Sikilta (Mayangna Sauni Bas): Historia y situación actual de los Mayangna Balna del Uliwas*. Managua: Impresiones Modernas.

The Nature Conservancy. (1997d). *Kipla Sait Tasbaika: Tradición oral y estudio socioeconómico de las comunidades indígenas del sector raudales*. Managua. Impresiones Modernas.

The Nature Conservancy. (1997e). *Miskitu indian tasbaika kum: Historia y situación actual de las comunidades Miskitu del alto Coco*. Managua: Impresiones Modernas.

The Nature Conservancy. (1997f). *Población mestiza en el suroeste de la reserva Bosawas: Estudio socioeconómico*. Managua: The Nature Conservancy.

Williamson, C. D. (1997). *Análisis Económico de la Inversión Extranjera Directa en la Costa Atlántica de Nicaragua 1969–1978*. Managua, Nicaragua: CIDCA-UCA.

World Bank. (2014). Poverty headcount ratio at $1.25 a day (PPP) (% of population).

World DataBank. (2012). GNI per capita, Atlas method (current US$).

ENVIRONMENTAL CAMPAIGNS IN ISRAEL AND THEIR SPATIAL IMPACTS

Benny Furst

ABSTRACT

The purpose of this study is to serve a political-cultural perspective on the environmental movement in Israel through an examination of the organizations' impact on the shaping of geographic space, the political system, and the general culture. The practical expression of political cultures is examined through an analysis of prominent organizations' activism as manifested in seven representative environmental campaigns waged over the last four decades. The study is qualitative in nature, and as such is based on in-depth interviews with numerous actors from the environmental arena – institutions, NGOs, the business sector, and academia. The originality of the study is to deeper the cultural aspects of the phenomenon.

Some of the findings regarding the effectiveness of the environmental campaigns are showing that there is significant gap between the political system and the civil society in Israel, relating the environmental issue as a whole: while the first still see it as a marginal sector, the Israeli public and local leadership has made a change in the attitude toward

Occupy the Earth: Global Environmental Movements
Advances in Sustainability and Environmental Justice, Volume 15, 137–169

ISSN: 2051-5030/doi:10.1108/S2051-503020140000015006

environmental issues. This cultural gap is due to the difference in cultural values, between the local and national level.

Keywords: Environmentalism in Israel; environmental movement organizations; environmental campaigns; political cultures

BACKGROUND

The global environmental movement is the last "survivor" among the multitude of large-scale social movements that emerged beginning in the 1960s (Rootes, 2003). Today, with the perspective of several decades, the environmental movement can be said to have undergone two major trends: institutionalization and professionalization. Institutionalization is manifested primarily in the assimilation of not only some of the formal organizations but also many environmental concepts and issues into the political establishments of developed countries. With respect to professionalization, there has been an increasing sophistication in everything related to the study of environmental problems, modi operandi and principally the infusion of science into the dynamics of the environmental system. The outcome of these two trends is that environmentalism comprises today an all-encompassing conceptual super-framework for the activities of thousands of NGOs dealing with environmental issues both broadly and specifically, which accordingly have come to be termed "Environmental Movement Organizations" or EMOs (Rootes, 1999). Scholar and environmental activist Paul Hawken argues that quantitatively speaking there are today approximately 30,000 NGOs dealing with environmental issues, and that if we add to them organizations dealing concurrently with issues of social justice and minority rights, their numbers reach more than a 100,000 (Hawken, 2007).[1]

In Israel, public activism in the environmental field has increased over the last two decades, both in terms of public discourse and the degree of impact on decision-makers. This finds expression in the number of organizations and bodies dealing with environmental issues, which has soared to well over a 100, as well as in greater sophistication of methods of action, as manifested in the professionalization of environmental knowledge, enhanced mobilization capacity, and inter-organizational coalition-building (Karassin, 2001; Tal, 2011).

Most of the relevant studies to date have dealt primarily with the action dimension of environmental activism, in a cause and effect sense, or with specific aspects, for example, political, communicational, public. What most of them lack is the broader, more abstract dimension that has the potential to help us better understand effects and behaviors beyond the context of legitimate resistance or protest. Wherefore the present study offers an additional theoretical dimension of environmental protest by presenting *culture* as an explicative factor for the phenomenon. It should be noted that even in empirical studies conducted on environmental campaigns in other parts of the world (Close, 1998; Della Porta & Rucht, 2002; Doyle, 2002), there is virtually no in-depth treatment of the cultural context and its repercussions on interactions between actors; hence the essentiality of investigating this novel topic.

RESEARCH DESIGN

The present study seeks to identify and characterize the political cultures of environmental organizations in Israel with the aim of examining their impact on the activities and campaigns the organizations engage in and on the manner in which they help to shape geographical space, the political system, and the general culture. Accordingly, the political cultures comprise the independent variable impacting the organizations' activity, while spatial construction is the dependent variable affected by the organizations' activity and the manner in which they manage and lead their campaigns.

The theoretical concept "political culture" refers to those values, norms, and symbols that shape human perception and in turn the modi operandi of a group of people. Methodologically speaking, "political culture" is an approach that examines the political structure of organizations and institutions as a function of the cultural structure, that is, the world of values, norms, symbols, and worldviews of which it is composed (Lane, 1992; Wildavsky, 1987). This research approach makes an important contribution since the majority of studies dealing with cultural movements generally and environmental organizations specifically have focused more on agents' activities, organizational structures or historical processes that generated results, and less on the impact of different cultural models that shape action and its outcomes. The novelty of this study, therefore, lies in its positing of the cultural aspect as explicative factor. Moreover, embracing Miller's (1994) assertion that the most significant contribution of social movements has been a cultural one, I argue that political cultures are a major factor shaping environmental organizations and their effectiveness. Empirically speaking, the political cultures of environmental organizations are examined by studying their behavior in seven prominent environmental campaigns.

Two primary research questions are explored:

1. Which political culture models characterize environmental organizations in Israel?
2. How do such political culture models impact the effectiveness of the organizations' work in terms of their effect on space?

Because this study is qualitative in nature, most of the sources used for the purpose of data collection are documents (newspapers, protocols, position papers, and internet publications) as well as in-depth interviews with key actors. The case studies selected represent different types of development projects in Israel from the last decade, as well as an array of environmental aspects. Below the seven case studies are concisely outlined (Furst, 2013)[2] (Fig. 1).

Case Study 1: The Campaign for Protecting Wildflowers

This was a public education campaign launched in 1964 by the Society for the Protection of Nature in Israel (SPNI) and the Israel Nature Reserves Authority (INRA). Its aim was to eradicate the then common practice of picking and selling wildflowers as this was leading to the destruction of natural habitats as well the disappearance of several species of flowers from the Israeli landscape. The educational campaign rested upon innovative legislation, "soft" educational enforcement, and the use of children and teenagers as social agents of change. The combination of these elements was remarkably effective, as within a few short years wildflower picking had become virtually non-existent in fields and wildernesses in Israel.

Case Study 2: The Campaign against the Plan to Establish the Settlement "Michal"

At the beginning of the 2000s, the Israeli government advanced a policy of establishing new settlements in peripheral areas out of security and ideological considerations. One of these settlements, which was given the biblical

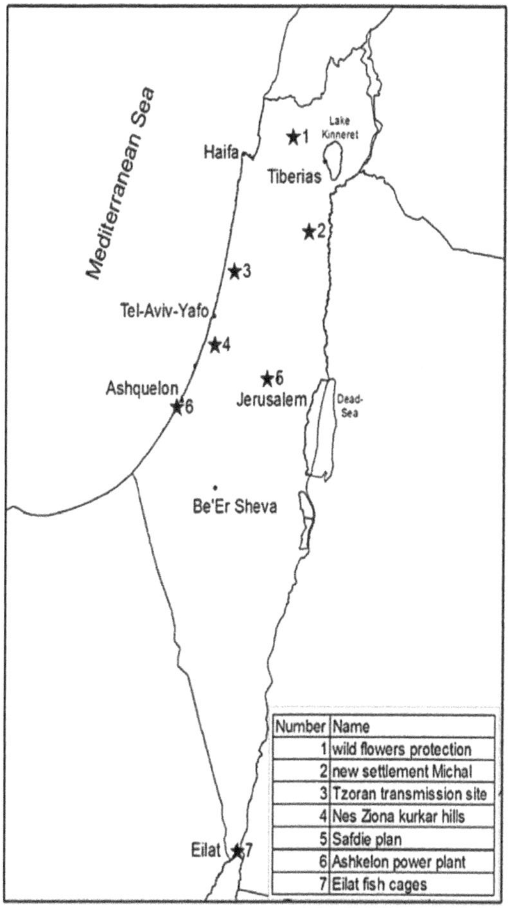

Fig. 1. Environmental Campaigns in Israel Case Studies.

name "Michal," was planned for the ridge of Mount Gilboa, in the middle of a nature reserve abundant in a long list of rare species of wildflowers topped by the Gilboa Iris. A broad-based public campaign led by a number of national and local environmental organizations brought about the plan's annulment, as the main argument being that existing settlements should be reinforced rather than establishing new centers of development in high nature value areas.

Case Study 3: The campaign against Development of the Kurkar Hills in Ness Ziona

This environmental campaign centered on residents' attempts to stop housing construction plans for the kurkar (Sandstone) hills near the city of Ness Ziona, based on the claim that they serve as an open country recreational area, and due to the wide variety of flower species on the hills. The developers, who presented legal deeds to the land, appealed to the Supreme Court to allow construction on the site. However, in 2008, the Supreme Court rejected their appeal on the grounds that acquisition of land does not entail a guarantee that the owner can build on it, and that the rare value of the area necessitate edits preservation as a wilderness.

Case Study 4: The Campaign to Close the Hillel Transmission Station in Tzoran

This campaign, which continued for eight years, was launched by the residents of Tzoran, a small town in Israel's Central District, who succeeded in proving that a nearby radio transmission station was emitting dangerously high levels of radiation causing severe health problems. The residents' campaign was waged vis-à-vis government officials and powerful financial bodies who insisted on the station's continued operation. The residents enlisted experts, formulated a strategy, and acted consistently over years until finally in 2003 authorities agreed to close the transmission station.

Case Study 5: The Campaign against the "Safdie Plan" to Expand the City of Jerusalem

The Safdie Plan, named after its chief architect, dealt with the expansion of the Jerusalem city limits by approximately a third of its size. The aim of the plan was to allow for building provisions and to bolster demographically the city's Jewish population. The areas designated for expansion were west of the city in uninhabited areas abundant in forests, natural groves, heritage sites, springs, and streams. Environmental organizations in Jerusalem united to form the "Coalition for the Preservation of the Jerusalem Hills," and launched a highly publicized campaign, waged over several years. In 2007, the Coalition succeeded in convincing the National Council for

Planning and Construction that there were already sufficient land provisions within the existing city limits, that the expansion plan could harm many natural and scenic assets, and that it was therefore gratuitous. The Safdie Plan was subsequently scrapped, and the campaign went down in history as one of the most significant achievements of the environmental movement in Israel.

Case Study 6: The Campaign to Remove the Fish Cages from the Gulf of Eilat

During the 1980s and 1990s, two successful enterprises operated in the Gulf of Eilat that engaged in the raising of sea bream in underwater cages. At the beginning of the 2000s, environmentalists and academics observed that the fish cages were producing organic waste that was harming the rare coral reefs in the gulf waters. Environmental organizations launched an aggressive and determined campaign, pressured for the establishment of a professional commission of inquiry, and mobilized public support for the cages' removal from the Red Sea. The campaign ended successfully in 2008 when the expert commission of inquiry and the government decided to end the operation of the cages notwithstanding findings that the cages' part in the pollution of the gulf was relatively minor.

Case Study 7: The Campaign to Prevent the Establishment of a Coal-Fired Power-Plant in Ashkelon

The need to meet electrical demands ensuing from the accelerated development of the Israeli economy led the Israel Electric Corporation (IEC) to present the government with a plan to add a production unit to its existing power-plant in the city of Ashkelon. The plant, which was to be coal-fired, aroused intense public opposition among area residents and national environmental organizations. After approximately a decade of public and legal struggles, it was decided in 2011 that the new plant would be erected but that it would operate exclusively on natural gas. This victory is an important reference point in the history of environmental activism in Israel in that a high environmental standard was set, which has subsequently been applied in all other power-plants in Israel.

METHODOLOGY – THEORETICAL MODEL

Mapping the Campaigns

In order to gain a more general picture of the case studies, I elected to map them along two axes (Hasson, 1993): the first the ideological dimension, or *orientation* (i.e., the political message and cultural code), and the second the behavioral dimension, or *action* (i.e., how and in what framework to act). The contribution of such mapping is to discern cultural variance among environmental organizations as manifested in their public campaigns.

The first pattern on the orientation axis, of a specific instrumental action, represents organizations acting rationally toward a clear and pragmatic goal, while the second entails broader action including scathing moral criticism directed primarily against the government and the political establishment or the financial sector, rather than merely against local centers of power.

The intersection of the orientation and action axes yields two additional action patterns, differentiated by preferred action strategy: inside or outside the establishment. The first pattern is marked by conservative public activism, such as establishing a lobby or political action group, conducting negotiations, intensive use of the mass media, supporter mobilization, and any other protest action that conforms to legitimate cultural codes. At the other pole, we find the extra-institutional activity typical of activist groups that sometimes employ violent protest techniques that may be considered illegitimate from a legal or moral standpoint. In this way, four types of environmental campaign political cultures have been arrived at: *targeted moderate campaign*; *broad-based moderate campaign*; *targeted radical campaign*; and *broad-based radical campaign.*[3]

Mapping the Organizations

In order to map the political culture models for the actors themselves (the organizations), I chose to refine the model presented above for the campaigns, and add the cultural dimension to it in the form of basic values and rhetoric. Here, I would like to argue that basic values are the main factors impacting orientation, while rhetoric is the primary instrument used by organizations to sway decision-makers, professionals, and the general public regarding their messages.[4] The centrality of rhetoric in the organizations' activity stems inter alia from the fact that being NGOs they effectively lack formal decision-making authority or the ability to advance legislation, and thus most of their power lies in making an extrinsic impact on the establishment and the public.[5] Since the way of doing this is, as aforesaid, through effective rhetoric, I will also examine environmental rhetoric here, as Taylor (2000) suggests. However, the main distinction between the organizations will be made according to Schwartz's (2006) theory of cultural valuations, which proposes two types of cultures: mastery and harmony. Mastery cultures are characterized by a greater need to lead exclusively, to dominate the conversation, and to transform reality according to one's own vision and interests. The basic values of this culture include ambition, success, daring, independence, self-assertiveness, competence, and ability to cope with adversity. By contrast, harmony cultures are marked by reconciliation with extant reality and fitting into it. The basic values of this culture include integration, serenity, unity with nature, protecting the environment, accepting one's portion, and an aspiration to the aesthetic.

Accordingly, two sub-cultures, mastery and harmony, are taken to exist within each of the four types of campaigns, thus yielding eight different types of environmental organizations. The empirical account below will help to illustrate this.

Examining Effectiveness

The organizations' effectiveness is examined through the concept of "cultural environment," which, according to Williams' (2004) includes the following two entities:

(A) *Boundedness*: the totality of legitimate behavioral codes, values, concepts, and terms employed by an organization ("jargon"). Empirically speaking this means the messages conveyed and the modes of persuasion deployed (rhetoric) by the organization. Boundedness may also be termed a "frame in thought."

(B) *Resonance*: the salience and power of the legitimate concepts included in the boundedness of the organization. Variance exists between these concepts in terms of their salience and resonance due to differences in personages (conveying the message), time (periods) and public receptivity toward (consumption of) the concepts.

Boundedness and resonance are cultural resources used by environmental organizations to make an impact and accumulate power: boundedness relates to public power and the desire to influence decision-makers, mainly

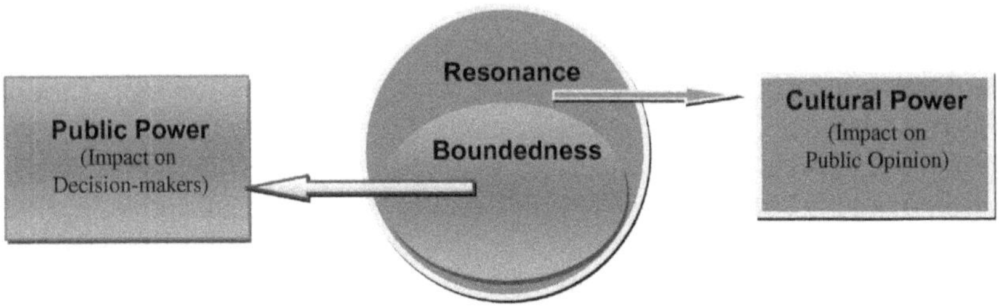

Fig. 2. "Frame Resonance" as Measure of Effectiveness.

by achieving legitimization over time, while resonance seeks to create cultural power and thereby make an impact on the public opinion. Empirically speaking we are talking about two interrelated concepts in that resonance is a sought-after product of boundedness. The integration of the two yields the concept of "frame resonance,"[6] which refers to the factors and conditions for the framing's success, which in turn determine the organization's effectiveness in mobilizing supporters (Fig. 2).

According to Bendford and Snow (2000), frame resonance can be empirically tested based on the degree to which three core tasks are adhered to in framing an organization's action:

- Diagnostic framing: Identifying the problem and assigning blame or responsibility.
- Prognostic framing: Suggesting solutions, strategies, and tactics for attacking a problem.
- Motivational framing: Providing a call to arms or rationale for action.

POLITICAL-CULTURAL MAPPING OF THE CAMPAIGNS AND ORGANIZATIONS

Based on the methodology presented above, the case studies may be categorized as follows (Fig. 3).

As I have shown in the theoretical model above, a more detailed characterization can be offered of environmental organizations' political cultures as reflected in the modi operandi within each campaign. In this mapping, the ideological approach of the organizations is ascertained as manifested in interactions with other actors and according to the types of rhetoric they employ.[7] The result is that each of the four types of campaign political

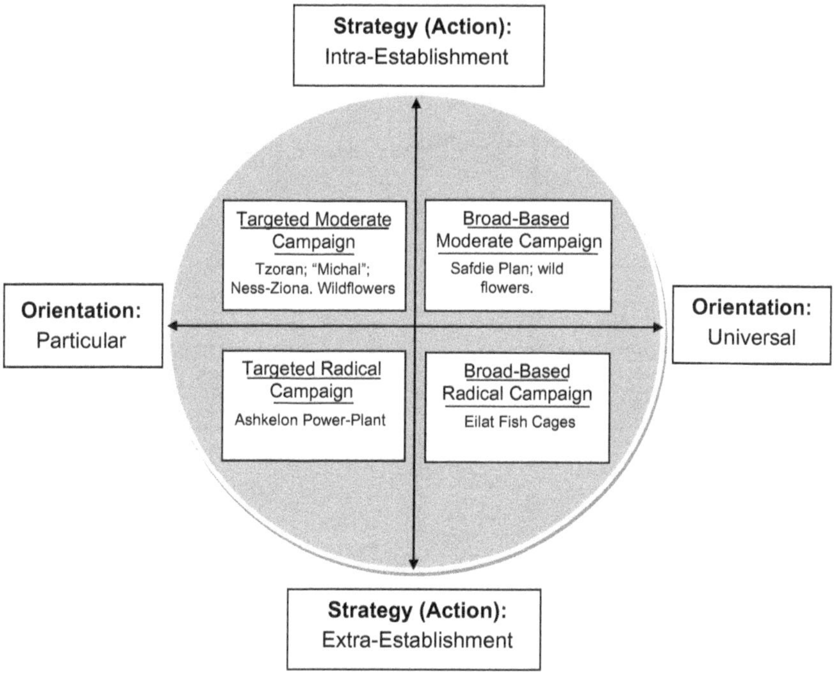

Fig. 3. Political Cultures of Environmental Campaigns – Four Types.

cultures splits into the two sub-cultures (mastery and harmony) that characterize the organizations operating in the context of each campaign (Fig. 4).

Now that the campaigns and organizations have been mapped, empirical descriptions can be presented of the four types of environmental campaigns, each one including the two organizational sub-cultures (mastery/harmony).

Targeted Moderate Campaign

The case of the *campaign to shut down the Hillel transmission station in Tzoran* (Furst, 2013) is a quintessential example of this type of political culture, where residents fought to close the transmission site out of public health considerations. With the exception of the threat to evacuate the town (which was not fully carried out), the Residents' Committee acted

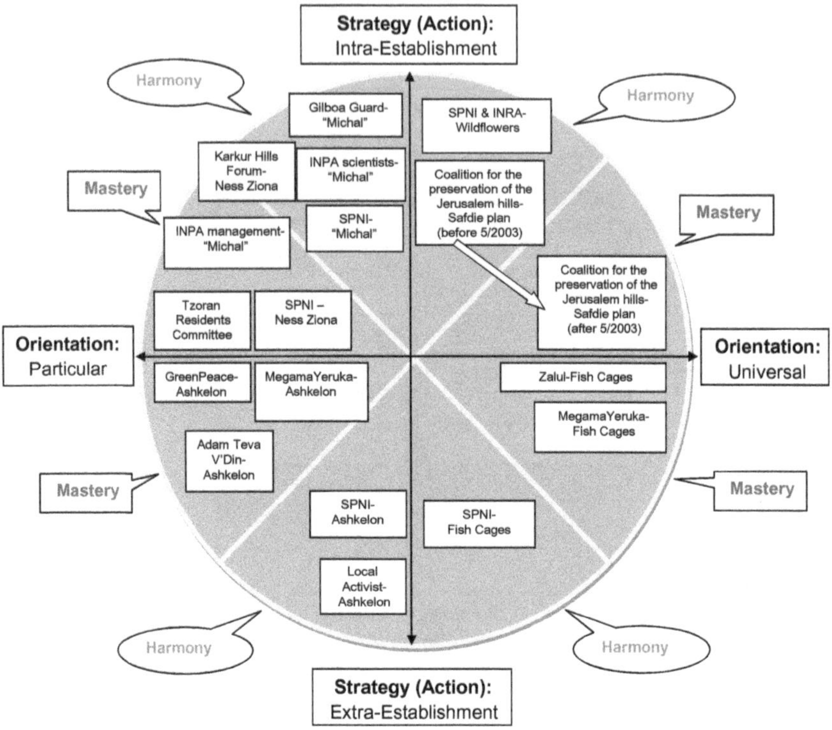

Fig. 4. Political Cultures of Environmental Organizations – Eight Types.

through accepted methods vis-à-vis the local authorities, government ministries, and Knesset committees. The interaction with establishment agents, in this case interested parties who sought to continue the operation of the site, was moderate and non-vehement in most instances, the main argument being that the transmission station was redundant. The Residents' Committee's tactics were to identify power sources within the establishment and to act in a targeted manner in relation to them. The broad use of the media and the Internet, particularly in order to place children on the frontline of the campaign as mortally threatened by cancer-causing radiation, brought into sharp focus the object of the campaign. The outcome of the Tzoran campaign, the transmission station's closure, created a sense of empowerment among both residents, who viewed it as a foundational struggle over the town's identity, and the local leadership, who found it to be an effective instrument for generating politico-electoral capital.

The sole actor in this environmental campaign, the Tzoran Residents' Committee, may be characterized as possessing a mastery political culture. The Committee acted with neither the cooperation nor the leadership of any other environmental organization, and did so exclusively based upon local initiatives and resources mobilized by the town's mainly upper-income residents. The residents' mastery culture was manifested in their unwillingness to accept any alternative to the station's closure and to removal of the antennas on the site. The Residents' Committee, which had lost almost all faith in "professionals" and the system, created its own scientific database (on radiation emitted by the antennas) and collected internal information on the transmission station and its operation. One of the heads of the Residents' Committee described this as follows.

> I set up an HQ and we began studying the issue. We collected material and made contacts. In the meetings we realized this thing [the antennas] was very dangerous, that there was a real danger, that this thing was lethal. For every negative newspaper article there were ten positive ones, because the state wasn't interested in changing the situation. We held a gathering of residents. The aim was to shut down the station. We weren't willing to live here ... the residents here are strong, smart, and enlisted in the campaign in the fullest sense of the word. we didn't stop for a second. This was our lives, the lives of all of us, and no one was going to stop us for a second.[8]

An attempt to understand the motives behind the Residents' Committee's uncompromising mastery approach leads to the conclusion that underlying it was a fundamental lack of confidence in the establishment, which stemmed from concealment of information and manipulations by those parties who had a vested interest in the continued operation of the station. In this case, the establishment, represented by bodies holding significant political power (the Ministry of Defense, the Israel Broadcasting Authority, Bezeq [the largest telecommunications group in Israel]) adopted a mastery approach evidenced in a patronizing attitude toward the public typified by invocation of arguments and values perceived as archaic from a public, geopolitical and technological perspective, including "military secrecy," "complex technological knowledge," and "Israeli public diplomacy toward the Arab world." In effect, the Residents' Committee ran into a closed establishment structure, which forced it into a clearly isolationist stance: non-collaboration with other environmental groups and expression of non-confidence in the establishment. Thus a situation developed in which the two sides became increasingly entrenched in their respective positions, the rift between them growing deeper and deeper. Ultimately, it took the intervention of two powerful actors with a harmony political culture, the Ministry of Environmental Protection and the State Control Committee, to open up channels of rational communication between the sides and lead to a solution.

The campaign reached its climax in the summer of 2003, when the residents' anger and frustration reached a boiling point. The residents and leaders of the campaign at this stage began to employ a rhetoric of irrationality, arguing that the station was politically superfluous, as other modes of transmission such as the Internet and cable and satellite television had rendered radio public diplomacy broadcasts archaic and irrelevant.

The case of *the campaign against the establishment of "Michal" settlement* (Furst, 2013) is also relevant to this type of political culture because most of the activity occurred within the planning establishment, mainly the National Council for Planning and Construction. The few protest actions held on Mount Gilboa were of a conservative character, and focused on informational and educational activities for teenagers and trekkers during the blooming season. The campaign was led by a number of well-known personalities identified with the Gilboa region and the Jezreel Valley, who dismissed what they regarded as an old and irrelevant Zionist value of establishing new settlements as a security imperative. The combined leadership of the local grassroots organization The Gilboa Guard ("Mishmar Hagilboa") and the veteran national organization, the Society for the Protection of Nature in Israel (SPNI), focused on the wider public interest of nature protection with a strong reliance on scientific arguments. The campaign's major outcome, cancellation of the plan, contributed greatly to enhancing the status of environmental organizations as public representatives, which had recently become anchored in law.[9] The campaign had a highly targeted orientation, whose goal was to prevent the establishment of the settlement, but also appealed to universal values of nature preservation and social justice. Most of the actors adopted a harmony ideological approach, as we will see below.

The SPNI was the dominant body leading the campaign against the "Michal" settlement, while its harmony worldview found expression primarily in the planning institutions. As in other environmental campaigns dealing with the protection of natural values threatened by development initiatives, a rhetoric of loss and disaster was employed, which relied on scientific arguments and research, such as the botanical survey carried out by the Israel Nature and Parks Authority (INPA), which proved that the number of rare irises in the area designated by the plan was greater than initially believed.

For the SPNI, the campaign was a sort of replay of a similar campaign in the late 1970s where it fought against an initiative to establish a new settlement in the Um al-Safa forest on the West Bank, an area declared a particularly unique ecological reserve. As in the case of Michal, the initiative by governmental bodies in Um al-Safa won the support of the INPA, which portrayed the SPNI as a group of extremists.[10] The similarity between the two cases allows us to better understand the interaction between the INPA and the SPNI: the former, as a professional establishment organization, operates based on considerations that often contradict the "classic" ecological worldview of romantic nature protection and opposition to development, while the latter regards itself as a watchdog over this organizational surface tension within the INPA.

The professional capabilities and harmony character of the SPNI, which vehemently opposed the proposed plan yet made realistic compromise offers, are noted by the special investigator appointed to review the plan:

> The Society for the Protection of Nature did its homework exceptionally well. There wasn't a thing they failed to raise as an argument regarding the case in question. They delivered it backed up with evidence, did truly splendid work. I read and delved into the material but they submitted excellent work ... They are believers. They have a sense of calling, even an extreme one, since they view their world through the green perspective.[11]

The SPNI's political-cultural power in this campaign was manifested in its success in negating three old political-cultural values that guided the establishment: erecting new settlements along borders as a military necessity (Prime Minister's Office), socioeconomic reinforcement by attracting higher-income populations (Regional Council) and appropriation of lands from nature reserves for the sake of establishing settlements (INPA). The mastery tactics employed by these three establishment bodies, all absolutely convinced they were right, and that they were the only ones ostensibly entitled to decide, were proven to be mistaken, as their mastery approach merely spurred the SPNI to take a compromising harmony approach of portraying the settlement scheme as illogical from a planning perspective and as opposed to social justice.

The second actor with a harmony political culture involved in this campaign was the Gilboa Guard, which acted as an "ad hoc" residents' organization. Its methods were highly conservative and included community activities and information evenings in area towns, treks, marches, and petition-signing. The messages operated mainly on the level of logical persuasion, with an emphasis on the claim that it was wrong to establish a new settlement along the separation barrier facing the Palestinian territories, and in the middle of a nature reserve, while two nearby veteran settlements (Ma'ale Gilboa and Meirav) were populated at only a fifth of their capacity.

The heads of the Gilboa Guard's acting committee stressed again and again that they did not oppose establishment of new settlements in principle but that it was wrong to do so at the proposed location on the ridge of Mount Gilboa:

> There were two weak spots to this plan: the rare species of Irises, and the security separation barrier. The proposed location seems rather silly as an "ecological" settlement ... It's absolute foolishness and perhaps even fraud: they're going and promising people an "ecological settlement with a view and quality of life," while they're building it right on the separation barrier. The message was that an injustice was taking place here, and the community got this.[12]

The third actor in this campaign was the INPA, in which two contradictory intra-organizational sub-cultures were observed: the scientists, who acted with a harmony approach, and the Authority's administration, which took a clear mastery approach. This cleavage stemmed from a professional ethical dilemma between the desire to expand the nature reserves in Gilboa and the anticipated harm that would be caused to a vital aesthetic and natural asset (the Gilboa iris). The non-opposition of the INPA to the plan and its compact with the developers and the Regional Council imparted it the image of a "property developer" and as we saw above, this drew harsh criticism from the SPNI. Thus two opposing political cultures were created within the INPA: one among to the Authority's administration, characterized by a patronizing and uncompromising attitude that was not based on a deep evaluation of the necessity of establishing the new settlement, and the other a harmony culture that developed among the professionals who sought to engage in a compromising dialogue that would lead to the cancellation of the plan out of scientific and rational considerations. One of the INPA chief scientists described this as follows:

> The whole process was wrong. The Nature and Parks Authority should have stood up to the Regional Council and said no, but here they didn't pay sufficient heed to the truth. Because this is a hierarchical governmental organization, it's very much influenced by who heads it ... The movement wasn't sufficiently directed to the basic value of pure nature protection, which is what the Nature and Parks Authority by law is supposed to be.[13]

The mastery approach of the Authority's administration in this case was manifested in the fact that its arguments were not based on an empirical appraisal of the value of the area. Only at the beginning of 2005, when the voices of the professionals within the INPA began to be heard, was a botanical survey of the irises in Gilboa conducted, whose data led to a turnaround in the position of the INPA, which swiftly led to the annulment of the plan altogether.

The third case that illustrates the targeted moderate campaign type is the *campaign to protect the kurkar hills in Ness Ziona* (Furst, 2013). In the political culture of this campaign, two actors with opposing ideological–cultural approaches were observed: the SPNI, which assumed a mastery approach, and the "kurkar hills Forum" which was marked by a harmony approach.

The mindsets evident in the activity of the SPNI in the Ness Ziona case reflect values of uncompromisingness and determination to a greater extent than in other campaigns. This stems possibly from the tension that developed during the campaign between the SPNI and the INPA, which, according to both the SPNI and the "kurkar hills Forum" did not act with adequate force to declare the hills a national park, and even refrained from opposing the development of the northern part of the site. A testament to the uncompromising determination of the SPNI can be observed in the following statement by the campaign coordinator:

> Our campaign was a preemptive one, to create a situation in which the courts wouldn't accede to them, the mayor or the Nature and Parks Authority wouldn't accede. I met with the district supervisor of the Nature and Parks Authority and we agreed that we wouldn't make any compromises involving construction on the hills. The campaign was entirely about nature preservation … Our approach from the get-go was that we wouldn't give up a single inch of the area. The way to preempt a threat is to offer something else. Sometimes you encounter a concrete threat, and you fight against it. But sometimes you act before the threat exists, and we come and make a proposal. And when you come with something prepared, that makes sense, you create a planning reality even before the threat is born.[14]

The Ness Ziona residents' organization ("the kurkar hills Forum") adopted a quintessentially harmonious ideological approach, which relied upon rational, non-vehement and non-threatening rhetoric with mainly positive and practical content, rather than just immediate opposition. Their key messages as conveyed on the Internet and in their protest activity were as follows:

> The residents of Ness Ziona have been blessed with a unique and rare natural treasure – the kurkar hills. At this very moment the hills are blooming with hundreds of species of plants, some extremely rare in Israel and the whole world … You are cordially invited to the dedication of the first segment of the kurkar hills trail, which will be held this Saturday.[15]

In contrast to other cases, the developers (land owners) themselves responded in a harmonious and compromising manner to the kurkar hills Forum's positive approach, and focused on the attempt to reach a land compromise of some kind, and at a later stage, on receiving just financial compensation for the loss of their building rights. And indeed, when the matter reached the Supreme Court, it was decided in a precedential ruling that land acquisition does not automatically accord building rights to the buyer, and that preservation of natural assets and the needs of the general public may sometimes override such rights. The long campaign waged by the SPNI and the kurkar hills Forum bore fruit, and today the unique area is in the process of being declared a national park.

Broad-Based Moderate Campaign

In this political culture category, we can place first and foremost *the campaign to prevent wildflower picking* (Furst, 2013), because the aims of the environmental organization's work were not only to protect wildflowers but also to forge new ethics that would shape relations between man and nature for many years to come. Actually, in the wildflowers case we are not talking about an environmental campaign in the usual sense of the term but rather action for cultural change based on "soft" public diplomacy and an ideology that reemphasized values of love of the land, nature preservation, and relations between people and the environment.

The campaign to prevent wildflower picking was distinguished by harmonious and effective collaboration between the SPNI and the Nature Reserves Authority,[16] as well as a high degree of assimilation within governmental and social institutions.[17] An examination of press coverage and archival documents shows that the overall ideological conception was not one of a campaign but rather "reeducation of the public." The rhetoric employed operated on simple logic by which if the public continued to pick flowers, the country would lose some of its beauty, and that this would harm everyone, in both the present and future. However, this was the first real challenge to the ethos of rapid development that then predominated in Israel. For the first time, a different kind of thinking emerged that attempted to place limits on development in the name of protecting natural assets. The following excerpt from the SPNI magazine illustrates this:

> In the Israeli press a new spirit has been blowing in recent months: nature protection concepts now seem obvious. The public has started to sense that even within the pace of development we must ask: development – what is it for? Is it only for financial profit? Or also for a pretty home and country that has both character and beauty?[18]

The wildflower case is marked by harmony not only between the two leading environmental organizations but also in their interaction with other actors, among both the establishment and the public. The cultural uniqueness of this case lies in the organizations' harmonious appeal to the general public calling for their enlistment in the cause of ending a widespread cultural custom (mass flower-picking) for the sake of a new and unfamiliar cultural value (ending the flower-picking). The organizations did not relate to the public as an "enemy" but rather found the right way to "softly" penetrate the collective consciousness. It should be noted that this form of activism does not aim to challenge the establishment but rather to partner with it and make intelligent use of its tools – legislation and enforcement.

The case of *the campaign against the "Safdie Plan"* (Furst, 2013) also falls under this type of political culture as it was a public campaign that focused on issues deliberated in planning institutions. The conservative nature of the field campaign carried over into the demonstrations and assemblies, and the messages were broader than mere opposition to the "Safdie Plan" per se. The campaign led by the "Coalition for the Preservation of the Jerusalem Hills" (CPJH) against the plan generated in effect various sub-debates on wider issues, including inter alia demographics, the dilemma between vertical construction and wilderness preservation, the limits of urban development, and environmental and social justice.

The ideological outlook of the CPJH in its campaign to stop the "Safdie Plan" was quintessentially one of mastery, that is, it was uncompromising and driven by a high degree of self-persuasion that the plan did not do justice to the city or its residents. This approach, it should be noted, was adopted by the Coalition only after the campaign got underway, as at the outset it took a harmony approach vis-à-vis the political and planning establishment, that is, it did not oppose the plan automatically and categorically but rather raised proposals for realistic development alternatives. However, this approach ran early on into a patronizing attitude from the developers, which forced the heads of the CPJH to change direction and launch an adamant and uncompromising campaign, as articulated in the following statements:

> Although throughout they advised us to go for a compromise, we resolutely did not go for a compromise. We took a hard, clear line, which showed that this plan was unnecessary and was harming the city, and therefore we weren't morally capable of going for a compromise.[19]

Thus a situation developed of mutual distrust between the CPJH and the establishment, and the desire be the one to decide what Jerusalem's real needs were became the central factor in the interaction between the sides. This led the CPJH to make use of every possible form of rhetoric: of disaster befalling natural assets, of irrational planning, of injustices being perpetrated on the city and its residence, etc., as emerges from the following:

> We had the slogan "Save the Jerusalem Hills," which is a warning of impending danger and disaster. Afterwards we called for the reinforcement of the city center, which is a rational argument: even though it's harder to build within the city, deal with the difficulty, gather resources, make legislative changes. We succeeded in convincing people that approving the plan, against all healthy common-sense, would turn all the city's residents into victims of development, because the residents' taxes are what would be funding the process. We were right not only as "Greens": we were right economically, socially and demographically. Justice in the Jerusalem hills and the city of Jerusalem triumphed.[20]

An interesting perspective on the CPJH and the SPNI is provided by the special investigator appointed to review the thousands of arguments put forward by opponents of the plan:

> The values the Coalition and the Society for the Protection of Nature presented were right ... I don't expect the SPNI to hand me what the significance of the matter is for the Jewish-Arab majority in Jerusalem ... They have a very rational and very business-like rhetoric. I take my hat off to them. It was a pleasure to listen to them.[21]

Effectively, the interaction throughout the deliberations between the special investigator and members of the Coalition embodied a deep cultural conflict between two opposing worldviews, which have virtually nothing in common except perhaps the fact that they have a long-term vision: nature preservation and avoidance of wide-scale development versus development that answers political needs and is not based on in-depth analyses. Ultimately, and notwithstanding the special investigator's recommendation, the National Planning Council decided in February 2007 to cancel the Safdie Plan, after having accepted the CPJH argument that there were sufficient land provisions for development within the existing Jerusalem city limits.

Targeted Radical Campaign

The case of *the campaign against the coal-fired power station in Ashkelon* (Furst, 2013) serves as the singular example here of a targeted radical political culture. The prominent organization in this campaign was Greenpeace, which teamed up with both establishment bodies (the city of Ashkelon) and Non-establishment bodies (Adam Teva V'Din, Megama Yeruka), while sticking to its signature non-violent direct action methods.[22] In addition, there were two other actors who took a harmony approach – the SPNI and a local activist. Below is an empirical account of this case.

The dominant organization in this campaign, Greenpeace, assumed an unequivocal and uncompromising mastery approach, that is, total opposition to the coal-fired power-plant. It is evident from Greenpeace's messages and methods of action that the organization is generally unwilling to compromise or accept proposals that are remote from its worldview, which it adheres to fiercely and seeks to impose by all legitimate means. The organization is well aware of its "tough" activist image and makes clever use of it, as the following remarks by the leader of Greenpeace's energy campaign reveal:

> It was very important to me not to make just any provocation. Provocation has to serve the message: we're stopping the plant's operation in order to literally stop the cranes from feeding coal into the plant. The smart thing to do is create a new attraction each time, to flood the public consciousness with the issue and to speak to it from a different angle each time. Greenpeace wants to represent the unheard voice.[23]

Greenpeace's uncompromising mastery mindset is manifested in the rhetoric of its messages as well with their emphasis on disaster and macabre pathos in order to attract media attention. The main rationale behind this rhetoric is to pin personal responsibility to some higher-up, in this case the Minister of National Infrastructure. The following is a press account of one of Greenpeace's demonstrations in this campaign:

> This morning approximately 30 Greenpeace activists gathered across from the home of Minister of National Infrastructure. Dressed as terminally ill patients and sitting in wheelchairs, they danced to the sound of people suffocating, mechanical ventilators and ambulance sirens. "We tried to simulate a field hospital to illustrate what state Ashkelon residents will be in when their health is damaged as a result of the pollution emitted by yet another coal-fired power-plant," explained the campaign director ..."We view the Minister of National Infrastructure as the main one responsible for the malicious and needless decision that prioritizes the interests of the Israel Electric Corp. over those of thousands of people."[24]

The same mastery values and disaster rhetoric that characterize Greenpeace may also be observed in Megama Yeruka, an organization that has pioneered its own local Israeli brand of Greenpeace-inspired activism. Megama Yeruka views the establishment's deliberate disregard of its efforts as contemptuous and patronizing, which only gives it added incentive to preserve its own mastery political culture and even intensify it by raising the bar of bold confrontation. This is a vicious cycle of mastery from which it is difficult to extract productive dialogue over time.

The third actor in this campaign with a mastery political culture was Adam Teva V'Din: Israel Union for Environmental Defense.[25] The basic values of this organization inform its policies, which espouse assertive legal action against and within the establishment, wielding universal arguments from the field of planning and building for the benefit of the general public and based on legal information and practice. Accordingly, for a public hearing held in Ashkelon, the organization prepared a legal document entitled "On the Senselessness of Building Another Coal-Fired Power-Plant in Ashkelon," which henceforth was used by all the environmental organizations involved in the campaign. This, however, effectively limited the possibility of harmonious dialogue that could have led to a compromise, as its blunt and emphatic style created a climate of warfare between the actors where each side attempted to impose their truth. This may be the reason the campaign dragged on for so many years.

As aforesaid, in this radical targeted campaign two other actors characterized as having a harmony approach were involved as well. The first, the SPNI, whose field activism was insignificant in this case, focused mainly on making a harmonious impact within the planning institutions. Thus, in its capacity as a full member of the National Infrastructure Committee, the SPNI displayed a position more interesting than one would normally infer regarding its worldview in recent years among environmental organizations, particularly regarding infrastructure projects:

> Our objective was to postpone the construction of the power-plant for as long as possible. Such postponement is meaningful, because as time goes on, technologies less harmful to the environment and more energy-efficient come into use … and in this period of time it's possible to bring about significant electrical savings …[26]

The second harmony actor in this campaign was *the local activist*.[27] Based on the present research, this actor's ideological worldview was quintessentially a harmony one. He saw himself as a mediator between the local establishment and the community, called for compromising dialogue, and offered his rhetorical and professional skills for action that occurred effectively within the establishment while simultaneously joining forces with those environmental organizations acting outside and against the establishment. The local activist's extra-institutional strategy found expression in his calling for the community to act against the (local and national) establishment, sometimes even to the point of incitement to rebellion. Regarding his harmony approach, distinguished by a strong sense of justice, sensitivity to others and capacity for giving, compromising dialogue we can learn a great deal from his words below:

> I was always the "native" who brought political power, who brought the community, because they couldn't … I always conveyed to them, and when in their company, I said to them, "This is right next to your home, to our home it's even closer. You're living

here right next to this, and if they put up another power-plant, we won't be able to breathe here." You start broadcasting this in every direction, and the residents internalized it. We showed them that they had soot on their balconies, and that this was harming their children. You always show them concrete things ... I learned cheap psychology – people want attention. Don't solve every problem for them. Relate to them, listen to them.[28]

The local activist had local power, which is compatible with a worldview that regards grassroots activism as an important element in the success of environmental public campaigns, an element not always present in national organizations.[29] The local activist channeled his personality and power into forging an authentic bond between the residents and the local authority by making the environmental issue a relevant and central one for everyone.

Broad-Based Radical Campaign

The case of *the campaign against the Gulf of Eilat fish cages* is illustrative of the broad-based radical type of political culture. The leading organization in this campaign was the Zalul Environmental Association, for whom fighting for justice is a basic value shaping its activity. Accordingly, it relates to developers as adversaries, holding the conviction that if their interests can be satisfied by alternative means, there is a greater likelihood of protecting the environment. The theme of social justice, which carries no less importance than environmental justice from the organization's perspective, is expressed in Zalul's messages:

We have proven that you can do the same things, only differently, that is, entrepreneurs can go on making profits without harming public resources. Before every argument of ours we say we don't want to hurt anyone's pocket, don't want anyone to lose their job. What will I have to be proud of then? People being laid off?[30]

However, Zalul's harmony approach did not have a chance to come into play because the owners of the fish farms expressed nothing but total contempt and animosity toward the organization, which they viewed as attempting to sabotage an economic success story and an original Israeli scientific development (marine agriculture). Actually, the combative interaction between the fish companies and Zalul only helped the organization to crystallize its identity as a fighter for universal justice. The expressive style of the fish farm owners toward Zalul was scathingly hostile:

The Greens have very strong mobilization and persuasion capabilities ... They deal in brainwashing. The battle was decided effectively in the media, rather than through

> methodical, in-depth deliberations. They're violent, they're bastards ... The Greens are a religion. And it's very difficult, if at all, to change a religion and religious ideas.[31]

Like Zalul, *Megama Yeruka* also had a mastery political culture in this campaign, with most of its expressions being articulated in a disaster rhetoric typical of radical environmental organizations. This is evidenced in such slogans as: "Painting the Sea Black," "The Red Sea is Turning Black," "Save the Gulf of Eilat!," "Liberating Eilat!," "The Coral are Crying 'Help!'" "Megama Yeruka's mastery approach is encapsulated in its notion of taking responsibility in the sense of if we don't take things into our own hands, no one is going to do it." This conception is a unifying factor relevant to a central basic value in radical organizations–sociability, according to which activists join such organization not only because of their socio-environmental ideology but also (and sometimes primarily) to satisfy their need for social affiliation with a group that lends the individual a strong sense of identity and purpose:

> There's a lot of social interest and student enthusiasm in Megama Yeruka, and most join the organization because of the social aspect rather than the environmental cause ... Another value is "optimism," but also "fatalism." ("If we don't act, there will be a disaster")[32]

Unlike the two aforementioned actors that took a mastery approach in this campaign, the SPNI acted harmoniously, mainly within the planning institutions. The fish cages campaign also relied upon extra-institutional elements such as legal consultation, mobilization of the local community, and consolidation of a political lobby. The SPNI's main message was distinguished by a rhetoric of loss couched in simple "black and white" terms: if the cages remain, the coral will die. The general tenor of the campaign was one of "life or death," with no shades of gray in between, while Eilat and its coral were constructed as a familiar and beloved brand ("The Gulf of Eilat is Finished!").[33]

EFFECTIVENESS ANALYSIS: THE SPATIAL IMPACTS OF ENVIRONMENTAL ORGANIZATIONS

Now that the political-cultural characteristics of the campaigns and actors have been covered, an analysis of their spatial impacts, that is, the protest actions' effectiveness, can be presented. In accordance with the theoretical model introduced at the outset and in light of the empirical account above, we are able to show the campaigns' effectiveness in a breakdown of the four political cultures characterized: targeted moderate campaign, broad-based moderate campaign; targeted radical campaign; broad-based radical campaign. Because, as we noted, each of the four types includes two organizational sub-cultures, harmony and mastery, I relate to the characteristics of these sub-cultures as well in the analysis (Table 1).

The effectiveness analysis yielded a number of key insights regarding the degree of variance in spatial impacts:

A. The most significant geographical spatial impact was made by the broad-based moderate campaigns, while the targeted radical campaigns had the least impact. This finding reinforces the argument that environmental conflicts are multidimensional and complex, and that the most effective way to impact one's immediate physical space is through moderate and patient dialogue rather than swift and militant confrontation.
B. Broad-based moderate environmental campaigns appear to have the profoundest impact on the general culture, while targeted radical campaigns have the least impact in this context. This likely stems from the fact that the public knows how to identify the motives, messages, and intentions of the actors and translate them into support when convinced of their (ideological) authenticity. Thus rational arguments presented in a diverse rhetoric by moderate protest leaders resonate more successfully among the public and enjoy greater receptivity that the disaster rhetoric of radical organizations perceived as extreme and unpersuasive.
C. In the types of campaigns with a universal orientation ("broad-based campaigns"), the degree of variance between the two types of actor sub-cultures was minimal. In other words, in universal campaigns there is no significant difference between the two ideological approaches of harmony and mastery. This is indicative, in my estimation, of the centrality of the organizations' causes, while the other elements of the campaign (organizational ego, the battle for credit, use of gimmicks) become marginal from their perspective. By contrast, in targeted campaigns significant variance is evident in the degree of impact made by harmony and mastery actors.
D. In the breakdown of bodies impacted, it appears that the political establishment was impacted the least relative to the other bodies, while the greatest impact occurred on the "articulating the problem" dimension. I am of the opinion that there is a relationship here, which leads to two conclusions that are highly significant for understanding the

Table 1. Summary of the Political Cultures' Spatial Impacts.

Spatial Impact		Geographical	Political (Boundedness)		Cultural (Resonance)		
Political culture		Outcome Criteria	Legitimization	Impact on the Establishment	Articulating the Problem	Formulating the Solution	Mobilization
Campaign	Organization						
Targeted Moderate Campaign	Harmony	(++)	(++)	(0)	(++)	(++)	(+)
	Mastery	(+)	(+)	(-)	(+)	(–)	(++)
Broad-Based Moderate Campaign	Harmony	(++)	(++)	(+)	(++)	(++)	(++)
	Mastery	(++)	(++)	(+)	(++)	(++)	(++)
Targeted Radical Campaign	Harmony	(0)	(+)	(+)	(++)	(0)	(+)
	Mastery	(+)	(–)	(-)	(++)	(+)	(0)
Broad-Based Radical Campaign	Harmony	(+)	(+)	(+)	(+)	(++)	(+)
	Mastery	(++)	(++)	(++)	(++)	(+)	(+)

Key:
(–) Very minor impact.
(-) Minor impact.
(0) Moderate impact.
(+) Major impact.
(++) Very major impact.

interaction between the environmental movement and the political establishment: firstly, environmental organizations have a high level of professional expertise on environmental issues, and this allows them to clearly articulate problems on the agenda. The second conclusion is that the political establishment does not treat environmental organizations seriously enough. While a few actors are exceptions, the environmental movement still runs into dismissal by government institutions, which tend to view environmental protection as a "niche" and marginal issue. These two conclusions point to a significant cultural schism between the political establishment and civil society on everything related to the environmental field in Israel.

CONCLUSION: THE SOCIAL AND POLITICAL SIGNIFICANCE OF ENVIRONMENTAL ACTIVISM

The survey of environmental activism presented above allows us to visualize the general structure of the environmental movement in Israel as three concentric circles representing three types of organizations: in the center, *the environmental establishment*, including the various branches of the Ministry of Environmental Protection;[34] in the middle circle, a limited number of non-governmental *core organizations* of a unique professional hue that operate on the national level;[35] in the outer circle, dozens of grassroots organizations some of which are "ad hoc," dissolving effectively at a campaign's conclusion.

Below I discuss the significance of environmental activism in Israel in three contexts: political, cultural, and global.

The Political Context

The study's empirical findings point to a cultural and ideological cleft between the political establishment and the environmental movement in Israel: environmentalists, and particularly those of a mastery persuasion, view politicians as corrupt, opportunistic, and untrustworthy, while the latter (in most cases) regard environmentalists as part of a passing fad with extreme religious attributes, and as being naïve and out of touch with reality. The result is a double missed opportunity: productive dialogue is precluded and alienation occurs from the political establishment, which is effectively the central arena for resolving environmental problems in a democratic society.

Hence, the abstention of most environmental organizations in Israel from engaging in politics in a broader sense shows that their degree of participation in the general political arena is far from complete. Notwithstanding heightened public awareness of the issue, environmentalism is still perceived as a niche because it dares not delve into macro-political issues such as the Israeli-Palestinian conflict and socioeconomic disparities in Israeli society.

In my estimation, until such time as environmental movement organizations make a significant cultural change, that is, start attacking macro-political issues, they will go on reinforcing the negative image that some in both the public and political sphere have of them as too biocentric (mainly radical mastery organizations) on the one hand or too anthropocentric (moderate harmony organizations) on the other. Only by taking the middle road, the path of assertive moderation, while presenting rational and well-established politico-environmental arguments, will the organizations overcome the sometimes contemptuous treatment by the establishment. Signs of coming closer to such a discourse have been seen in recent years in an ideological joining of forces with the financial and business sector, which until less than a decade ago suffered from a very negative image among environmental organizations in Israel.

The Cultural Context

The data and event analysis in this study have shown that identity, belonging to an ideological symbolic framework, and the individual's need for self-realization all have a significant place in environmental activism in Israel. Thus in many cases the organizations serve for local activists as a forum for self-expression and cultivating social or political leadership. This derives from the inclusive image of the term "environment," which is perceived as clean, pure, and within the general consensus.

The present study, which has highlighted the quest for identity, shows how culture serves as a motivating factor for action in the context of a social order defined by competition and conflict. This idea, articulated by Veugelers (1994) in relation to Touraine, Habermas, and Gramsci, is supported by the current data, based on the following explanation: environmental organizations in Israel have acted over the last two decades within the context of a competitive culture in which each organization vies for its place in political-cultural space and for the vital resource that is its raison d'être – public support. Because this resource is "limited," that is, individuals tend to support no more than one or two organizations, the organizations attempt to differentiate themselves culturally by means of distinctive rhetoric, symbols or ideas. Thus competition rages that sometimes devolves into a veritable culture war. Evidence of such competition in this study would be the organizational tension between the SPNI and the INPA, and between Greenpeace as an international organization and Megama Yeruka as a local one.

In reference to Castells' (1997) suggestion that the global environmental movement's biggest contribution has been the creation of a new culture, the "Green Culture," as it has imparted new values to everything concerning the relationship between people/society and the natural environment, one could certainly say that a significant green culture has burgeoned in Israel. Like Castells, who draws a distinction between "ecological" organizations and "environmental" organizations, I have shown here the transition that generally moderate organizations have made from an ecological ethics (i.e., a focus on "classic" environmental problems such as air pollution, noise pollution, water pollution, and hazardous materials) to environmental ethics based on broader social principles among which the environment is a central but not an exclusive one.

The Global Context

Christopher Rootes, well-known claim about the global environmental movement being the last "survivor" among social movements since the 1960s raises certain difficulty regarding how to characterize the Israeli case, as the development of the environmental movement in Israel occurred in virtually no connection with the activity of other local social or political movements.

Looking at a timeline of events, it appears the environmental movement in Israel has had a slow evolution in which organizations were established generally as a response to previous campaign failures (the wildflower campaign after the Hula Lake campaign, the "Michal" campaign after the Separation Barrier campaign), catastrophic events (the Maccabiah bridge collapse in 1997, revelations regarding pollution of the Kishon River in 2002) or challenging situations (the 1973 oil crisis, absorption of waves of immigration from the former Soviet Union and Ethiopia between 1989 and 1991). Here the question arises as to whether there actually exists an effective environmental movement in Israel comparable to environmental movements in other developed countries, or whether we merely are talking about sporadic activity by a hodgepodge of different organizations acting outside of any overriding ideological framework. The research findings enable us to confidently assert that that former is the case – that in Israel there does exist an environmental movement that can be defined not only quantitatively and organizationally but also through observation of its significant role in Israeli society and impact on the culture. This determination accords with the definition of a "social movement" as suggested by Snow, Soule, and Kriesi (2004, p. 11): "collectivities acting with some degree of organization and continuity outside of institutional or organizational channels for the purpose of challenging or defending extant authority, whether it is institutionally or culturally based, in the group, organization, society, culture, or world order of which they are a part."

In light of the above, the environmental movement in Israel can be seen as part of the global environmental movement, given not only the similarity in organizational and ideological structure but also the interaction with international environmental organizations and the capacity to assimilate global ideas within the local Israeli context. The clearest indication that environmental organizations in Israel are also part of a discernable social movement is the profound cultural and ideological pluralism within the movement, as evidenced in a broad diversity of actors, interests, and values (harmony and mastery), thus creating an effect of cultural transformation. All the environmental campaigns presented here comprise turning points in terms of not only the specific issue they addressed but also the broader spatial, political, and cultural transformations they brought about. The fact that this has been a consistent, ongoing, and long-term trend entailing a process of institutionalization shows that environmental organizations in Israel are part of a defined movement and not just a random repertory of actors.

NOTES

1. Hawken claims further that it is the largest social movement in history, whose full scale remains unknown. Particularly interesting is the subtitle of Hawken's bestselling book *Blessed Unrest: How the Largest Movement in the World Came into Being and Why No One Saw It Coming*.

2. The description of all seven case studies is based on deep interviews with major relevant actors (activists, decision-makers, business, and private sector), media and official protocols from planning administration discussions. All these sources were held by the author as part of his PhD dissertation (Furst, 2013).

3. The terms "moderate" and "radical" were chosen in order to draw a distinction between action patterns and the action spaces in which organizations operate. Accordingly, "radical" does not refer here to the use of extreme violence but rather to a higher degree than usual of daring and sacrifice, initiation of out-of-the-ordinary actions, etc.

4. According to Aristotle, rhetoric is a human art, a situation in which a person try to get an agree from others, that is the art of persuading.

5. An exception is the Society for the Protection of Nature in Israel, whose representatives sit as members at the National council for Planning, representing those organizations concerned with environmental protection.

6. The concept of "frame resonance" appears in Snow and Bendford (1988).

7. In each of the boxes there appears the name of the environmental organization and the relevant campaign, for example: Zalul – The Fish Cages. It should be stressed that the mapping of the organizations is based on their activity in the specific case studies rather than on a generalized organizational-studies-type characterization.

8. Interview, Mr. Givon, leader of activists in Zoran, July 2011.

9. Representation of Environmental Public Bodies Law, 2002. The law established for the first time the official status if environmental organizations in planning institutions and granted them the right to vote.

10. The Um al-Safa campaign bore fruit as in 1979 it was decided to establish the new settlement (Halamish) southwest of the forest rather than within it. See Tal (2002).

11. Interview, Mrs. Ringer, special interrogator of the national council for planning, December 2010.

12. Interview, Mrs. Gilboa-Ron, an activist of "Mishmar ha-Glboa," November 2010.

13. Interview, Prof. Dayan, scientific division, Nature and Parks Authority, June 2010.

14. Interview, Mr. Pearlmuter, coordinator of SPNI, June 2011.

15. Local media advertisement and local website, 2004.

16. The SPNI and the Nature Reserves Authority cooperated fully on the campaign, and the ideological and rhetorical identity between the two was strong. Accordingly, I relate here to the two organizations essentially as a single unity of analysis.

17. The division of functions between the two organizations was that the Nature Reserves Authority would lead legislative initiatives and coordination with government bodies (the Ministry of Education, the court system, the Israel Police, the IDF) as well as commercial ones, while the SPNI would supply the network of volunteers and fieldwork.

18. "Nature Protection," SPNI magazine, October 1962.

19. Interview, Mrs. Elyashar, coordinator of CPJH, June 2010.

20. Interview, Mrs. Zur, head of CPJH, July 2010.

21. Interview, Adv. Vitkon, special interrogator of the national council for planning, June 2010.

22. Non-violent direct action refers to protest methods that are not illegal and that are aimed mainly at attracting media attention by staging theatrical

presentations, demonstrating dangers and engaging in physical acts of sacrifice. A key principle of direct action is that it is held at those physical sites that are the object of protest. For more on these tactics, see Furst (2012).

23. Interview, Mrs. Grossman, coordinator of Clean Energy campaign, GreenPeace, August 2011.

24. News item, www.ynet.co.il, August 2007.

25. Adam Teva V'Din literally translated is Man, Nature and Law.

26. Words of Mrs. Hann, coordinator from SPNI, news item, HAARETZ newspaper, 13 June, 2008.

27. By the term "local activist" I mean an individual person who initiates a campaign but does not establish an organization, and with whom, notwithstanding collaboration with other bodies and organizations, the campaign is identified with personally.

28. Interview, Mr. Shefer, environmental coordinator, Municipality of Ashkelon, September 2011.

29. Exceptional in this regard are two organizations: the SPNI, which partners with local organizations, and Adam TevaV'Din, which regards itself as a legal clinic providing inter alia aid to local campaigns.

30. Interview, Mr. Fishler, deputy manager, "Zalul," March 2010.

31. Interview, Mr. Tondovsky, general manager, "Ardag," March 2010.

32. Chat on "Megama Yeruka" website, www.green.org.il, July 2005.

33. In many senses there is a similarity between the success of SPNI's coral campaign and the wildflower campaign in the 1960s: in both cases there was a valued, aesthetic and well-loved object under threat, for which it is relatively easy to mobilize sympathy while presenting the "enemy" as a bully or entrepreneur that wishes them harm.

34. Including the Nature and Parks Authority, the National Parks and Nature Reserves Council, environmental units, river authorities, and two government-owned corporations.

35. Within this circle fall Life and Environment – The Israeli Union of Environmental NGOs as a representative umbrella organization, the SPNI as a planning and educational body, and Megama Yeruka and Greenpeace as radical activists.

REFERENCES

Bendford, R., & Snow, D. (2000). Framing processes and social movements: An overview and assessment. *Annual Review of Sociology*, *26*, 611–639.

Castells, M. (1997). *The power of identity*. Oxford, UK: Blackwell Publishers.

Close, D. (1998). Environmental NGOs in Greece: The acheloos campaign as a case study of their influence. *Environmental Politics*, *7*(2), 55–77.

Della Porta, D., & Rucht, D. (2002). The dynamics of environmental campaigns. *Mobilization: An International Journal*, *7*(1), 1–14.

Doyle, T. (2002). Environmental campaigns against mining in Australia & the Philippines. *Mobilization: An International Journal*, *7*(1), 29–42.

Furst, B. (2012). *"Think Global-Act Local": A descriptive analysis of environmental protest organization – The case of Greenpeace Israel.* Research Paper No. 7. Institute for Israel Studies, University of Maryland.
Furst, B. (2013). *Political cultures in environmental organizations in Israel: Typology, effectiveness and meaning.* PhD dissertation, submitted at the Hebrew University of Jerusalem (in Hebrew).
Hasson, S. (1993). *Urban social movements in Jerusalem: The protest of the second generation.* Albany, NY: State University of New York Press.
Hawken, P. (2007). *Blessed unrest: How the largest movement in the world came into being and why no one saw it coming.* London, UK: Penguin Books.
Karassin, O. (2001). Organizations acting for quality of the environment and life in Israel. *National priorities in the environmental field in Israel: Position document 2.* Haifa, Israel: Samuel Neaman Institute for Advanced Studies in Science and Technology. (In Hebrew).
Lane, R. (1992). Political culture: Residual category of general theory. *Comparative Political Studies, 25*, 362–387.
Miller, J. (1994). *Democracy in the streets: From Port Huron to the siege of Chicago.* Cambridge, MA: Harvard University Press.
Rootes, C. (1999). *Environmental movements: Local, national and global.* London, UK: Frank Cass.
Rootes, C. (2003). *Environmental protest in Western Europe.* Oxford, UK: Oxford University Press.
Schwartz, S. (2006). A theory of cultural value orientations: Explication and applications. *Comparative Sociology, 5*(2–3), 137–182.
Snow, D., & Bendford, R. (1988). Ideology, frame resonance and participation mobilization. *International Social Movement Research, 1*, 197–217.
Snow, D., Soule, S., & Kriesi, H. (2004). Mapping the terrain. In D. Snow, S. Soule, & H. Kriesi (Eds.), *The blackwell companion to social movements.* Oxford: Blackwell Publishing.
Tal, A. (2002). *Pollution in a promised land.* Berkeley, CA: University of California Press.
Tal, A. (2011). *The environmental movement in Israel: Trends, needs and potential.* Research report submitted to the JMG Foundation, Ben Gurion University, Beer Sheva, Israel (In Hebrew).
Taylor, D. E. (2000). The rise of the environmental justice paradigm. *American Behavioral Scientist, 43*(4), 508–580.
Veugelers, J. (1994). Social movements. In J. Curtis & R. Richardson (Eds.), *Social world.* Toronto, Canada: McGraw-Hill Ryerson.
Wildavsky, A. (1987). Choosing preferences by constructing institutions: A cultural theory of preference formation. *American Political Science Review, 81*(1), 3–21.
Williams, R. (2004). The cultural contexts of collective action: Constraints, opportunities, and the symbolic life of social movements. In D. Snow, S. Soule, & H. Kriesi (Eds.), *The blackwell companion to social movements.* Oxford, UK: Blackwell Publishing.

EXPLORING IDEOLOGY AS A 'RESOURCE' FOR ENVIRONMENTAL JUSTICE ACTIVISM: REFLECTIONS FROM THE ANTI-GMO MOVEMENT IN FRANCE

Darren McCauley

ABSTRACT

Environmental justice activism is increasingly globalized, multi-faceted and multi-scaled (Bickerstaff & Agyeman, 2009; Walker, 2009a, 2009b). The existence or perception of injustice triggers the development of social activism in increasingly diverse contexts. The present contribution seeks to assess the explanatory value of resources in understanding activism (Freeman, 1979). In place of justice, the under-studied social movement theory of resource mobilization is explored as a complementary and partly oppositional account of justice activism. The highly controversial anti-GMO movement in France is selected as an invigorating context for evaluating activism. The perceived injustice of lifting restrictions on

Occupy the Earth: Global Environmental Movements
Advances in Sustainability and Environmental Justice, Volume 15, 171–193

ISSN: 2051-5030/doi:10.1108/S2051-503020140000015007

the importation of GM maize into France inspired the mobilization of a nationwide movement. In sharp contrast to existing literature, ideology is considered as a resource that effectively promotes or hinders social activism. Significant conclusions are developed for environmental justice activism research around emphasizing instability, heterogeneity, cultural sensitivity and above all, the limitations of agency-centric arguments.

Keywords: French society; French politics; environmental justice activism; social movements; resource mobilization theory

The scope of environmental justice has grown considerably since its inception in the United States (Agyeman, 2002). The range of socio-environmental concerns has, indeed, widened alongside understandings of its definitions and constituents (Davies, 2006; Kurtz, 2004). Researchers have been challenged to move beyond exploring proximity and scale (Bickerstaff & Agyeman, 2009; Walker, 2009a). The environmental justice frame has effectively globalized beyond US preconceptions to new locations, contexts and concerns (Walker, 2009b). This chapter explores environmental justice activism within the context of the French anti-GMO movement. The lifting of a national moratorium in 2003 on the importation of GM maize is pinpointed below as a key mobilizing injustice moment resulting in a nationwide anti-GMO movement. However, the study of activism involves more than the exploration of justice frames and the resulting development of justice theory. This chapter concentrates, rather, on the contribution of existing (under-studied) social movement theories for understanding environmental justice activism (Freeman, 1979; Obershall, 1973; Tilly, 1978). Due to word restrictions, the resource mobilization theory (only) is explored as a conceptual framework for explaining activism.

The anti-GMO movement in France offers an intriguing backdrop for the assessment of resource mobilization. The movement is intimately linked to the rebirth of French social mobilization in the 2000s as the cornerstone to the national anti-globalization movement. The iconic site of Larzac became the symbol of French anti-GM sentiment, inspiring waves of nationwide civil disobedience, known as *Les Faucheurs Volontaires* (FV). The movement is often associated with its charismatic (and formerly imprisoned) Presidential candidate Jose Bove. This chapter reveals, however, that the anti-GMO movement is much broader than the personality of Jose Bove, or the actions of 'les faucheurs'. It explores below a wide range of interests, spanning from environmental and agricultural to consumerist and human rights. It argues that shifting ideology is central to a movement's ability to respond to changing opportunity structures.

The use of biotechnology in food has generated substantial research into its potentially harmful effects for human beings. I concentrate specifically on the 'release of genetically modified organisms into the environment'. The particular issue of 'release' is highly emotive among French citizens. Interest in France is derived from exceptionally consistently negative opinion polls towards the release of GMOs. As a result, French policy-makers have attempted to translate such research into legislation and, more generally, policies. The relative novelty and inter-departmental nature of biotechnology in food have provided a serious challenge for consecutive French administrations (Toke, 2007). Moreover, a peculiarity of social mobilization in France rests in its relationship with the state. I explore below how French social movements have become defined by an overbearing and exclusionary state apparatus. Rather than suppressing social action, this history continues to inspire the emergence of new activities, personalities and movements.

THE FRENCH STATE, IDEOLOGY AND SOCIAL MOBILIZATION

The country case study of France in mobilization literature is synonymous with the emergence of new social movements in the late 1960s. 'May 1968' became known as an ideological watershed moment that led to the replacement of conservative morality by more liberal values. New movements consisted of 'those whom traditional institutions had forgotten or excluded' (ecologism, antinuclearism, feminism, consumerism and postmaterialism). The traditions of individual nation-states are integral to structuring social movements (Dryzek, Downes, Hunold, Schlosberg, & Hernes, 2003). France is essentially classified as a *strong* and *passively exclusive* state. A *strong* state makes important political decisions without much recourse to various interests. With little regard to variation between policy areas, resulting national policy styles dominate the entire political system (Kriesi, 2004). As *passively exclusive*, movements are excluded from traditionally corporatist forms of national decision-making. The exclusion is termed as passive because the state does nothing to undermine 'outside' movements, interests or associations (Dryzek et al., 2003, pp. 1–20).

As an idea that can be traced back as far as Jean-Jacques Rousseau, social movements have been dominated by a centralized ideology of exclusion. Following the traditional French conception of democracy, Cohen-Tanugi (1991) underlines that lobbying or groups represent an attempt to prioritize particular interests to the detriment of the general public interest. In the French Republican tradition, the state has been held to be superior to the total of competing interests: groups exist in a subordinate relationship with the state. Effectively banning civil society representation after the French Revolution, le Chapelier Law of 1791 was the legislative expression of the political culture of Jacobinism in France. The traditional role of the French state is therefore classified as Jacobin, which stipulates that elected governments are mandated with the will of the people directly, without the mediation of other interests (Hazareesingh, 2002).

The recent history (post 1970) of the environmental movement in France is wrongly characterized by over-reliance on state funding, impotent campaign management and an ultimately futile Green party representation. As the figurehead of environmental campaigns, the anti-nuclear movement mobilized more protests in France than in any other country. French environmental groups (*FFSPN, la société nationale de protection de la nature*) sprang up in the 1970s, initially in an attempt to stop the construction of nuclear plants. However, the catastrophic failure of budding environmental groups to campaign successfully resulted in the fortification of an overbearing centralized state and the subsequent absorption of environmental representation by mainstream parties. The environmental movements have, as a result, become marginalized and devoid of a substantial support base. It is claimed that the initial success, and recent failure, of *Les Verts* has marked the demise of a national environmental movement.

Such a gloomy diagnosis neglects, however, to take into account changing patterns of environmental, and more generally, social mobilization. It ignores the networks established among movement actors in France on policy-specific and ideologically shared issues (Ollitrault, 2004). Emerging values of 'global justice' have challenged environmental and non-environmental movement actors to become involved in a combined struggle. Fougier (2004) maintains that environmental actors are an integral part of what he calls '*la mouvance contestaire*' (the sphere of protest). The case study (below) on the anti-GMO movement demonstrates how environmental actors, alongside agricultural, forestry and specialist associations, were equally vital in protests against GMOs. The French anti-globalization movement is, therefore, considered as a space or aggregation of *interests* united in the common goal of 'environmental justice'.

The continued ideological development of direct action movements in France alludes to a long-standing inclination towards protest behaviour among French citizens and groups (Wilson, 2008). Many high profile cases have managed to successfully influence government policy. Sustained pressure from a wide range of anti-waste incinerator interest groups, for example, led to the abandonment of several waste-to-energy plants throughout France (McCauley, 2009). There has also been a long list of protests from more excluded agriculture groups against certain objectives of the Confederation Agricole Paysanne (CAP) and the decision of the EU and the government to allow market-driven priorities to transform the countryside (Thompson, 2003). Most recently, the anti-GMO movement has become the focus for renewed protest activity (McCauley, 2011, 2013). I set out below the conceptual framework of resource assessment before exploring the rationale and methodological approach of assessing the anti-GMO movement in France.

RESOURCE EXPLANATIONS OF ACTIVISM

The resource mobilization theory emerged with Olson's 'Logic of Collective Action', in response to the shortcomings of classical collective behaviour theories. It sought to develop an understanding of what makes collective action possible. Although it does not systematically evaluate different resources, Olson's work introduces the relationship between the group size and its overall effectiveness (Olson, 1965). The resource mobilization theory developed further as a distinctive approach to the study of social movements. Obershall (1973) used this theory to identify and define potential resources for individual actors in reacting against the state. Tilly (1978) introduced the importance of internal organization for social movements. However, this chapter emphasizes that these fundamental works (particularly those of Obershall and Tilly) on the resource mobilization theory are heavily dependent upon a rather narrow material-based understanding of resource mobilization. As a result, it argues that a study of social movements and its constituent parts demands a more systematic assessment of individual resource bases – in this case even ideology.

Sufficient levels of resources are needed for initial and sustained social mobilization (Freeman, 1979). Focusing on the resource mobilization theory essentially prioritizes the study of movement organizations and activists over movements (Chazel, 1993). It concentrates on the rationality of movement groups through emphasizing a range of 'resources'. The principal hypothesis maintains that the activity of any group is increased when it acquires more resources. As a result, the groups with more resources attempt to exert more effort through a wide range of largely non-confrontational strategies. Poorer resourced movements are more likely to perform more confrontational activities (Dalton, Recchia, & Rohrschneider, 2003, p. 756). The current underdevelopment of the resource mobilization theory requires urgent attention from social movement scholars. Too often consigned to financially driven causal arguments, resource mobilization research designs lack an appreciation of new trends in social-psychological driven movement studies.

This chapter upholds indeed that 'the tendency of resource mobilization to marginalize ideological issues may deflect critical attention from one of the most vital issues for movement formation' (Buechler, 1997, pp. 198–199). Collective action by individuals and organizations is equally determined by their ideologies and values. Alongside Touraine (1977), Eyerman and Jamison (1991) criticized resource mobilization studies for avoiding issues of values and ideologies in the mobilization of activists. Diani (2000) and Zald (2000) first introduced this idea in response to critics of the 'value-free' nature of resource mobilization through the development of 'ideologically structured action'. This chapter identifies (and limits to) three core issues in exploring ideology 'as a resource' for mobilization, namely, *collective solidarity*, *shared goals* and *individual beliefs*.

Doherty, Plows, and Wall (2003) underlined the pivotal role of a common philosophy and ideology to the mobilization of direct action in Britain. A sense of solidarity among activists ensured the long-term survival of loosely formed organizations or movements. Moreover, their shared belief structure became as important as other resources in stimulating collective action. La Goutte d'eau campaign and the Bergama environmental movement are examples of how sustained mobilization can be ensured through shared ideology (Doidy, 2004). La Goutte d'eau is a veritable symbol of how mobilization can be ensured through shared ideology alone. This is a mountain ranch that was occupied by Eric Petetin on the site of a proposed European road project. It became the venue for concerts, festivals, public debates and general assemblies arranged by local environmental groups. Throughout a 10-year period, la Goutte d'eau created a loosely knit network of environmental groups, ecologists and sympathizers.

Coban (2004) underlines that the successful mobilization of the Bergama environmental movement was largely due to shared belief towards a common goal. A common environmental philosophy was forged between individuals (peasants, villagers, engineers, architects, doctors) and groups (local and national associations, German-based FoodFirst, Amici della Terra, Washington-based Mineral Policy Centre). It was their shared belief in opposing mining programmes that sustained a successful mobilization of the environmental movement. A common philosophy is, in this way, a resource that can create a more cohesive and effective organization. Fillieule (2001) demonstrates the essential role of philosophy in the recent boom of 'solidarity' associations in France. Essentially, his study shows that the philosophies held by individual members are an equally important resource to the association as the collective ideology. The motivations and determinants of increased activism may, therefore, be explained through individual philosophical considerations.

Solidarity, shared goals and individual beliefs can significantly influence the form of social mobilization (Polletta & Jasper, 2001). Dalton et al. (2003) demonstrated that self-identified ecologist groups tended to use more unconventional methods of political activity. Conservation groups focused, in contrast, on limiting their activities to consultative and lobbying practices. This difference in collective ideology resulted, therefore, in two unique strategies based on a range of action repertoires. It is, of course, acknowledged that treatment of ideology as a resource category cannot cover such issues in any real depth. As a result, attention is strictly focused on examining the core issues of collective solidarity, shared goals and individual beliefs in determining a movement actor's strategy selection.

CASE SELECTION AND METHODOLOGY

The particular issue of GMOs is highly emotive among French citizens. France has witnessed the growth of a distinctive societal response to the release of GMOs. Unique from other EU member states, the activities of French activists have turned progressively more aggressive and on a national scale. Before 2003, anti-GM campaigns were essentially structured around sub-national resistance based on demonstrations, marches and even sabotage. At a time when similar movements across Europe dwindled, FV campaign in 2003 marked a unique period of widespread and radical national level 'civil disobedience' (Hayes, 2007). The main catalyst – and perceived injustice – for this shift was the lifting of a national moratorium (1999–2003) on the importation of GM maize. The anti-GMO movement has encompassed a wide range of interests, including environmental, consumer, counter-globalization, agriculture and specialist.

I concentrate on the four most active social movement organizations in the anti-GMO movement: environmental interest group 'Greenpeace-France' (referred below as Greenpeace), trade union 'Confédération Paysanne' (CP), counter-globalization actor 'Association pour la Taxation des Transaction pour l'Aide des Citoyens' (ATTAC) and specialist group 'Inf'OGM'. Greenpeace was the first environmental association to establish an anti-GMO campaign in France. Until the FV campaign in 2003, Greenpeace was recognizably the leader of the anti-GMO movement. The Larzac 'altermondialiste' festival in August 2003 marked a significant transformation in leadership. An axis built upon la CP–ATTAC emerged as the new leading partnership. Moreover, the controversial issue of GMOs in France prompted the establishment of specialist associations. Inf'OGM specializes in monitoring the GMO debate. Similar to CP, this association multiplied its activities after participation in the 'Larzac' festival. I examine below the main strategies employed by each association throughout the post-FV period of 2004–2011.

I assess below the main strategies employed by each association throughout the post-FV period of 2004–2011. A triangulated data collection methodology includes semi-structured interviews with key social movement actors, detailed policy documentary analysis and an examination of newspaper coverage. Document analysis provided the main data collection method for ascertaining the variety of strategies employed by the selected groups. A newspaper archival search outlined the range of actions undertaken by each group throughout the lifetime of the post-2004 period in question. Newspaper articles are often cited as a major resource for research on social movement mobilization and environmental activism (Fillieule, 2003). A mixture of both semi-structured interviews and document analysis was, therefore, employed in the collection of data on resources, in this case being 'ideology'.

THE FRENCH ANTI-GMO MOVEMENT

Centred on anti-globalization activities, anti-GM food protests have become active in the French 'sphere of protest' (see Fougier (2004) above). This sphere is also not limited to the activities of José Bové, the CP or ATTAC. In reality, it encompasses a range of large-scale interest groups, social movements and new protest groups. The interest groups derive from an assortment of sectors advocating the protection and promotion of developing countries, solidarity, free trade, human rights, feminism, ethnic minorities and environmentalism. For example, a new trade union has emerged directly from this movement, 'Solidaires, Unitaires and Democratiques', which epitomizes emerging forces and contradictions at play (Le Queux & Sainsaulieu, 2010). Its activities allow a range of groups to identify certain elements of their particular cause that are associated with the larger anti-globalization movement (Fougier, 2002, 2004). I set out below the main anti-GMO activities of such interests. The chapter then concentrates on the four key organizations of Greenpeace, CP, Inf'OGM and ATTAC.

Movement Leaders: Greenpeace-France and Confédération Paysanne

Since its establishment in 1977, Greenpeace has been known as the poor relative in the global Greenpeace 'business'. The failure of the anti-nuclear campaign in the 1970s was only the prelude to its greatest crisis, the Rainbow Warrior fiasco.[1] In contrast to other countries, Greenpeace suffered from negative implications in the affair, culminating in the closure of their offices in 1987 (explored below under 'organizational resources'). With a permanent office of 46 staff, Greenpeace is now the fastest growing environmental group in France in the last 10 years.[2] Throughout this period, Greenpeace strategically constructed a process of professionalization and social integration. The former refers to the restructuring of their offices, campaigns and overall operational structure and procedures. The latter attempted to regain both the support of the French public and the environmental community. Greenpeace created a series of networks with other actors while establishing a system of regional offices to bring their work 'closer to the people'.

Greenpeace was the first environmental association to establish an anti-GMO campaign in France. At first, it remained essentially at the regional level. This campaign was triggered in late 1996 when Libération reported that a shipment of GM soybean was to arrive. In a well-known media event, Greenpeace set up a blockade on the arrival of the soybean in France. Throughout the proceeding five years of the campaign, Greenpeace was recognizably the leader of the anti-GMO movement. The Larzac 'altermondialiste' festival in August 2003 marked a significant transformation in leadership. An axis built upon *CP–ATTAC* emerged as the new leading partnership. This shift in power was symptomatic of 'wholesale' changes in the movement's scope, focus and activities. Within this context, José Bové developed into a symbolic figurehead of this shift in leadership.

Indeed, CP is essentially a peasant farmers' association that can be traced back to an opposition movement to the Gaullist programme of agricultural intensification in the 1950s. Traditionally in the shadow of the Federation Nationale des Syndicats d'Exploitants d'Agricoles (FNSEA)/CNJA,[3] CP structured its increasingly political movement alongside the urban proletariat in its pursuit of a class struggle. Officially established in 1987, CP completed its transformation from a movement to a fully structured agricultural trade union. It has recently become the second most popular agricultural trade union in France (behind FNSEA). As will be discussed in greater detail, its activities became synonymous with its charismatic leader Bové. His decision to stand for the 2007 Presidential Elections has particularly thrust the CP's GMO campaign into the spotlight.

Specialist and Counter-globalization Groups

The controversial issue of GMOs in France notably prompted the establishment of specialist associations: Inf'OGM. Guided by both environmentalist and consumerist principles, Inf'OGM is an officially not-for-profit association that specializes in monitoring the GMO debate. Under the leadership of Thierry Raffin, Inf'OGM puts pressure on national policy-makers to offer transparent information to the public. After an international conference in India in late 1998, Isabelle Tarradellas officially created the association to provide the French public with information on the issue. In terms of activities, the group publishes articles, press statements and a monthly bulletin. Intensifying its networks with other associations (ATTAC, FNE (France Nature Environnement), FRAPNA, *Nature et Progrès*) since 2003, Inf'OGM has been involved in joint publications and campaigns on GMO issues. Similar to CP, this association multiplied its activities after participation in the 'Larzac' festival. Nonetheless, their activities concentrate uniquely on information collection and dissemination, usually in the wake of larger high profile Greenpeace media campaigns.

ATTAC was, at the same time, founded in 1998 as the first major 'counter-globalization' association in France. ATTAC is present in 34 countries throughout the world. ATTAC-France has a membership base of 30,000 with a complex network of regional associations. It particularly concentrates on the promotion of 'alternative' Left-wing politics. Its interests often position this association within mainstream political debate. The close connections between the FV campaign and CP would largely demote ATTAC to a secondary partner in the leadership of the anti-GMO movement. Nevertheless, the close association of José Bové to these associations maintains a distinct confrontational approach to the GMO issue.

In addition to Greenpeace, CP, Inf'OGM and ATTAC, there are also various other actors that are active in anti-GMO activities that can be categorized into four groups: environmental, consumer, agricultural and specialist. First, *les Amis de la Terre* (AdT), FNE and *Agir pour l'Environnement* have been consistently involved in collective actions against the release of GMOs. AdT has multiplied its links with ATTAC on the GMO dossier, although their activities in this area are on an ad hoc and often irregular basis. FNE initiated a series of high profile legal cases on the GMO dossier between 1998 and 2003. However, their involvement diminished significantly throughout the more aggressive proceeding period. *Agir pour l'Environnement* became active in the anti-GMO movement in 2002 in response to the proposed lifting of the EU moratorium. Essentially, they adopted a role as coordinator of collective events for the movement.

Second, the main consumer group UFC-*Que Choisir* has consistently monitored the GMO debate. Their strategy has focused almost uniquely on information dissemination. Third, there are a variety of agricultural groups that partake in anti-GMO activities. *la Coordination Rurale* (CRUN) is a smaller version of CP in terms of size and impact. Founded more recently (1991), CRUN is an organization of peasant farmers that participates in demonstrations and collective campaigns against GMOs. *Fédération Nationale pour l'Agriculture Biologique* (FNAB) federates over a hundred local and regional groups representing the majority of organic farmers. After joining the anti-GMO movement in 2000 after risks of contamination, their activities have been largely limited to petitions for better regulations. *Mouvement pour le Droit et le Respect des Générations Futures* (MDRGF) is an association for non-intensive farming. It has been active in national campaigns and demonstrations against the release of GMOs. Ecoropa is also involved in alerting the French public about any dangers in the field of genetic engineering.

Finally, *OGM Dangers* is the other (in addition to Inf'OGM) particularly active specialist group in this area. OGM Dangers provides a space for ideological reflection on the GMO debate. Established in 1998, this association is primarily an organizer of events, conferences and seminars. The background to this association is intimately connected to the personnel and ideology of ATTAC and Ecoropa. OGM Dangers places the GMO debate within wider philosophical questions on sustainable agriculture. In this way, a series of speakers are invited to sponsored events to debate the various facets of releasing GMOs into the environment. In addition to these activities, members are involved in national and local demonstrations. It is an official signatory to a series of local demonstrations under the 'Pas d'OGM dans ma commune'[4] campaign that includes 20 other anti-GMO associations. OGM Dangers also publishes a series of online articles and 'reflective essays'.

ANTI-GMO ACTIVISM AND STRATEGIES

The period (2004–2011) is selected as the main time frame for exploring the strategies of four key movement actors. The year 2004 marked the beginning of widespread national protest in response to the lifting of a national ban on GMOs. It is clear during this phase that the associations strategized on three distinct lines: civil disobedience, lobbying and media campaigns. First, it is revealed that CP and ATTAC led a successful expansion of the FV campaign across France. This campaign has encompassed wider interests since 2008, including Greenpeace activists. Second, the initial leadership of Greenpeace and Inf'OGM in lobbying government was usurped by CP and ATTAC. Third, Greenpeace and Inf'OGM (in contrast to CP and ATTAC) have focused their attention on lobbying businesses on the labelling and tracing of their produce.

Civil Disobedience

The FV campaign was officially launched at the famous Larzac counter-globalization rally of August 2003. Activists converged on Larzac to express their dismay at the consequences of current globalization pressures and the decision of France to lift the moratorium on GMOs. Even though anti-GMO protesters were involved in crop destruction before 2003, the FV campaign marked a distinctive period of widespread and radical 'civil disobedience' on a national scale (Hayes, 2007). As an emerging movement leader during this period, José Bové (associated officially with both CP and ATTAC) called on individuals to destroy crops and biotechnology research centres throughout France. Links between CP and Les Verts were evident as the latter actually called for citizens to partake in 'la désobéissance civile' (civil disobedience). CP and ATTAC developed an unambiguous strategy of civil disobedience from early 2004 onwards, firmly rooted in counter-globalization ideology.

Although Greenpeace and Inf'OGM were present at Larzac, they had largely avoided involvement in any aggressive actions as part of the FV campaign. There has been a clear distinction between the non-aggressive activities of Greenpeace and Inf'OGM as opposed to the often aggressive activities of CP and ATTAC. Greenpeace and Inf'OGM have focused their attention on lobbying government, especially the Environment Ministry, since the late 1990s. Both groups struggled with the regionalization of lobbying activities (including legal cases) due to the government's increasingly pro-GMO stance. However, both Greenpeace and CP have become increasingly embroiled in protest activities. In 2008, activists from both Greenpeace and CP were arrested for destroying suspected GM crops in South West France. Statements from those arrested underlined the need to 'resist Europe's wish to pollute France' with GM maize (Moreau, 2010). Moreover, José Bové (currently MEP for the party Europe Ecologie and CP activist) and seven others were found guilty of destroying GM crops in late 2011.

Lobbying for Moratoria on GMOs

Greenpeace and Inf'OGM were instrumental in lobbying for a first moratorium (1999–2003) on GMOs. Greenpeace was particularly instrumental in 2000 by communicating statistics to the Brussels based office on the location and extent of GMO crops in France. First, Greenpeace attempted to exploit well-established links with the Environment Ministry on the GMO dossier. Members were involved in monthly (largely ad hoc) meetings with officials. Similarly, CP lobbied connections with the Agriculture Ministry. Its sway with agricultural policy-makers was inferior to the less active (on the GMO dossier) FNSEA. Second, Greenpeace and Inf'OGM kept regular contact with Les Verts (and other Left Parties Parti Communiste Francais (PCF) and Lutte Ouvriere). They were particularly involved in the 1999 European Parliament elections with a consistently anti-GMO campaigner and now Green MEP (Daniel Cohn-Bendit). In contrast, the bulk of CP's lobbying activities were located at the sub-national level of government.

In contrast, CP and ATTAC played a decisive role in lobbying government to impose a second moratorium (post 2008). CP's status as a trade union, of course, ensured a minimum of such political activity. However, three significant factors have propelled the anti-GMO stance of CP and ATTAC further into national political debate. First, José Bové (closely associated with CP and ATTAC) decided to stand for the 2007 presidential elections and the European Parliament election in 2009 for Europe Ecologie. He firmly vowed to (re-)instate a moratorium on all GMO activities. Reinforced by a suspended prison sentence for anti-GMO activities, Bové undoubtedly promoted the CP and ATTAC campaign against the use of biotechnology in food. Second, the candidates for the Socialist Party for the presidential elections (2007 Ségolène Royale and 2012 Francois Hollande) had created favourable links with CP (particularly Royale) and a distinctly anti-GMO stance. Third, the uncertainty surrounding a proposed new bill on GMOs created a vacuum for political debate on the issue.

Campaigns against Companies

Greenpeace and Inf'OGM have maintained a campaign against retailers in France since 2004. They established a 'black list' of those retailers selling products that contained GMOs. Without undertaking any scientific research, they simply concentrated on those products containing soya or maize. On compiling the list, the name of any retailer that did not guarantee the absence of GMOs in their products was published on their website. This issue was central to their overall agenda on GMOs. From this perspective, Greenpeace was campaigning for a more transparent and regulated process of labelling and tracing GMOs in products. This would prove to be the backbone of their anti-GMO campaign, particularly post 2004. In their view, European regulations and national government should not have established a minimum requirement of 0.9% to be considered genetically modified.

As a result, their campaign uncovered any produce that they considered to contain any trace of GMOs. In 2004, Greenpeace created a website 'Détectives OGM' dedicated to publishing regular updates on the current state of retailers in France. Inf'OGM placed similar research online under the heading 'OGM et la santé' (GMOs and health). Greenpeace attacked a well-known meat produce in 2008, *Boeuf de Bazas*, for allegedly containing GMOs. In this way, Greenpeace demonstrated a sophisticated ability to use media and advertising pressure on businesses. They highlighted that European and national law did not govern the existence of GMOs as a foodstuff in animal rearing. Moreover, Greenpeace distributed a guide on the presence of GMOs in produce to consumers throughout France at stores of the supermarket chains *Géant* and *Carrefour*. Carrefour decided in late 2010 to label 300 of its own products 'GMO free' with a green and white sticker in response to the Greenpeace-led campaign (Durand, 2010).

IDEOLOGY AND THE ANTI-GMO MOVEMENT

This section explores a resource based account of 'ideology' in terms of why each group selected and modified their strategic outlook as explored above. The tendency of the resource mobilization theory to 'marginalize ideological issues may deflect critical attention from one of the most vital processes for movement formation' (Buechler, 1997, pp. 198–199). In addition to organizational and material factors (outside the scope of this chapter), collective action by individuals and organizations is equally determined by their ideologies and values. This section attempts to reinforce the importance of ideology in social movement behaviour through exploring the extent to which ideologically structured action is observable (Diani, 2000; Zald, 2000).

Individual Philosophies and Shared Beliefs on GMOs

CP and ATTAC were essentially the ideological leaders of the 'altermondialiste' movement in France (Agrikolansky, Fillieule, & Mayer, 2005, p. 23). In political terms, these groups endeavoured to pilot a revitalization of the French Left. This desire is epitomized by the decision of their campaign leader José Bové to challenge in the 2007 French Presidential and 2009 European Parliament elections. CP originates essentially from two principles based on the 'refusal of corporatism' and 'international solidarity'. The former refers to a belief that small to medium farmers are increasingly exploited by large food processing companies. In this way, the peasant has evolved into the symbol of a new proletariat. The latter concentrates on demonstrating solidarity with third world farmers. CP considers the unwanted imposition of 'Western' agricultural practices in such countries as the source for the plight of third world peasants (Agrikolansky et al., 2005; Herman, 2002). Their guiding philosophies endeavour to position the GMO debate in a larger (often aggressive) movement for an alternative form of globalization and sustainable agriculture for small farmers (see Table 1). This factor has been crucial to the politicization of the CP-ATTAC political strategies post 2006.

Table 1. Philosophies and Shared Beliefs.

Philosophies/Group	Greenpeace	CP	ATTAC	Inf'OGM
Altermondialisme	X	X	X	X
Refusal of corporatism		X	X	
Sustainable development	X			

As a specialist association in GMO policy, Inf'OGM is guided simply by an overarching mission to offer information to the public on GMOs. Its links with ATTAC through Aurelien Bernier reveal a motivation for mobilizing anti-GMO sentiments among its members/readers. Greenpeace International and other national offices (particularly the United Kingdom and Germany) are major actors in counter-globalization activities. While Greenpeace-France remains an active player in the movement, the CP-ATTAC axis has largely led the philosophical and activism-based challenges in France. Greenpeace campaigns cover a wider remit than CP-ATTAC under the guiding umbrella of sustainable development – in this case a notable prioritization of encouraging green business/industry in society. As the main founding issue to Greenpeace, their campaign against nuclear power remains a major philosophical undercurrent. Above all, a distinguishing factor of Greenpeace is the willingness to use media in promoting their campaigns through large-scale non-aggressive activities (Auger & Ferrante, 2004, pp. 11–32). This was most evident in their successful campaign against retailers.

The personal ideology of individual members can also determine the strategies employed by a particular organization (Snow, 2004). In sharp contrast to Greenpeace, CP and ATTAC have had recourse to a major charismatic figure in the anti-GMO movement, José Bové. He has been the leader of CP and remains a member of the 'Collège des fondateurs' of ATTAC-France. His involvement in both CP and ATTAC-France since 1998 embodies the often cited 'CP–ATTAC axis' in anti-GMO activities. Famously presented as a simple farmer from the Midi, he has positioned himself as a protector of peasant farmers from the rise of liberal economics and globalization. From this perspective, Bové finds a natural home in the anti-productivist CP and the counter-globalization ATTAC-France. His leadership essentially allowed these two organizations to move the anti-GMO movement towards direct action through local and national 'civil disobedience' (Martin, 2003).

Lepault (2005, pp. 25–39) argues that Bové has imposed his personality on the GMO issue to such an extent that a debate on the potential advantages of GMOs is no longer possible in France. The role of individuals at Inf'OGM has been, in contrast, limited for understanding strategy selection. Christophe Noisette (1998–2002) and then Eric and Aurélien Bernier (2002+) have concentrated on creating links with Greenpeace and to a lesser extent ATTAC. However, there is a notably high turnover in personnel in Greenpeace. According to a former representative on the GMO dossier, this has significantly hampered the emergence of a Greenpeace leader on the issue of GMOs. As a result, individuals (such as José Bové) do not have the opportunity to develop. He also maintains that it would be very difficult to compete with the high profile (and often illegal) actions of José Bové.

Ideology vis-à-vis the State and GMOs

An association's ideology towards the state significantly affects the strategies undertaken to influence policy-making (Edwards & McCarthy, 2004; Zald, 1992). WWF-France (as part of an initially larger sample of organizations) believes that cooperation with institutions and parties is the most effective relationship with the state. From this perspective, state actors are targeted as power players through their status as policy-makers. Greenpeace, Inf'OGM, CP and ATTAC profess, in contrast, that societal confrontation with state institutions is the necessary consequence of unpopular policy decisions (see Table 2). Movement actors accordingly mobilize popular support as the most productive response to, in their view, an impenetrable overbearing state. Greenpeace and Inf'OGM perceived their role as instigators of popular support against state policies (GMOs in this instance) through high profile media campaigns and the dissemination of relevant information. CP and ATTAC claimed to simply represent undercurrents in society that are mobilizing against state policies through spontaneous actions of civil disobedience.

Table 2. Ideologies *vis-à-vis* the State.

Relation with State/Group	Greenpeace	CP	ATTAC	Inf'OGM
Cooperation				
Confrontation	X			X
Aggression/disobedience		X	X	

Table 3. Ideologies *vis-à-vis* Successive Governments and the EU.

Organization	*Vis-à-vis* Governments		*Vis-à-vis* the EU	
	Left/Right	Intervention	Relations	Intervention
Greenpeace	Indifferent	More	Confrontation	More
CP	Left	More	Confrontation	Less
ATTAC-France	Left	More	Confrontation	Less
Inf'OGM	Left	More	Confrontation	Less

The political ideology of these associations largely determines their response to successive French governments. There are important party political affiliations among the associations (see Table 3). As a result, their reaction to governments differs in proportion to the respective government in power (i.e., Centre-Right government from 2002 to 2012). All four associations are unsurprisingly sympathetic to the political objectives of Left governments. Greenpeace declared the least interest in party politics in France. This became debilitating in the increasingly politicized issue of GMOs (2006+). CP, ATTAC and indirectly Inf'OGM are proactively involved in the promotion of their own brand of Left-wing politics. The leader of the CP-ATTAC axis stood for the 2007 French Presidential and 2009 European elections. In his first step into mainstream politics, Bové pledged to reinvigorate the Left by highlighting ecologist themes and representing the so-called 'provocateurs'.[5]

It is striking how little attention was paid to the EU level by all associations in their strategies on the GMO issue. Greenpeace and, more recently, ATTAC officially recognize the growing weight of EU decision-making by establishing permanent offices in Brussels. All four associations were ideologically 'confrontational' *vis-à-vis* the EU on the GMO dossier (see Table 3). Above all, these associations expressed a particular desire for greater state intervention on the issue (Hayes, 2007). A CP representative concluded that adhering to an EU-wide suspension of the GMO moratorium post 2003 was symptomatic of government 'selling out on their Republican values'. In stark contrast to the other associations, Greenpeace maintained that the EU should also have more control on decision concerning GMOs in France. It believed that only strict zero tolerance EU-wide regulation on GMO in food could secure the long-term objective of GM-free produce. It is argued, therefore, that this ideological difference resulted in only Greenpeace lobbying EU officials, until Bové was elected to the European Parliament in 2009.

CONCLUSION

It is acknowledged that the understanding of the 'justice activists/movement actors' in this chapter is in stark opposition to much of the justice (Kurtz, 2004; Walker, 2009a, 2009b) and social movement literature (Coban, 2004; Doherty et al., 2003; Fillieule, 2003). A group's mobilization is usually considered less reactionary and more impulsive. In following, a movement does not respond to its resource base or shifting opportunity structures. The 'ideological' resource type is indeed capable of only modest contributions to understanding psychological reasons for mobilization. First, it does not provide a detailed account of the collective moral imperative driving the overall movement. Second, ideology as a 'resource' only concentrates on individual or organizational (rather than movement-wide) specific accounts of resource and opportunity exploitation. However, the chapter has demonstrated how resource accounts of environmental justice activism can move beyond financially driven arguments. Similar to financial clout, ideology can partly define what a movement actor will, and will not, do.

It is argued that ideology was a significant 'resource' for the associations in focus throughout this chapter. All four associations based their activities, under the overall banner of 'altermondialisme'/counter-globalization, on varying degrees of confrontation (not cooperation) with the state as a notion, consecutive French governments and the EU. This 'ideological resource' allowed all four associations to respond to the excluding opportunity structures within the GMO issue. Moreover, ideals of non-aggression (Greenpeace and Inf'OGM) and aggression (CP and ATTAC) limited each association's capacity for action. Under closer examination, the ideological approach of each association is uniquely different vis-à-vis strategy selection. Greenpeace displayed no interest in becoming politically involved on the GMO issue. Moreover, it was less confrontational in comparison with the three other associations. CP, ATTAC, and to a lesser extent, Inf'OGM became politically active in the promotion of alternative Left-wing politics in France. Finally, it is argued that Greenpeace lobbied supranational institutions as the only association calling for greater EU intervention. More generally, the confrontational approach of all associations to the EU hampered their ability to strategize on the supranational stage.

Environmental justice activism in France is formed around – and in opposition to – the centralist and all imposing notion of the *state* in its many guises. The traditional role of the French state is classified as Jacobin, which stipulates that elected governments are mandated with the will of the people directly, without the mediation of other interests (Hazareesingh, 2002). The collective ideological stance of all four organizations against the state and its policies on GMO issues encouraged the development of multi-level mobilization strategies. In contrast to US accounts of justice activism, the primary 'frame' (albeit not explicitly explored here) for French activism remained, therefore, the state – rather than industry and corporate actors. Moreover, civil disobedience represented a focal point for justice activist strategies in France. The emergence of the FV campaign united a wide range of interests around the notion of nationwide 'crop destructions'.

The existence or perception of injustice remains the birthplace of activism. However, first, social movement theories remind us that activism is rarely a stable homogeneous phenomenon. In support of Bickerstaff and Agyeman (2009), justice activism is based on highly unstable and fragile networks of actors. Resource mobilization underlined, moreover, the heterogeneity of justice activism through emphasizing conclusions based more firmly on the rationality of movement actors. The less confrontational and political bias stance of Greenpeace allowed it to focus more heavily on lobbying local, national and EU organizations, whilst CP capitalized the reshaping of the anti-GMO movement trajectory towards civil disobedience and Left-wing political leadership. And yet, Greenpeace activists became increasingly embroiled in civil disobedience activities post 2004. Similarly, CP and ATTAC used links with José Bové to apply pressure in the European Parliament against GMOs. Second, a resource account of activism demonstrated that environmental justice activism is often too wedded to the politics of '*localism*' rather than scale more generally. Extensive research in justice activism has pointed to the reframing of injustices at 'higher' scales (Bickerstaff & Agyeman, 2009; Davies, 2006).

Indeed, the iconic meeting at Larzac in 2003 sparked a major shift in the French anti-GMO movement. However, emphasis remains too rooted on the 'local' as the key point for distributional or procedural injustices to occur. For activists, the lifting of a national moratorium in 2003 was the key catalyst for mobilization – resulting in the meeting in Larzac. Institutionalized centralist notions of the state in France reveal a more pronounced cultural and historical expectation of injustice occurring at a national level – in this case the importation of GM maize into France. Movement actors in this instance were rather ideologically 'wedded' to the perceived distributional and procedural injustice wielded by national governments and the 'state' – rather than a long history of civil rights struggles as for US environmental justice activism (Agyeman, 2002). Third, resource explanations of activism emphasize the limitations of justice activism. The resource mobilization theory effectively prioritizes a 'rational institutionalist' viewpoint on movement trajectories. In other words, rational resource based inadequacies effectively constrain activism 'frames'. In following, agency must work within the confines of structures, no matter how far-reaching a perceived injustice may actually be.

NOTES

1. 'Le Guerrier de l'arc-en-ciel', or the more commonly known (even in French) 'Rainbow Warrior' was a symbolic Greenpeace boat that was used in order to protest against French nuclear testing in 1985. An operation was carried out by the French military (including Gérard Royal, brother of Ségolène Royal) to destroy this boat. However, Greenpeace-France was the victim of a 'disinformation campaign' by a government in denial, and more specifically France Inter. It was claimed that the Rainbow Warrior was actually a 'spy boat' that was in collaboration with the 'East', that is, communist Russia. This resulted in the closure of Greenpeace-France.
2. The total number of members throughout France is currently 86,000.
3. Fédération National des Syndicats d'Exploitants Agricole/Comité national des jeunes agriculteurs.
4. Not in my backyard!
5. Bové refers to 'les provocateurs' as the people involved in crop destructions (and other direct action activities). More broadly, he claims to represent a wider public that is increasingly disillusioned with mainstream parties. In his view, this disillusionment originates from ineffective environmental programmes and oppressive corporatist philosophies.

REFERENCES

Agrikolansky, E., Fillieule, O., & Mayer, N. (2005). *L'Altermondialisme en France: La longue histoire d'une nouvelle cause*. Paris: Flammarion.

Agyeman, J. (2002). Constructing environmental (in)justice: Transatlantic tales. *Environmental Politics*, *11*, 31–53.

Auger, P., & Ferrante, J. (2004). *Greenpeace: Controverses autour d'une ONG qui dérange*. Paris: La Plage.

Bickerstaff, K., & Agyeman, J. (2009). Assembling justice spaces: The scalar politics of environmental justice in north east England. *Antipode*, *41*(4), 781–806.

Buechler, S. (1997). Beyond resource mobilization? Emerging trends in social movement theory. In S. Buechler & F. Cylke (Eds.), *Social movements: Perspectives and issues* (pp. 193–211). London: Mayfield Publishing Company.

Chazel, F. (1993). *Action collective et mouvements sociaux*. Paris: Presses Universitaire de France.

Coban, A. (2004). Community-based ecological resistance: The Bergama movement in Turkey. *Environmental Politics*, *13*(2), 438–460.

Cohen-Tanugi, L. (1991). Légitimité et Réglementation: Les Réticences Françaises. *Problèmes Politiques et Sociaux*, *662*, 54–57.

Dalton, R., Recchia, S., & Rohrschneider, R. (2003). The environmental movement and the modes of political action. *Comparative Political Studies*, *36*(7), 743–771.

Davies, A. (2006). Environmental justice as subtext or omission. *Geoforum*, *37*, 708–724.

Diani, M. (2000). The relational deficit of ideologically structured action. *Mobilization: An International Journal*, *5*, 17–24.

Doherty, B., Plows, A., & Wall, D. (2003). 'The preferred way of doing things': The British direct action movement. *Parliamentary Affairs*, *56*(4), 669–686.

Doidy, E. (2004). Une lecture inquiète de l'activité militante: Épuisement et effondrement dans deux mobilisations contemporaines en France. *Modern and Contemporary France*, *12*(1), 63–74.

Dryzek, J., Downes, D., Hunold, C., Schlosberg, D., & Hernes, H. (2003). *Green states and social movements: Environmentalism in the United States, United Kingdom, Germany and Norway*. Oxford: Oxford University Press.

Durand, E. (2010). Nourri sans OGM: l'étiquette carrefour bien recue. Ouest-France, *5*, 7.

Edwards, B., & McCarthy, J. (2004). Resources and social movement mobilization. In D. Snow, S. Soule, & H. Kiesi (Eds.), *The blackwell companion to social movements* (pp. 116–152). Oxford: Blackwell Publishing.

Eyerman, R., & Jamison, A. (1991). *Social movements: A cognitive approach*. Cambridge: Polity Press.

Fillieule, O. (2001). Dynamics of commitment in the sector known as 'solidarity': Methodological reflections based on the case of France. In M. Guigni & F. Passy (Eds.), *Political altruism?: Solidarity movements in international perspective* (pp. 51–67). Oxford: Rowman & Littlefield Publishers.

Fillieule, O. (2003). Local environmental politics in France: Case of the Louron valley, 1984–1996. *French Politics*, *1*(3), 305–330.

Fougier, E. (2002). *La Contestation de la Mondialisation: Une Nouvelle Exception Française?* Paris: Institut Français des Relations Internationales.

Fougier, E. (2004). Le mouvement altermondialistes en France: Les groupes contestaires Français. *Problèmes Politiques et Sociaux*, *897*, 25–28.

Freeman, J. (1979). Resource mobilization and strategy: A model for analyzing social movement organization actions. In M. Zald & J. McCarthy (Eds.), *The dynamics of social movements: Resource mobilization, social control, and tactics* (pp. 167–190). Cambridge: Winthrop Publishers.

Hayes, G. (2007). Collective action and civil disobedience: The anti-GMO campaign of the faucheurs volontaires. *French Politics*, *5*(4), 293–314.

Hazareesingh, S. (2002). *The Jacobin legacy in modern France*. Oxford: Oxford University Press.

Herman, P. (2002). *Confédération Paysanne: Changeons de politique agricole*. Paris: Mille et Une Nuits.

Kriesi, H. (2004). Political context and opportunity. In D. Snow, S. Soule, & H. Kiesi, (Eds.), *The blackwell companion to social movements* (pp. 67–90). Oxford: Blackwell Publishing.

Kurtz, H. (2004). Reflections on the iconography of environmental justice activism. *Area, 37*(1), 79–88.

Le Queux, S., & Sainsaulieu, I. (2010). Social movements and unionism in France: A case for revitalization? *Labor Studies Journal, 35*(4), 503–519.

Lepault, S. (2005). *Il faut désobéir à Bové*. Paris: La Martinière.

Martin, J. (2003). José Bové, un activiste sans projet? *Modern & Contemporary France, 11*(3), 307–321.

McCauley, D. (2009). Wasting energy? An analysis of campaigns against waste-to-energy plants in France. *Environmental Politics, 18*(6), 917–938.

McCauley, D. (2011). Bottom-up Europeanization exposed: The anti-GMO movement in France. *Journal of Common Market Studies (JCMS), 49*(5), 1019–1045.

McCauley, D. (2013). Protest, politics and produce. *Local Environment: The International Journal of Justice and Sustainability, 18*(6), 223–247.

Moreau, F. (2010). Les Faucheurs volontaires d'OGM ont 'eu raison avant l'heure'. *Le Monde, 14*, 4.

Obershall, A. (1973). *Social conflict and social movements*. Englewood Cliffs, NJ: Prentice Hall.

Ollitrault, S. (2004). Des plantes et des hommes de la défense de la biodiversité à l'altermondialisme. *Revue Française de Science Politique, 54*(3), 443–463.

Olson, M. (1965). *The logic of collective action*. Cambridge, MA: Harvard University Press.

Polletta, F., & Jasper, J. (2001). Collective identity and social movements. *Annual Review of Sociology, 27*, 283–306.

Snow, D. (2004). Framing processes, ideology, and discursive fields. In D. Snow, S. Soule, & H. Kiesi (Eds.), *The blackwell companion to social movements* (pp. 380–413). Oxford: Blackwell Publishing.

Thompson, D. (2003). Protest politics and violence in the French countryside in the 1990s: An expression of euroscepticism. *Modern and Contemporary France, 11*(3), 293–306.

Tilly, C. (1978). *From mobilization to revolution*. Reading, MA: Addison-Wesley.

Toke, D. (2007). *The politics of GM food*. London: Routledge.

Touraine, A. (1977). *The voice and the eye*. Cambridge: Cambridge University Press.

Walker, G. (2009a). Beyond distribution and proximity: Exploring the multiple spatialities of environmental justice. *Antipode, 41*(4), 614–636.

Walker, G. (2009b). Globalizing environmental justice: The geography and politics of frame contextualization and evolution. *Global Social Policy, 9*(3), 355–382.

Wilson, F. (2008). *Interest-group politics in France*. Cambridge: Cambridge University Press.

Zald, M. (1992). Looking back to look forward: Reflections on the past and future of the resource mobilization research program. In A. Morris & C. Mueller (Eds.), *Frontiers in social movement theory* (pp. 326–349). London: Yale University Press.

Zald, M. (2000). Ideologically structured action: An enlarged agenda for social movement research. *Mobilization: An International Journal, 5*(1), 1–16.

THE ART OF CONTESTATION AND LEGITIMACY: ENVIRONMENT, CUSTOMARY COMMUNITIES, AND ACTIVISM IN INDONESIA

Keith Bettinger, Micah Fisher and Wendy Miles

ABSTRACT

*Indonesia is one of the world's "megadiverse" countries, providing ecosystem services that accrue at the global scale. However, control over access to and use of natural resources has historically been a source of tension between the central government and local communities, with the latter usually being marginalized by the former. Since the fall of the authoritarian Suharto regime in 1998, however, a grassroots movement supports the revitalization of customary communities and their traditional systems of social organization (*adat*). A major part of this quest for legitimacy is the portrayal of indigenous people as environmentally benign. This chapter describes how indigenous systems have been influenced by political processes over time. We then describe how the changing political–administrative landscape has given rise to a national indigenous rights movement. We also analyze international factors that have contributed to the emergence of the indigenous movement before*

Occupy the Earth: Global Environmental Movements
Advances in Sustainability and Environmental Justice, Volume 15, 195–224

ISSN: 2051-5030/doi:10.1108/S2051-503020140000015008

discussing potential challenges facing the movement in the future. This chapter seeks to get beyond the simplistic conflation of indigenous peoples and environmentalism by understanding the strategic articulation of indigeneity and environmentalism.

Keywords: Indonesia; conservation; indigenous movement; *adat*; indigenous environmentalism; AMAN

INTRODUCTION

> For the Oma'lung tribe in Setulang village, near the Malay-Indonesian border, who apply the Dayak code restricting access to the forest and its resources, the death of the forest is the demise of their promise to their ancestors. Their ancient indigenous knowledge guides them to manage their natural landscapes sustainably by shifting cultivation to agro-forestry that does not harm nature, while humans can still benefit from it. This is part of their cultural heritage. (Davary, 2012, p. 18)

> According to traditional beliefs, forests are places to seek spiritual discipline where one can cut oneself off from the world and be close to nature and the Almighty. In so doing, one might hope to find inspiration about new ways of tackling things that need to be done in the community. (Kleden, Chidley, & Indradi, 2009, p. 304)

> One good solution to mitigate climate change is to change current high carbon production and consumption patterns to low carbon ones, such as those practice by indigenous peoples. With their traditional wisdom, indigenous peoples have proved they can sustainably maintain their forests and the carbon stock contained within them. ... to them, forests are inherited from their ancestors and their responsibility to take care to maintain. Should they not take care of them, their ancestors will get angry and bad luck will come. (Syarif, 2010, pp. 116–117)

These images and accounts describe indigenous peoples in Indonesia in a certain light: as natural environmentalists. The excerpts depict their inherent connection to the land and its importance in the maintenance of their livelihood, as well as their conservation ethic in which "sustainability" has necessarily played an important role for generations. Respect for the land is implied to be inherent in their cosmogonies, and they have a spiritual responsibility to safeguard the planet. According to these framings, indigenous knowledge has utility not only in terms of local resource management but also for global environmental governance. Images such as these help to support the *articulation* (Hall, 1996; Li, 2000) of indigenous rights and environmentalism in Indonesia, where recognition for the rights of customary societies as well as their role in environmental governance have emerged as a major political issue over the past 15 years.

Masyarakat adat, or "customary communities," have taken on a new significance in terms of environmental politics in Indonesia since the downfall of President Suharto in 1998, whose authoritarian *New Order* government[1] ruled the country for more than 30 years. The official discourse during the Suharto years was that Indonesia had *no* indigenous people (Li, 2000), and so *adat* structures were marginalized and their role in local governance and resource management was minimized. Now, though, Indonesia has entered a new era of decentralization and democratization (known by the expression "*reformasi*") and *adat* has emerged as an important discourse and political vehicle not only for those wishing to rejuvenate and revitalize local cultural practice, but also for those aiming to contest state claims over land and natural resources such as forests. At the same time, *adat* is perceived by international conservation organizations as having great potential as a structure through which conservation programs can be implemented. As the aforementioned accounts indicate, an important part of both of these discourses is the environmentalism inherent in *adat* codes.

The implications of this are important not just in Indonesia, but for the world in general. Indonesia is a key battleground for environmentalists because of the myriad forces threatening the country's ecological resources, which provide ecosystem services at a global scale. The 17,000 island archipelago has the third largest tropical forest cover in the world, which has been recognized as a critical site for carbon sequestration. Indonesia is also home to approximately 20% of the world's biodiversity. It has been recognized by Conservation International as one of the world's most "megadiverse" countries. It is home to 18 of the World Wildlife Fund's "Global 2000" ecoregions and contains two dozen of Bird Life International's "Endemic Bird Areas" (United Nations, 2014). At the same time, most areas within the archipelago are still primarily rural and agrarian, and a significant percentage of the population is considered to be "forest dependent."[2] Hence, for countless millions of Indonesians, farming and the utilization of forest products are the primary means of making a living, and the most realistic way for getting ahead, be it to pay school fees, medical expenses, or a new cell phone, is through agricultural expansion. Moreover, Indonesia's macroeconomy has long been dependent on the exploitation of the archipelago's natural resources; extractive industries have played a key role in economic development throughout the post-colonial era, and the country's leaders have identified the expansion of tree crops such as oil palm and mineral extraction to be key economic drivers of the future (Gellert, 2010).

Thus, it should not be surprising that over the past few years, the country has experienced one of the world's highest rates of deforestation

(Hansen et al., 2013). Related to this is the fact that Indonesia is one of the world's largest emitters of greenhouse gases, the majority of which are caused by deforestation, forest degradation, and peatland conversion (Brockhaus, Obidzinski, Dermawan, Laumonier, & Luttrell, 2012). Massive fires set to clear land for expanding oil palm plantations annually blanket the entire region in smoke, creating tensions with Indonesia's international neighbors. As a result, Indonesia has become a key target for new conservation technologies seeking to address issues of deforestation for the sake of slowing down the increasingly rapid march of global warming induced climate change. For example, the United Nations' global framework for addressing greenhouse gas emissions through market structures, REDD+ (Reducing Emissions through reduced Deforestation and forest Degradation), has been the impetus for dozens of programs aimed at compensating local people for providing ecosystem services that accrue at the international scale. At the same time, since the fall of Suharto, indigenous communities have been framed as guardians of the environment and potential partners for international actors.

The *adat* community in Indonesia has played a key role in publicizing and protesting the abuses of Indonesia's extractive economy, which has for decades sacrificed local livelihoods for the broader goal of economic development (Moniaga, 2007; Neidel, 2006). They claim that they have long been victims of the predatory and corrupt resource management practices of the Suharto regime, and that a restoration of the rights of indigenous peoples to manage their land and natural resources is the first and most important step in putting Indonesia on the path to sustainability and environmentally friendly economic development. Since the fall of the long-ruling authoritarian regime in 1998, new space has opened up to challenge government resource management policies and their negative impacts on the environment. However, *adat* is about more than just the environment, and its new found success is due in part because it reflects the priorities of a range of different movements and interests operating across scales. The resurgent *adat* movement brings together elements of democratization and decentralization, indigenous identity and rights issues, land tenure reform, and a backlash against what has been referred to as the "global land grab" (Hall, 2011). Moreover, *adat* has been perceived by outside actors as a channel through which they can work to further their environmental, political, and economic interests. This chapter will show how the *adat* movement in Indonesia has benefitted from a confluence of interests that have come together to support indigenous rights and management of resources. We show how the *adat* movement has also recognized the utility of allying with conservation and environmental interests, and this is consistent with a global trend whereby indigenous rights activists have recognized that environmentalism is a useful tool for articulating claims to land and natural resources (Dove, 2006). We delve into the politics of environment and indigenous identity to understand how environmental discourses and performances have been deployed in the service of achieving official recognition for *adat* rights.

This chapter is divided into five sections. In the section "Introduction," we have introduced the concept of *adat* and how it is being linked with environmental discourse in Indonesia. In the section "*Adat* and the Control of Natural Resources Over Time", we describe how the definition of *adat* has evolved over time, demonstrating that the very notion of *adat* can indeed be considered a colonial innovation, which was deployed in the service of seeking to consolidate control over natural resources. We describe the significance of *adat* during the struggle for independence, and then how *adat* communities were marginalized by the authoritarian Suharto regime. This short history helps to explain how the post-authoritarian revitalization of *adat* has been susceptible to a range of agendas and actors working both within Indonesian and at the international level. In the section "The Downfall of the Authoritarian State and the Emergence of Environmental *Adat*", we describe how decentralization and democratization reforms enacted after the fall of Suharto have created legal and discursive space for the reemergence of *adat*. We then provide three case studies, which illustrate different aspects of the articulation of customary communities and environmentalism. In the section "International Influences on Indonesia's Indigenous Movement," we analyze three global currents of discourse and practice that have helped to create space within Indonesia for the emergence of a resurgent indigenous movement, and how the emerging indigenous movement has been able to draw on these currents for support and legitimacy. In the next section, "Into the Future: Challenges and Opportunities for Adat in Indonesia," we describe several emerging issues that could potentially frustrate the marriage of indigenous rights and environmentalism in Indonesia before some concluding remarks.

ADAT AND THE CONTROL OF NATURAL RESOURCES OVER TIME

The term *adat* is often loosely translated in the popular media and grey literature as "customary law." However, most scholars agree that this rendering does not sufficiently express the multiple dimensions of the term. Davidson and Henley (2007, p. 3) rather define *adat* as a "complex of rights and obligations which ties together three things – history, land, and law – in a way that appears rather specific to Indonesia," whereas von Benda-Beckmann and von Benda-Beckmann (2011, p. 168) assert that "*adat* in Indonesia has become a generic term to indicate an often undifferentiated whole constituted by the morality, customs, and legal institution of ethnic or territorial groups." For the purposes of this chapter, we refer to *adat* collectively as locally rooted systems of social organization, since this term suggests customs and codes of governance, conflict resolution, and resource management devised through time in response to local conditions rather than in reference to broader norms and canons existing at the national or international scales.

The concept of *adat* law emerged out of the friction (Mahadi, 1987; Tsing, 2004) arising from the contact of an expanding Dutch colonial hegemony and the local inhabitants of the thousands of semi-independent villages and other locally rooted socio-political units spread across the hundreds of populated islands of the Indonesian archipelago. This friction engendered a dilemma as to how the colonial masters should administer their new subjects. Beyond the island of Java, the Dutch were committed to a strategy of indirect rule, working through already existing systems of administration and jurisprudence. However, the nature of the colonial endeavor is one of control and coordination (Burns, 2007), and so the Dutch needed a way to apprehend and thus manipulate these various and diverse systems of social organization. The solution would come from the ethnographic efforts of Dutch scholars.

Adat was first codified by Cornelis van Vollenhoven (1874–1933) and his students at the University of Leiden in the Netherlands during the late 19th and early 20th centuries (Burns, 2007). According to the interpretation of these colonial legal scholars, *adat* is produced at the local scale and is the product of the interaction of people with one another, outsiders, and their environment. It is contingent on geography and local traditions and reflects local cosmologies and dynamics of power. van Vollenhoven and his followers viewed local systems as an under-appreciated treasury of knowledge and felt that local people were best governed by their own systems with as little external modification as possible, favoring a dualistic legal arrangement in which *adat* regimes prevailed at the local level and colonial law at the national level (Lev, 1962). Though van Vollenhoven is almost universally regarded as having respect for locally rooted systems of social organization and as a scholar who viewed his efforts as helping to ensure the integrity and independence of local communities, the Dutch colonial authorities had more practical uses for his meticulous inventory of *adat* systems in mind. *Adat* systems became convenient structures through which the Dutch could enact indirect rule, governing through extant hierarchies, councils, and chiefs (Li, 2001; Tyson, 2011). At the same time, those local officials privileged by this system were compensated for their complicity in the endeavor. Thus, the work of van Vollenhoven and his students created the category of *adat*, allowing it to be used by the state throughout the colonial and post-colonial periods for the maintenance, projection, and consolidation of power (Antlov, 2003; Benda, 1966; Burns, 2007; Jaspan, 1965). The interaction between the Dutch colonial apparatus and local elites permanently altered the character of *adat* systems, in some cases perverting them to such an extent that local people had to be compelled to follow supposedly customary rules (Eghenter, 2000; Galizia, 1996; Zerner, 1994).

During the struggle for independence (1945–1949), the role of *adat* was largely symbolic, as a number of prominent nationalist leaders had been trained at Leiden under van Vollenhoven and hence viewed *adat* as a uniquely Indonesian feature that could be the basis of a distinct socio-legal identity rooted in Indonesian ideals and traditions, in contradistinction with those originating from the West (Bourchier, 2007). Thus *adat* was used to bolster nationalists' assertions that the people of the archipelago had governed themselves in the past and could do so in the future. This role for *adat* was recognized in the Indonesian Constitution, but in the years following independence, the practical realities of establishing a dualistic legal system featuring *adat* principles at the local level had become apparent, and by the late 1950s, pressure was mounting to establish a unified system of national law encompassing private, public, criminal, and procedural aspects (Arizona & Cahyadi, 2013; Lev, 1962). By 1959, *adat* courts were being dismantled all over the country and *adat* law was being relegated to secondary status behind international law in the national legislature (Bourchier, 2007; Jaspan, 1965). At the same time, the young nation's leaders were struggling with regional rebellions, calling into question the future of a unified Indonesia. Emphasizing and celebrating regional diversity would have been contrary to the goal of establishing an Indonesian *nation* (Anderson, 2006), and so rather than formally recognizing *adat* communities, the government, in a move that would have far-reaching implications, reprised the Dutch colonial dualism of "native" (*pribumi*) and "non-native" (*non-pribumi*) Indonesians, with the latter term referring mainly to Indonesians of Chinese descent. By the mid-1960s, President Sukarno's failed adventures in economic autarky and international neutrality had isolated Indonesia, but this classification scheme would be instrumental in establishing the next chapter in the Indonesian political history.

When General Suharto came to power following a coup in 1965, he inherited an economy in turmoil and a nation in chaos. After permanently sidelining Sukarno in 1967, now President Suharto charted a course to putting the country back on solid footing, which involved opening the country up to foreign investment in exploiting and extracting the country's rich natural resources. The primary project of the three decades of Suharto's New Order government would revolve around centralization of administration and the consolidation of power in a patrimonial, top-down hierarchy. The primary framing used by the Suharto government to justify its policies was a discourse of developmentalism (Escobar, 1995; Li, 2007; Scott, 1998) in which the technocratic state would bestow the fruits of modernization upon the rural populace, facilitating the social and economic transformations needed for a true democracy in the spirit of the 1945 Constitution to take shape. The state's modernization project involved an unquestioning populace embodied in the concept of the "floating mass" (*massa mengambang*), whereby the people were envisioned as politically neutral, tasked with the duty of working to fulfill the state's development goals (Crouch, 1979; Samson, 1973). This depoliticization of the citizenry was accompanied by the extension of a standardized bureaucracy, the insertion of the military into the politics, and administration of all levels of the country.

Suharto's strategies were based on the exploitation of the archipelago's abundant natural resources, and so the first step to achieving his goals was to establish and enforce state control over forests and mineral resources, which was accomplished by legislation (Act 5 of 1967, generally referred to as the Basic Forestry Law) that declared that virtually all forested areas across the archipelago were the property of the state. A major upshot of this law was that land and forests, which had previously been locally regulated as common property resources, became the property of the state. Moreover, small-scale and smallholder forestry as well as swidden agriculture were perceived to be inefficient uses of land, which would impede the state's developmentalist agenda (Lindayati, 2002). The national-level Ministry of Agriculture, which until 1983 administered state forests, was granted authority over more than 120 million hectares of forests, comprising more than 70% of the land area of Indonesia (Peluso, Lee, & Rachman, 2008). This astonishing usurpation by fiat enabled the state to award logging and mining concessions, and in most cases, these lucrative permits were distributed on a patronage basis to enterprises headed by people with close ties to the regime, or in other cases they went to military and former military officials as rewards for loyalty (Barr, 1998). By 1995, concessions had been awarded for more than 60 million hectares, which was at the time approximately half of the remaining forested area and slightly less than one-third of Indonesia's total land area (Arnold, 2008).

Key legislation was also passed to nullify local claims to forests and dismantle and marginalize *adat* systems. The existence of *adat* posed two potential threats to the New Order. The first was as a challenger to the state's claim to resources. The second potential threat was as a source of knowledge and legitimacy opposed to the unquestioned wisdom of the centrally driven modernization project. As a result, *adat* was removed from the official vocabulary as it related to local government and governance. This annihilation by omission allowed the state to avoid any positive connotations associated with *adat* still lingering in the public imagination, a process Neidel (2006, p. 4) refers to as "the metaphorical erasure of local history." Instead, *adat* became virtually synonymous with backwardness and isolation, and many communities were officially labeled "isolated communities" (*masyarakat terasing*), making them targets of state-led development projects (Arnold, 2008; Sangaji, 2007).[3] *Adat* was also officially conflated with communism, which had been constructed as a mortal enemy of the regime, and hence, indigenous communities were subjected to more intense scrutiny on the part of the state (Davidson & Henley, 2007). In other words, the potential of *adat* communities to challenge the state's claim over forest resources, as well as the state's prioritization as to how those resources would be used, was almost completely neutralized. *Adat* was instead essentialized as a regime-friendly symbol of diversity, characterized by variations in dress and architecture (Spyer, 1996).

By the late 1970s, state-sponsored logging of the forests was in full swing, but an increasing consciousness within the global community of the negative consequences of unrestricted logging in the tropics as well as the development of an international environmental movement led the Indonesian government to establish a network of national parks. However, like the expropriation of land and forests for commercial exploitation, the creation of protected areas also entailed the marginalization of local communities. Smallholders across the archipelago had their agroforestry plots sequestered within the newly established parks, and in some instances, entire villages were enclosed within protected areas. This had a significant impact on local livelihoods as new regulations prevented farmers from expanding the area under cultivation and legally obliged them to abandon swiddens and agroforestry plots. Local people continued to be marginalized, while the exploitation and management of natural resources continued to be directed by a technocratic elite based in Jakarta. Throughout this period, local people played virtually no role in management and conservation, and they derived very little benefit from the exploitation of forests (Lindayati, 2002; McCarthy, 2006; Peluso, 1992).

THE DOWNFALL OF THE AUTHORITARIAN STATE AND THE EMERGENCE OF ENVIRONMENTAL *ADAT*

This picture changed dramatically in 1998 when, amidst the Asian Financial Crisis and increasing violent and vocal street protests, President Suharto resigned. The Indonesian parliament immediately took steps to dismantle the vertically articulated authoritarian apparatus of government. Perhaps the most influential piece of reform legislation was Act 22/1999 on Regional Government, which not only devolved vast powers to the district level governments,[4] but also nullified Suharto-era laws that had imposed a uniform structure of village government throughout the country.[5] This opened the door for subnational levels of government to pass laws reviving traditional systems of spatial organization and administration, and in many places they did just this (Bedner & van Huis, 2008). Now local voices have started to reassert themselves in the breakdown of central control and the reorganization of power in the post-Suharto era. *Adat* has become a vehicle to articulate claims to land and resources, since in 1999, the Agriculture Ministry issued rules for registering *adat* lands. More and more laws have been passed since 1998 that have strengthened the position of recognized *adat* groups. However, recognition of *adat* has not been universal amongst Indonesia's sectoral ministries, and the Ministry of Forestry has long remained a stubborn holdout against reforming statutes and procedures to allow for a greater role for *adat* groups in the management of forest resources.

One of the main movers behind the resurgent *adat* movement has been the Alliance of Adat Communities of the Archipelago (*Aliansi Masyarakat Adat Nusantara*), or AMAN. AMAN was founded in 1999 by a coalition of activists, many of whom had been part of the movement that brought down Suharto (Bourchier, 2007). As is evidenced by the oft-quoted motto of this first congress ("If the state will not recognize us, then we will not recognize the state!"), AMAN's early orientation revolved around securing broader recognition for indigenous issues in Indonesia, and the formation of AMAN as a national umbrella organization was soon followed by the establishment of regional coordinating bodies wherein previously there existed no communication or cooperation between indigenous communities and activists (Avonius, 2003). During the first four years of its existence, the organization made significant progress toward these ends and was particularly successful in developing linkages with international organizations, and so the emphasis shifted at the second congress of indigenous peoples (2003) to focus more on issues of implementation (Acciaoli, 2007). AMAN has succeeded not only in broadening the traditional definition of indigeneity in Indonesia but also in emphasizing the environmental credentials of *adat* communities. During the New Order years, the number of people officially classified as indigenous was between one and two million (Li, 2000). However, AMAN's definition allows for a significant expansion in the number of indigenous people: "a group of people who, based on ancestral origin, live in a specific geographical area, have a distinct value and sociocultural system, sovereignty over their land and natural resources and control and take care of their survival by means of customary laws and institutions" (Syarif, 2010, p. 66). This definition omits any reference to ethnic or religious identities, which is part of a strategy to cast as wide a net as possible. Now AMAN counts more than 2,000 communities among its members. AMAN claims that there are 50–70 million *adat* people in Indonesia, which is a far greater number than any prior estimate, and this is due to the more catholic definition of *adat* that AMAN has used.

In a sense, then, what was once understood as generically rural is now understood as specifically indigenous (Moniaga, 2007). There is within the public imaginary in Indonesia an image of indigenous people as tied to the land, harboring an innate conservation ethic. This image has been encouraged to a certain extent by AMAN, but more central to its propagation has been the numerous media accounts that have surfaced in the post-Suharto era, portraying rural people as indigenous stewards of the land. Li (2000) uses Hall's (1996) concept of "articulation" to describe this dual-faceted framing in which local issues are first expressed in a way amenable to outside interests and then linked to larger discourses of indigenous rights and environmentalism. Accordingly, Li argues that a group's indigeneity has to be produced and packaged; it is a "positioning which draws upon historically sedimented practices, landscapes and repertoires of meaning, and emerges through particular patterns of engagement and struggle" (2000, p. 151). Indeed indigenous groups provide favored fodder for features in Indonesia's national daily newspapers (*Kompas, Java Pos*) and weekly glossy news magazines (*Tempo, Gatra*). This includes extolling the virtues of locally held traditional ecological knowledge and its value in forest management (see Yodono, 2010), the sublime connection between local people and the divine manifested through local nature (see Kristanto, 2012), the threat to local culture and arts posed by uncontrolled deforestation (see Indriasari, 2011), and the exotic customs of the few remaining hunter-gatherer groups inhabiting the country's forests (see Fatris, 2013). Through these portrayals, indigeneity is produced for and passively consumed by a growing urban, middle class who for the most part already placed a fairly high value on Indonesia's biodiversity resources. These media accounts link forest dwelling people to the forests as an essential(ized) part of the ecosystem and as "expert holders of communal land rights … the ideal performers of community based forest management" (Hauser-Schaublin, 2013, p. 14). As Bourchier (2007, p. 123) notes, the success of AMAN "was also due to the fact that their representations of *adat* as wise, socially harmonious, communalistic, and in tune with nature chimed with one of the favorite themes in Indonesian political thinking: the idea that *adat* is inherently good, pure, and authentic." In the paragraphs that follow, we describe three cases in which indigenous rights discourses have been linked with environmentalism in an effort to contest the state's control over land and natural resources, and what the outcome of these linkages has been.

Case 1: The Suku Anak Dalam versus Plantations in Jambi, Sumatra

In the province of Jambi, a group known as SAD113 has become a *cause célèbre* for its efforts to thwart plans by an agricultural conglomerate to develop an oil palm plantation on land they claim as their customary inheritance. SAD stands for *Suku Anak Dalam*, or "tribes descended from the interior." This term refers to a group of proto-Malay people inhabiting the forested interior of Jambi province, some of whom still adhere to hunter-gatherer livelihood strategies.[6] The "113" refers to the number of SAD families originally involved in the conflict. Like many other land conflicts across the archipelago, this one is rooted in Suharto-era resource policies. In 1971, a company was awarded a timber concession of approximately 70,000 hectares of which a significant amount overlapped SAD customary territory. Over time, this concession was logged out and the rights to use the land switched hands several times, and by the early 2000s, it had ended up in the hands of a plantation development firm. Also by this time, the repressive Suharto regime had ended, and so like elsewhere in Indonesia, SAD communities were freer to critique government policies and articulate their claims to land and resources. In 2003, the SAD113 group coalesced around an effort to recover 3,550 hectares of land from the plantation company. Although the territory in question was part of the original concession in the 1970s, a provincial survey conducted in the 1980s indicated that the area was already being actively cultivated by the local people, and fallowed fields and gravesites further indicated long-term habitation and use. Between 2006 and 2011, the SAD113 group and the plantation company worked through an NGO-led mediation process to resolve the dispute, but no satisfactory solution was reached (IPAC, 2014). Now the conflict continues to fester, and the SAD113 group has led marches and other demonstrations in the provincial capital, seeking to enlist the government's support in revoking the plantation company's concession.

The conflict in Jambi has brought together three different activist agendas as the SAD113 reached out to advocacy groups to build alliances to increase the perceived legitimacy of their claims: environmental, land reform, and indigenous rights. The original SAD113 group has attracted support from a regional political party as well as a national agrarian rights organization, both seeking to use the issue to further their agendas. Moreover, the ranks of those seeking to "recover" land have swelled into the hundreds, as the movement's leaders have encouraged land-seeking migrants to join the cause. The supposed innate environmental ethic of the forest dwellers has played a role in the controversy, as the plantation company, Wilmar, has a history of being a sort of environmental villain in the past, drawing fire from international groups like Greenpeace due to environmental damage caused by their investments elsewhere (Steinebach, 2013). However, this does not accurately reflect the original aims of the movement as the local people have long been open to a solution that includes some form of benefit-sharing from oil palm cultivation (IPAC, 2014). Still, the Suku Anak Dalam became a symbol in a Greenpeace campaign to press Wilmar to halt its environmentally destructive activities (Butler, 2014; Greenpeace, 2013).

This case illustrates how indigenous rights have been voiced and articulated with other reform agendas, and how each of these agendas draws support and legitimacy from one another. Indigenous groups seeking greater control over land and natural resources have adeptly adopted an element of environmental discourse, and as a result, they have found willing supporters at the national and international levels. However, the SAD113 case demonstrates how tenuous these connections can ultimately be, and how such a broad-based coalition can obscure the original message of the subaltern group. As this chapter goes to press, the SAD113 have not received official recognition of their claims, while landless migrants seeking to take advantage of the situation continue to move into the area, potentially "crowding out" the original SAD group. Moreover, Wilmar recently agreed to stop cutting old growth forests, and so Greenpeace is no longer active in the area. The case also shows how, even though the Suharto regime has been replaced, its resource management policies and practices have left a lasting impact that Indonesia's leaders and local communities are struggling to cope with.

Case 2: Conflicts Over Land and Lifeways in West Kalimantan

Since the 1970s, Kalimantan has been viewed by the central government as a frontier for plantation agriculture and transmigration to relieve population pressures on Java, but the Suharto-era policies oriented around these goals have led to the loss of a significant amount of land for the local Dayak people (Achwan, 2012). At the same time, many Dayak communities were targeted by programs by the Department of Social Affairs in the 1970s and 1980s, including efforts to resettle them, and so their traditional *adat* systems also came under assault from the Suharto regime (Syahruji, 2009). Because of the oppressiveness of the Suharto government, though, visible and outward acts of protest were rare and most expressions of discontent were manifested as "hidden resistance" (Kerkvliet, 2005) utilizing "weapons of the weak" (Scott, 1985), taking the form of subtle acts of sabotage and non-compliance. Much of this resulted from the incompatibility between the state's vision of orderly plantations and the swidden agriculture and agroforestry employed by many local communities and seen to be an important part of their culture. However, when the Suharto regime ended, communities almost immediately began acting out against plantation companies, and there were numerous instances of demonstrations, local people burning company offices and camps, seizing heavy equipment, and reoccupying land and destroying oil palm crops (Potter, 2008).

Images and discourses portraying indigenous environmentalism have been a major part of Dayak activism. Much of this effort has focused on the depiction of Dayak *adat* as a valid but fundamentally different form of resource management that is incompatible with the developmentalist agenda, an inconsistency that has caused it to be marginalized by the government. For example, the land ownership system is said to be based on trust and agreement within the *adat* framework and so there has been no need for written records describing land ownership (Syahruji, 2009).

However, in many cases, smallholders complain that they have been pushed into the cash economy under unfavorable terms, which has led to the erosion of indigenous community resilience. Representatives from the communities also argue that there is a strong connection between the community and the forest and that *adat* rules protect that relationship. Dayak culture is said to be based on five basic principles: sustainability, solidarity, biodiversity, subsistence, and compliance with *adat* law. So, "applied properly, these principles will bring about a form of sustainable development which is economically viable, environmentally friendly, and culturally sensitive" (Syahruji, 2009, p. 127).

The propagation of this green narrative has been facilitated by the emergence of a regional *adat* association, the *Aliansi Masyarakat Adat* (AMA, not to be confused with AMAN), which was formed in 1998, just two months after the fall of Suharto. AMA was one of the founding members of AMAN, and so it played a role in guiding the agenda of the national organization. This also provided exposure and a platform from which appeals for support have been made to the national media as well as international environmental and indigenous rights groups. These NGO linkages have also helped to establish a pan-Dayak consciousness that is able to speak, through AMA and AMAN, with a unified voice. Now the dozens of disparate groups that had been virtually powerless against the predations of the Suharto state have become a politically important group, and for the first time in 35 years, a Dayak has been elected governor of West Kalimantan. However, the province continues to struggle with sectarian tensions, and there are concerns that *adat* is and will continue to be used as a tool of identity politics.

Case 3: A Constitutional Challenge to the State's Ownership of Forests

Perhaps the most heralded development relating to resources and *adat* groups has been a decision handed down by the Constitutional Court of Indonesia in 2013.[7] The decision stems from a case brought by AMAN along with two indigenous groups, the Kuntu from Riau Province on Sumatra and the Kasepuhan Cisitu community from Banten Province on Java. The three plaintiffs argued that the wording of the Basic Forestry Act of 1999, which is the fundamental legislative framework for the management of the state's forests, was unconstitutional, since, while it recognized the existence of *adat* forests, it classified them as state, rather than privately held forests. This legal distinction meant that the Ministry of Forestry, as the government institution responsible for the administration of the forests, was the ultimate arbiter of the kinds of activities that would and would not be allowed in customary forest areas. The Constitutional Court agreed that the wording of the law was in fact in violation of a provision in the Constitution that required the state to recognize the rights of *adat* communities, and the word "state" was stricken from the law: "Customary forest are forests located in indigenous peoples' territories." Moreover, the Court ruled that the government, acting through the Ministry of Forestry, was in violation of its constitutional obligation to manage the country's natural resources for the maximum prosperity of the people of Indonesia, alluding to the years of mismanagement and irresponsible awarding of logging concessions. The court's implication was that not only had the Ministry's management of forest resources not been in the best interests of the citizenry, but it also had deprived people of their constitutionally guaranteed rights to livelihoods and a healthy environment (Butt, 2014).

In a statement released after the verdict was handed down, Abdon Nabadan, the secretary general of AMAN, declared that the decision would facilitate "the shift from centralized, authoritarian rule to a more sustainable and equitable Indonesian society." The outcome was a huge victory for AMAN and *adat* communities. The environmental angle always was and continues to be an important aspect in marshaling support for the indigenous position, and the centrality of the image of indigenous people as responsible stewards of the environment was apparent in the Court's decision. However, AMAN has continued working to build a strong coalition with national and international environmental groups in the wake of the decision. Part of the reason for this is that the Constitutional Court's verdict carries little weight as the court has no enforcement powers and has in the past relied on moral suasion and the acquiescence of the executive and legislative branches for its decisions to be implemented (Butt, 2014).[8] So far, the Ministry of Forestry has signaled officially that the decision is in line with its policies, issuing a statement that:

> ... the Ministry of Forestry considers indigenous peoples living in a certain area as being part of the forest itself. They cannot be separated. If there's an issue with forest management, indigenous communities will have their own mechanisms to deal with it. We're sure that these communities are environmentally friendly, concerned with sustainable economic practices and devoted to environmental protection. (Pasandaran, 2013)

However, the Ministry has always jealously guarded its control over the country's forests, and has fought the decentralization movement at every step in the past. Indeed, it has been argued that the Ministry may even be "structurally incapable" of relinquishing its power over the forests. Thus, moving forward, AMAN will need to maintain a strong coalition of support in order to pressure the legislature and the Ministry of Forestry to reform both resource management laws and practices. The best chance for securing real changes in resource management regimes is the Draft Law on the Recognition and the Protection of the Rights of Indigenous People, which would spell out the rights and responsibilities of *adat* communities in a legally binding way; as the name suggests this legislation is still under debate in parliament (Arizona & Cahyadi, 2013). Thus, while promising, the Constitutional Court's decision is still essentially a symbolic victory. This case demonstrates that while AMAN and the *adat* revitalization movement have scored some impressive achievements, the realization of the movement's true objectives is still a long way off.

INTERNATIONAL INFLUENCES ON INDONESIA'S INDIGENOUS MOVEMENT

AMAN has succeeded in becoming a national aggregator for indigenous issues in Indonesia, and has become adept at presenting conflicts over land in terms sympathetic to the public as well as international donors. AMAN owes part of its success to broader trends operating at the international level, which have provided discursive space for the emergence of the unified indigenous movement in Indonesia. At the same time, AMAN's leadership has proven proficient at reading the contours of the international discursive landscape, taking advantage of currents and trends. There are at least three international currents coming together to provide a propitious political niche for AMAN and the organization's portrayal of *adat*.

The first of these is a long-standing fetishization of the local within the conservation establishment, which has taken place in the context of a more general global trend towards the support of indigenous rights (Nadasdy, 2005). Davidson and Henley (2007, p. 4) argue that this is part of "international anti-imperialism as well as a 'cultural turn' in the intellectual life of what used to be the political left, leading to a shift in practical politics towards the defense of cultural diversity." In other words, support for indigenous groups has become an important channel through which anti-hegemonic activists can operate. This is particularly important amidst the decline in the 1990s and 2000s of more explicitly Marxist critiques and the advent of neoliberalism as a dominant economic structure throughout the world (von Benda-Beckmann & von Benda-Beckmann, 2011). This led to a "privileging of customary law, community rights, and relationships to conservation and sustainable development throughout the developing and developed world" (Zerner, 1994, p. 1081). At the same time, indigenous communities have been perceived as potential partners by more mainstream actors in conservation and development as well. The "local turn" began to be manifested in conservation discourse and practice in the 1980s, when the notion that environmental well-being could not be achieved without a concomitant improvement in the livelihoods of those communities and sectors of society most dependent upon the exploitation of natural resources.

This confluence has deep roots in Indonesia, since during the authoritarian Suharto years, strict governmental control limited the space in which civil society organizations could flourish. However, the Suharto regime's hostility toward the mere idea of indigeneity as a politically operative concept limited the role of *adat* in local conservation initiatives as well as the discursive space available to *adat* activists. Would-be activists found that one relatively "safe" area to work for change against the system was environmentalism, and so those activists interested in working for change on land and indigenous issues learned to couch their agendas in terms of environmental discourse, which was more acceptable to the regime (Peluso et al., 2008). In the post-authoritarian era, though, the empowerment of *adat* as a discourse has allowed AMAN to articulate a discourse of environmentalism while rejecting all forms of state-managed community-based conservation that do not explicitly recognize the sovereignty of indigenous people over their land and forests. This position is fortified by AMAN's networking with international organizations, which have long supported the articulation of indigenous rights and environmentalism.

The second international current of discourse and practice supporting the emergence of AMAN and indigenous rights in Indonesia is neoliberal conservation. Neoliberalism refers to a set of political-economic assumptions and practices and "revolves around the restructuring of the world to facilitate the spread of free-markets" (Igoe & Brockington, 2007, p. 433). Neoliberalism emerged as the dominant paradigm of economic orthodoxy in the 1980s and is guided by the general tenet that government involvement in the economy is virtually always counterproductive. Neoliberal prescriptions for addressing economic woes include deregulation, privatization, and improving access to markets. Conservation strategies based on neoliberalism began to take hold in the late 1980s and early 1990s and have involved privatization of protected areas, commoditization of biodiversity and genetic material, and payments for ecosystem services (PES). These approaches to conservation are constantly evolving, and AMAN has positioned itself to be a pivotal player in the unfolding of the most recent neoliberal innovations aimed at addressing environmental problems. These new developments represent an interesting change in terms of the potential agency afforded to local groups in the implementation of conservation activities, as in the past, neoliberal conservation strategies have been judged harshly based on their negative impacts on local and indigenous communities (Igoe & Brockington, 2007).

Over the past decade, there has been a discursive shift towards identifying market-based approaches to addressing environmental issues such as climate change, and PES strategies are becoming a new norm in international conservation circles. One major PES framework that has started to be implemented in Indonesia is REDD+, which has become a major component in Indonesia's national greenhouse gas reduction and conservation strategies. A major departure from the implementation of earlier neoliberal interventions is the potential role for *adat* communities and the appearance of AMAN as an influential player on the national stage. There is a remarkable philosophical convergence between AMAN and advocates for market-based conservation approaches. Whereas the latter argues that most environmental problems are rooted in a failure to properly value environmental assets and ecosystem services as well as a flawed relationship between local people, the market, and nature (Igoe & Brockington, 2007), the former asserts that this flawed relationship is the result of marginalization caused by the natural resource management policies of the Suharto era. Planning for the implementation of REDD+ has thus brought debate over the formalization of customary land rights back to the political agenda (Hein, 2013). The reason for this is that the decentralized, community-oriented characteristics of *adat* are perceived by many to be ideal structures though which REDD+ programs could operate, but the complexity of designing individualized programs for each *adat* community is far beyond the resources of even the most ambitious of REDD+ programs. Thus, AMAN has the potential to serve as an aggregator of *adat* characteristics, and could potentially function as a sort of middleman between individual communities and REDD+ program managers, minimizing transaction costs while providing businesses and governments a relatively easy path to obtaining their "indigenous cred." In other words, *adat* communities are consistent with market-oriented interventions because they can be considered corporate service providers. However, in virtually all places where intact *adat* systems have been identified, the community holds no formal title to lands claimed to be community property. This frustrates the implementation of REDD+ because it relies on private or corporate ownership of forests and land.

The attention focused on REDD+ and the potential for it to become an additional source of revenue for Indonesia has provided an opportunity for AMAN to engage with the responsible government authorities in discussion concerning access to forest land (Hein, 2013), since international actors have exerted their influence to ensure that local people are incorporated into discussions over REDD+. For example, the Letter of Intent between Norway and Indonesia whereby Norway would provide US$1 billion to Indonesia in exchange for action to protect carbon sinks calls for participation from all stakeholders, including indigenous groups and local communities (Hein, 2013). As a result of these pressures, the national REDD+ task force has announced plans to work with other government agencies to improve tenure security and to institutionalize Free, Prior, and Informed Consent for REDD+ and other resource management activities (Hein, 2013).

The third thread of international discourse and practice that supports the emergence of *adat* as a local unit of governance and resource management is ideologically related to the second: the neoliberal norms pushing political–administrative decentralization.

The most significant political event in Indonesia over the past quarter century has been the shift from the highly centralized New Order government to the current era of *reformasi.* These sweeping decentralization and democratization reforms mirror a global trend of decentralization that began in the 1970s and has been increasingly championed by international financial institutions, bi- and multilateral aid agencies, and non-governmental organizations worldwide. There are many supposed benefits to decentralization, including improved government service delivery, greater participation and representation in government, taking advantage of local knowledge, and increasing transparency and accountability. When Indonesia entered the *reformasi* era, it was hoped that all of these benefits would accrue. The way that *reformasi* has in fact unfolded has greatly impacted *adat* communities as well as the shape *adat* discourse has taken (Tyson, 2010). These impacts on *adat* and *adat* communities have been both indirect and direct.

Indonesia moved almost overnight from being one of the world's most centralized countries to being one of the world's most decentralized countries. Although decentralization was pushed for and supported by major donors, it was clear in the early stages of the reform movement that the "big bang" (Silver, 2003) nature of Indonesia's decentralization was going to be problematic because there had been virtually no capacity building to prepare lower levels of government for their expanded powers and responsibilities. Moreover, while decentralization was successful in dismantling the upwardly accountable patronage structures that were so critical to Suharto's hold on power, the legacy of the Suharto regime made it difficult to create functioning structures of downward accountability. "Elite capture" (Hadiz, 2004) was identified as a major problem of decentralization, as local and regional actors that had previously been part of the Suharto regime repackaged themselves, drawing on name recognition and informal (and corrupt) support networks to seize power.

As has been noted, though, decentralization also allowed for the reemergence of *adat* as a politically relevant category. However, due to the chaotic nature of decentralization in Indonesia and the disarray in which the New Order left *adat* structures in many areas, there have been numerous instances of politically ambitious local elites employing *adat* discourses as a form of identity politics (Syahruji, 2009; Tyson, 2010). In extreme cases, this contributed to shocking violence against minorities and migrant communities (McCarthy, 2004). On the other hand, this indirect opening of political space has greatly benefitted the *adat* movement in general, and it is highly unlikely that AMAN would have been established without decentralization reforms.

More directly, AMAN and other organizations have benefitted from the "indigenous turn." From an early stage, it was hypothesized that recognition of *adat* rights would help the implementation of decentralization and democratization (Patlis, 2004), and this stance has slowly been incorporated into the general policies of international lenders and donors. For example, the World Bank's operational directives currently explicitly call for the empowerment of indigenous peoples in decentralization activities (Ehrentraut, 2011). The Bank continues to be a major funder of decentralization-related activities in Indonesia and requires the government entities it partners with to agree to respect the rights of indigenous people (World Bank, 2010). Although Indonesia has thus far not created a unified legal framework to recognize and protect the rights of indigenous groups, the World Bank and other international organizations have created a climate of expectation as there are increasing pressures stemming from multilateral institutions, including the Asian Development Bank and the World Bank, to recognize *adat* communities on a basis consistent with international norms as enshrined in agreements such as the International Labor

Organization's Convention 169 on the rights of indigenous and tribal peoples (Bedner & van Huis, 2008).

INTO THE FUTURE: CHALLENGES AND OPPORTUNITIES FOR *ADAT* IN INDONESIA

AMAN and other organizations have succeeded in creating a unified national movement capable of challenging state-led resource management regimes. As we have shown, a key strategic asset for AMAN and the *adat* movement in general lies in its ability to position itself not only as a movement of the subaltern masses, but also as a more environmentally friendly alternative to the state's failed resource management endeavors (Fay, 2009). However, the *adat* movement is still developing; it began as a somewhat radical movement opposed to the central government policies and has evolved into a powerful organization now capable of wielding influence in the formation and implementation of policies relating to land management, subnational governance, and environmental issues (Acciaioli, 2007). As the movement's role in national politics and policy increases, its leaders and visionaries will need to address emerging issues related to the role of *adat* in society. The broad-based appeal of AMAN and the *adat* resurgence, which has hitherto been one of its greatest strengths, could gradually become a weakness, since diverse organizations often fall victim to mission creep and infighting amongst various agendas within the organization. In this section, we analyze some challenges facing the indigenous movement in Indonesia.

Challenge 1: Continuing Conflicts Over Natural Resources

Timber-based industries have long ranked as Indonesia's largest export after oil and gas (Arnold, 2008). In the future, timber will continue to play a key role in Indonesia's economic fortunes. However, the country's fastest growing crop is oil palm, which has a central part in the national economic development blueprint. In fact, Indonesia's national biofuel development strategy calls for an additional four million hectares of oil palm to be planted between 2016 and 2025. Most palm oil expansion is driven by large plantation corporations, who have emerged as a strong competitor with local communities for land resources in many cases. Disputes over land are frequent; Steinebach (2013) notes that there were 30 conflicts reported to authorities in 2010 in Jambi province[9] alone.

Moreover, the structure of decentralization laws has empowered and encouraged district governments to prioritize the expansion of natural resource-based industries including oil palm in an effort to increase tax revenues. In many cases, this disadvantages indigenous people because natural resources are often found on their land (Duncan, 2007). Related to this is the fact that the authority to officially recognize *adat* communities rests in many cases with the district government. Thus, the same officials that have been incentivized to expand extractive industries also have the power to approve or reject applications for recognition of *adat* communities, creating the potential for significant conflicts of interest.

The way forward should include ensuring transparency, impartiality, and accountability for determining the validity of indigenous systems. There is currently no government ministry or department tasked with investigating *adat* claims and promoting the rights of indigenous peoples. As a result, *adat* recognition is often subject to local political exigencies, and local elites have been known to exploit indigenous issues for political gain (Sangaji, 2007). The creation of an oversight body may help to separate the issue of rights recognition from local political processes, thereby insulating vulnerable *adat* communities that have not yet been recognized from being manipulated by politically ambitious elites. Additionally, external recognition for *adat* communities could allow them to function as an effective check against environmentally destructive extractive activities at the district level.

Challenge 2: Lack of Capacity/Political Will to Enforce Adat *Rights*

Related to the previous issue is the unwillingness or inability at all levels of government to adequately investigate and adjudicate conflicts between local communities and extractive industries (Tyson, 2010). As noted above, there have been hundreds of reported conflicts over land and access to resources between local people and concessionaires since the end of the New Order government. There are many causes for these conflicts, including encroachment on the part of corporations, squatting on the part of local people, and the issuance of extractive permits to land that is already in use by local people on the part of the government.

This is a complex problem and there is no easy answer. However, part of addressing these issues would revolve around guaranteeing *adat* communities secure tenure to their land. This would give communities and individuals legal standing to challenge corporate concessions in court, and would provide the institutional incentives for subnational governments to exercise greater care in terms of issuing permits and concessions. Moreover, it has been argued that equipping local communities with secure tenure as well as full ownership rights to *adat* forests would contribute to conservation goals. Lastly, *adat* activists argue that local people need to feel secure enough in their tenure arrangements as well as their livelihood security so that there is no pressure to overexploit commonly held resources such as *adat* forests (Abubakar, 2009).

Challenge 3: Realistic Expectations

As AMAN flexes its muscles and develops international partnerships to support programs in Indonesia, it will be essential to maintain realistic expectations as to what *adat* is and what it is capable of. As we have shown in this chapter, there is always a temptation to attribute positive environmental characteristics to indigenous communities and their systems of organization, and advocates for the empowerment of *adat* have in many instances contributed to the maintenance of the image of the "ecologically noble savage" (Redford, 1991). These framings have helped to enhance the image of *adat* in the public imagination; however, adat scholars are quick to point out that these indigenous systems are not and never have been inherently egalitarian (Davidson & Henley, 2007; Hauser-Schaublin, 2013). Moreover, it is important to recognize that *adat* has changed over time and that it will and should be expected and allowed to continue to change in the future (McCarthy, 2005; Potter & Badcock, 2004). Davidson and Henley (2007, p. 4) warn that "the informal, uncodified character of most 'adat law' makes it vulnerable to political manipulation, as does the idealization of order and stability with which adat is associated." Bearing these cautions in mind, it will be important in the future to establish mechanisms to ensure that *adat* institutions are imbued with values such as equity and fairness, and safeguards to ensure that *adat* systems do not become tools to serve the ends of local elites need to be established.

Moreover, virtually all rural communities in Indonesia are now exposed to and connected with international markets, and so the ideal of localized production systems is simply not realistic. There will always exist both pressures and opportunities stemming from external markets and economic dynamics (Sangaji, 2007). *Adat* systems and the expectations placed upon them in terms of resource management and environmental outcomes need to take these broader processes into consideration. *Adat* has the potential to provide a framework of flexibility and resilience for local people (Abubakar, 2009; Potter & Badcock, 2004), but it also has the potential to be undermined by exploitative pressures.

CONCLUSIONS

This chapter has described the historical formation of the concept of *adat* as well as its role in successive political-economic contexts in Indonesia. We have also shown how the *adat* revitalization movement has emerged as a result of a political transformation, and that this emergence has been shaped by internal and external factors. It is clear that AMAN draws upon international discourses for legitimacy, but also provides a channel through which international actors and agendas can operate. Because of this, AMAN's influence on the national stage has grown significantly since the movement was founded in 1999, and they now have the means to pursue broader projects, which generally speaking is the contestation of control over resources.

On the surface, the cases discussed in this chapter are about control over forests; but we have shown that the environment is merely an arena for larger struggles over indigenous identity and the division of power between the central government and local communities. More broadly, the articulation of indigenous discourses with the environmental movement is of interest because it represents a move away from state-led development and high-modernism, and so the emergence of AMAN is emblematic of the sea-change of attitudes and assumptions in development and conservation over the past 20 years. What remains to be seen is how the articulation of indigeneity and neoliberalism will play out in Indonesia, and how this will be manifested in terms of environmental outcomes.

NOTES

1. Suharto used the term "New Order" (*Orde Baru*) to distinguish his government from the "Old Order" (*Orde Lama*) that preceded his presidency.
2. One source estimates the number to be as high as 80 million, or approximately 38% of the national population (Ituarte-Lima, McDermott, and Mulyani, 2014).

3. Duncan (2007) describes one such project, the Program for the Development of Social Prosperity of Isolated Communities (*Pembinaan Kesejahteraan Sosial Masyarakat Terasing*), a five-year plan aimed at modifying the social, economic, and cultural characteristics of "isolated communities" to be more in line with the national philosophy of *Pancasila*. Common characteristics of these programs included forced resettlement, altered religious beliefs, and the adoption of new agricultural techniques.

4. Districts are the middle level in Indonesia's five-tiered administrative hierarchy: national–provincial–district–subdistrict–village.

5. The Village Act of 1979 invalidated local political–administrative structures throughout the county and replaced them with the Javanese *desa* system. In many areas, this led to the disempowerment of *adat* leaders, as their authority was usurped by village heads appointed by the central government. Numerous scholars provide accounts of the social and ecological impacts this had on local communities (Abubakar, 2009; Arnold, 2008; Kato, 1989; McCarthy, 2002).

6. The term *Suku Anak Dalam* has been used more loosely over the past 15 years, likely, at least partly, due to this issue. Steinebach (2013) describes how SAD *adat* allows for both men and women to marry into the group and thus become SAD themselves, and an IPAC (2014) report discusses how self-attribution of SAD has been used for political purposes.

7. The Constitutional Court (Mahkamah Konstitusi) was established in 2003 and has several special powers, including reviewing the constitutionality of laws.

8. Butt (2014) further notes that a corruption scandal revolving around one of the Court's most well-respected judges may damage this "soft power" ability.

9. Jambi province in central Sumatra is a major target for oil palm expansion.

REFERENCES

Abubakar, D. H. (2009). The Guguk indigenous community, Jambi: Protecting customary forests with local regulations. In E. O. Kleden & Y. I. Liz Chidley (Eds.), *Forests for the future: Indigenous management in a changing world* (pp. 63–99). Jakarta: AMAN.

Acciaioli, G. (2007). From customary law to indigenous sovereignty: Reconceptualizing Masyarakat Adat in contemporary Indonesia. In J. S. Davidson & D. Henley (Eds.), *The revival of tradition in Indonesian politics: The deployment of adat from colonialism to indigenism* (pp. 295–318). Abingdon, UK: Routledge.

Achwan, R. (2012). Hybridising state and ethnicity in an Indonesian region. *Asian Social Science*, *8*(6), 94–103.

Anderson, B. (2006). *Imagined communities: Reflections on the origin and spread of nationalism* (p. 240). New York, NY: Verso.

Antlov, H. (2003). Village government and rural development in Indonesia: The new democratic framework. *Bulletin of Indonesian Economic Studies*, *39*(2), 193–214.

Arizona, Y., & Cahyadi, E. (2013). The revival of indigenous peoples: Contestations over a special legislation on masyarakat adat. In B. Hauser-Schaublin (Ed.), *Adat and indigeneity in Indonesia: Culture and entitlements between heteronomy and self-ascription* (p. 240). Gottingen: University of Gottingen.

Arnold, L. L. (2008). Deforestation in decentralized Indonesia: What's law got to do with it? *Law, Environmental and Development Journal*, *4*(2), 77–101.

Avonius, L. (2003). Reforming adat: Indonesian indigenous people in the era of reformasi. *Asia Pacific Journal of Anthropology*, *4*(1/2), 123–142.

Barr, C. M. (1998). Bob Hasan, the rise of APKINDO, and the shifting dynamics of control in Indonesia's timber sector. *Indonesia*, *65*, 1–36.

Bedner, A., & van Huis, S. (2008). The return of the native in Indonesian law: Indigenous communities in Indonesian legislation. *Bijdragen tot de Taal-, Land- en Volkenkunde*, *164*(2/3), 165–193.

Benda, H. (1966). The pattern of administrative reforms in the closing years of Dutch rule in Indonesia. *Journal of Asian Studies*, *25*(4), 589–605.

Bourchier, D. (2007). The romance of adat in the Indonesian political imagination and the current revival. In J. S. Davidson & D. Henley (Eds.), *The revival of tradition in Indonesian politics: The deployment of Adat from Colonialism to Indigenism* (p. 377). Abingdon, UK: Routledge.

Brockhaus, M., Obidzinski, K., Dermawan, A., Laumonier, Y., & Luttrell, C. (2012). An overview of forest and land allocation policies in Indonesia: Is the current framework sufficient to meet the needs of REDD+? *Forest Policy and Economics*, *81*(1), 30–37.

Burns, P. (2007). Custom, that is before all law. In J. Davidson & D. Henley (Eds.), *The revival of tradition in Indonesian politics: The deployment of adat from colonialism to indigenism* (pp. 68–86). Abingdon, UK: Routledge.

Butler, R. (2014). *Clash with palm oil company leaves one indigenous community member dead in Sumatra.* Retrieved from http://news.mongabay.com/2014/0306-asiatic-persada-palm-oil-death.html. Accessed on May 20, 2014.

Butt, S. (2014). Traditional land rights before the Indonesian constitutional court. *Law, Environment and Development Journal*, *10*, 1.

Crouch, H. (1979). Patrimonialism and military rule in Indonesia. *World Politics*, *31*(4), 571–587.

Davary, B. (2012). Islam and ecology: Southeast Asia, adat, and the essence of keramat. *ASIA Network Exchange*, *20*(1), 12–22.

Davidson, J., & Henley, D. (2007). *The revival of tradition in Indonesian politics: The deployment of adat from colonialism to indigenism.* London: Routledge.

Dove, M. (2006). Indigenous people and environmental politics. *Annual Review of Anthropology*, *35*, 191–208.

Duncan, C. (2007). Mixed outcomes: The impact of regional autonomy and decentralization on indigenous ethnic minorities in Indonesia. *Development and Change*, *38*(4), 711–733.

Eghenter, C. (2000). What is *Tana Ulen* good for? Considerations on indigenous forest managment, conservation, and research in the interior of Indonesian Borneo. *Human Ecology*, *28*(3), 331–357.

Ehrentraut, S. (2011). Decentralization and the promise of indigenous peoples' empowerment: The case of the World Bank in Cambodia. *The Pacific Review*, *24*(1), 89–113.

Escobar, A. (1995). *Encountering development: The making and unmaking of the third world.* Princeton, NJ: Princeton University Press.

Fatris, M. (2013). *Into the wild.* Retrieved from http://en.tempo.co/read/news/2013/11/26/240532462/Into-the-Wild. Accessed on May 20, 2014.

Fay, C. (2009). An international overview: Indigenous natural resource management systems at the crossroads. In E. O. Kleden, L. Chidley, & Y. Indradi (Eds.). *Forests for the future: Indigenous management in a changing world* (pp. 277–294). Jakarta: AMAN.

Galizia, M. (1996). Village institutions after the Law no 5/1979 on village administration: The case of Rejang-lebong in South-Western Sumatra. *Archipel*, *51*(1), 135–160.

Gellert, P. K. (2010). Extractive regimes: Toward a better understanding of Indonesian development. *Rural Sociology*, *75*(1), 28–57.

Greenpeace. (2013). *Palm oil giant Wilmar caves to public pressure, commits to end forest destruction*. Retrieved from http://www.greenpeace.org/international/en/press/releases/Palm-oil-giant-Wilmar-caves-to-public-pressure-commits-to-end-forests-destruction/. Accessed on May 20, 2014.

Hadiz, V. (2004). Decentralization and democracy in Indonesia: A critique of neo-institutionalist perspectives. *Development and Change*, *35*(4), 697–718.

Hall, D. (2011). Land grabs, land control, and Southeast Asian crop booms. *Journal of Peasant Studies*, *38*(4), 837–857.

Hall, S. (1996). On postmodernism and articulation: An interview with Stuart hall. In D. Morley & K. Chen (Eds.), *Stuart hall: Critical dialogues in cultural studies* (pp. 131–150). London: Routledge.

Hansen, M. C., Potapov, P. V., Moore, R., Hancher, M., Turbanova, S. A., Tyukavina, A., & Townshend, J. R. G. (2013). High-resolution global maps of 21st century forest cover change. *Science*, *342*(6160), 850–853.

Hauser-Schaublin, B. (2013). The power of indigeneity: Reparation, readjustments and repositioning. In B. Hauser-Schaublin (Ed.), *Adat and indigeneity in Indonesia: Culture and entitlements between heteronomy and self-ascription* (pp. 5–16). Gottingen, Germany: University of Gottingen.

Hein, J. (2013). *Reducing emissions from deforestation and forest degradation (REDD+), transnational conservation and access to land in Jambi, Indonesia* (p. 33). Goettingen, Germany: Georg August University.

Igoe, J., & Brockington, D. (2007). Neoliberal conservation: A brief introduction. *Conservation and Society*, *5*(6), 432–449.

Indriasari, L. (2011). *Kerusakan Hutan Ancam Budaya Masyarakat*. Retrieved from http://oase.kompas.com/read/2011/09/16/23012932/Kerusakan.Hutan.Ancam.Budaya.Masyarakat. Accessed on May 20, 2014.

Institute for Policy Analysis of Conflict. (2014). *Indigenous rights vs. Agrarian reform in Indonesia: A case study from Jambi*. Jakarta: IPAC.

Ituarte-Lima, C., McDermott, C. L., & Mulyani, M. (2014). Assessing equity in national legal frameworks for REDD+: The case of Indonesia. *Environmental Science and Policy*, *44*, 291–300.

Jaspan, M. (1965). In quest of a new law: The perplexity of legal syncretism in Indonesia. *Comparative Studies in Society and History*, *7*(3), 252–266.

Kato, T. (1989). Different fields, similar locusts: Adat communities and the Villate Law of 1979 in Indonesia. *Indonesia*, *47*, 89–114.

Kerkvliet, B. (2005). *The power of everyday politics: How Vietnamese peasants transformed national policy*. Ithaca, NY: Cornell University Press.

Kleden, E. O., Chidley, L., & Indradi, Y. (2009). *Forests for the future: Indigenous management in a changing world* (p. 312). Jakarta: AMAN.

Kristanto, T. A. (2012). *Suku Dayak Anggap Pelestarian Alam Tanggung Jawab Manusia*. Retrieved from http://regional.kompas.com/read/2012/10/10/22474035/Suku.Dayak.Anggap.Pelestarian.Alam.Tanggung.Jawab.Manusia. Accessed on May 20, 2014.

Lev, D. S. (1962). The supreme court and adat inheritance law in Indonesia. *American Journal of Comparative Law*, *11*(2), 205–224.

Li, T. M. (2000). Articulating indigenous identity in Indonesia: Resource politics and the tribal slot. *Comparative studies in society and history*, *42*(1), 149–179.

Li, T. M. (2001). Masyarakat adat, difference, and the limits of recognition in Indonesia's forest zone. *Modern Asian Studies*, *35*(3), 645–676.

Li, T. M. (2007). *The will to improve: Governmentality, development, and the practice of politics* (p. 375). Durham, NC: Duke University Press.

Lindayati, R. (2002). Ideas and institutions in social forestry policy. In C. Colfer, & I. Resosudarmo (Eds.), *Which way forward? People, forests, and policymaking in Indonesia* (pp. 36–59). Washington, DC: Resources for the Future.

Mahadi. (1987). Islam and law in Indonesia. In R. Kipp & S. Rodgers (Eds.), *Indonesian religions in transition*. Tucson: University of Arizona Press.

McCarthy, J. F. (2002). Turning in circles: District governance, illegal logging, and environmental decline in Sumatra, Indonesia. *Society and Natural Resources*, *15*(10), 867–886.

McCarthy, J. F. (2004). Changing to gray: Decentralization and the emergence of volatile socio-legal configurations in Central Kalimantan, Indonesia. *World Development*, *32*(7), 1199–1223.

McCarthy, J. F. (2005). Between adat and state: Institutional arrangements on Sumatra's forest frontier. *Human Ecology*, *33*(1), 57–82.

McCarthy, J. F. (2006). *The fourth circle: A political ecology of Sumatra's rainforest frontier* (p. 353). Stanford, CA: Stanford University Press.

Ministry of National Development Planning. (2011). Masterplan for Acceleration and Expansion of Indonesia Economic Development 2011–2025 (MP3EI). Coordinating Ministry for Economic Affairs, Republic of Indonesia.

Moniaga, S. (2007). From bumiputera to masyarakat adat: A long and confusing journey. In J. S. Davidson & D. Henley (Eds.), *The revival of tradition in Indonesian politics: The deploymnet of adat from colonialism to indigenism* (pp. 275–294). Abingdon, UK: Routledge.

Nadasdy, P. (2005). Transcending the debate over the ecologically noble Indian: Indigenous peoples and environmentalism. *Ethnohistory*, *52*(2), 291–331.

Neidel, J. (2006). *The garden of forking paths: History, its erasure and remembrance in Sumatra's Kerinci Seblat National Park*. Ph.D. thesis, New Haven, CT: Yale University Press.

Pasandaran, C. (2013). *Constitutional court annuls government ownership of customary forests*. Jakarta Globe. Retrieved from http://www.thejakartaglobe.com/news/constitutional-court-annuls-government-ownership-of-customary-forests/. Accessed on May 20, 2014.

Patlis, J. M. (2004). *A rough guide to developing laws for regional forest management* (p. 31). Bogor: CIFOR.

Peluso, N., Lee, S. A., & Rachman, N. F. (2008). Claiming the grounds for reform: Agrarian and environmental movements in Indonesia. *Journal of Agrarian Change*, *8*(2–3), 377–407.

Peluso, N. L. (1992). *Rich forests, poor people: Resource control and resistance in Java* (p. 336). Berkeley, CA: University of California Press.

Potter, L. (2008). Dayak resistance to oil palm plantations in West Kalimantan, Indonesia. Paper presented at 17th Biennial conference of Asian Studies Association of Australia, Melbourne, July 1–3.

Potter, L., & Badcock, S. (2004). Tree crop smallholders, capitalism, and adat: Studies in Riau Province, Indonesia. *Asia Pacific Viewpoint*, *45*(3), 341–356.

Redford, K. (1991). The ecologically noble savage. *Cultural Survival Quarterly*, *15*(1), 46–48.

Samson, A. A. (1973). Indonesia 1972: The solidification of military control. *Asian Survey*, *13*(2), 127–139.

Sangaji, A. (2007). The masyarakat adat movement in Indonesia: A critical insider's view. In S. D. Jamie & H. David (Eds.), *The revival of tradition in Indonesian politics: The deploymnet of adat from colonialism to indigenism* (p. 377). Abingdon, UK: Routledge.

Scott, J. C. (1985). *Weapons of the weak: Everyday forms of peasant resistance*. New Haven, CT: Yale University Press.

Scott, J. C. (1998). *Seeing like a state: How certain schemes to improve the human condition have failed* (p. 464). New Haven, CT: Yale University Press.

Silver, C. (2003). Do the donors have it right? Decentralization and changing local governance in Indonesia. *Annals of Regional Science*. 37(3), 421–434.

Spyer, P. (1996). Diversity with a difference: Adat and the new order in Aru (Eastern Indonesia). *Cultural Anthropology*, *11*(1), 25–50.

Steinebach, S. (2013). "Today we occupy the plantation – Tomorrow Jakarta": Indigeneity, land and oil palm plantations in Jambi. In B. Hauser-Schaublin (Ed.), *Adat and indigeneity in Indonesia: Culture and entitlements between heteronomy and self-ascription* (pp. 63–79). Gottingen, Germany: University of Gottingen.

Syahruji, A. (2009). The Kiyu Dayak indigenous community, Meratus, South Kalimantan: Strengthening alliances to campaign for forest protection. In E. O. Kleden, L. Chidley, & Y. Indradi (Eds.), *Forests for the future: Indigenous management in a changing world* (pp. 101–141). Jakarta: AMAN.

Syarif, A. R. (2010). Exploring REDD+ implementation in Indonesia's state policies for indigenous peoples. In W. Alangui, G. Subido, & R. Tinda-an (Eds.), *State of forests, policy enviroment & ways forward: Indigenous peoples, forests & REDD plus* (pp. 61–138). Baguio, Philippines: Tebtebba.

Tsing, A. L. (2004). *Friction: An ethnography of global connection* (p. 376). Princeton, NJ: Princeton University Press.

Tyson, A. (2010). *Decentralization and adat revivalism in Indonesia: The politics of becoming indigenous* (p. 210). Abingdon, UK: Routledge.

Tyson, A. (2011). Being special, becoming Indigenous: Dilemmas of special adat rights in Indonesia. *Asian Journal of Social Science*, *39*, 652–673.

United Nations. (2014). *Convention on biological diversity Indonesia country profile*. Retrieved from http://www.cbd.int/countries/profile/default.shtml?country=id#facts. Accessed on May 20, 2014.

von Benda-Beckmann, F., & von Benda-Beckmann, K. (2011). Myths and stereotypes about adat law: A reassessment of van Vollenhoven in the light of current struggles over adat law in Indonesia. *Bijdragen tot de Taal-, Land- en Volkenkunde*, *167*(2–3), 167–195.

World Bank. (2010). *Technical guidelines for environmental management, land acquisition and/or resettlement and indigenous peoples*. Jakarta: World Bank.

Yodono, J. (2010). *Kalteng Manfaatkan Adat Lestarikan Hutan*. Retrieved from http://oase.kompas.com/read/2010/01/27/20135724/Kalteng.Manfaatkan.Adat.Lestarikan.Hutan. Accessed on May 20, 2014.

Zerner, C. (1994). Through a green lens: The construction of customary environmental law and community in Indonesia's Maluku Islands. *Law and Society Review*, *28*(5), 1079–1121.

FROM DISPARATE ACTION TO COLLECTIVE MOBILIZATION: COLLECTIVE ACTION FRAMES AND THE CANADIAN FOOD MOVEMENT

Rebecca Schiff and Charles Z. Levkoe

ABSTRACT

Academic and popular literatures have addressed growing concerns about the ways we produce, harvest, distribute, and consume food; manage fisheries and inputs to agriculture; and deal with waste. Throughout the 20th century, a series of issue-specific frames emerged that explicitly addressed issues of social justice, the environment, and human health in the food system. During the mid-1990s that comprehensive master frames were established in attempts to bring disparate ideas and actions together into a more inclusive food movement. *In this chapter, we explore the development of these collective action frames and turn to Canada as a case study to examine the key moments that have brought together diverse actors through collaborative networks to assert their place within a broader social movement. We argue that recognizing the*

Occupy the Earth: Global Environmental Movements
Advances in Sustainability and Environmental Justice, Volume 15, 225–253

ISSN: 2051-5030/doi:10.1108/S2051-503020140000015009

increasing development of food networks and making these relationships visible opens new theoretical and practical possibilities for food system transformation.

Keywords: Alternative food networks; Canada; community food security; food movement; food sovereignty; master frames

INTRODUCTION

There is an increasing recognition of the myriad of challenges related to the food system. Academic and popular literatures have addressed growing concerns about the ways we produce, harvest, distribute, and consume food; manage fisheries and inputs to agriculture; and deal with waste. There are also concerns about power differentials that leave the control of food – a basic human need – under the purview of large corporations and free market mechanisms. People around the globe are calling for change: change that favors social, economic, and environmental justice over self-interested profit-making, and change that respects the integrity and interdependence of humans, natural systems, and all species on earth. With increasing rates of land disposition, rising food prices, pervasive food-related health problems, ecological devastation, and growing inequality, there is an increasing demand for public control over our food systems.

The emergence of concepts such as *food security* and *sustainable agriculture* in the 1940s indicated the beginnings of a new global discourse centered on food systems issues. The following decades saw the evolution of issue-specific mobilizations adopting a multitude of food-related frames such as *fair trade*, *organics*, and *local food.* Alternative food networks associated with these frames developed across the globe throughout the 20th century. In the Global South, struggles of rural peasants focused on issues such as agrarian reform and the damaging impact of capitalist markets, corporate agribusiness, and global financial institutions on local communities and ecosystems. While some farmers and urban activists in the Global North were sympathetic to these issues, mobilization primarily focused on the development of alternative food initiatives (AFIs) that ranged from establishing innovative marketing opportunities (e.g., farmer's markets and community shared agriculture projects) to venues for social and political citizen engagement (e.g., community food hubs and food policy).

Many of these efforts focused on local issues and remained isolated with a limited scope of analysis and impact to articulate the challenges of the food system, their ideals, and activities in response. Isolated initiatives can be weak, as they tend to have a narrow understanding of broader structural inequities and limited analysis and impact. Further, they are easily incorporated into the logic of the dominant system. For example, while "organic" and "local" have become household concepts in respect to making sustainable purchasing choices, they are also major profit motivators for corporate retailers interested in selling product. This can be seen to compromise the integrity of those labels. It was not until the mid-1990s that *community food security* (CFS) and *food sovereignty* were introduced as master frames that attempted to bring disparate ideas and actions together into a more comprehensive food movement.

In this chapter, we explore these developments and look to Canada as a case study to examine recent efforts among food initiatives to come together through collaborative networks and assert their place within a broader social movement. In Canada, there have been a number of attempts to build a food movement across sectors, scales, and places and to link local initiatives regionally, nationally, and globally. We identify key moments where Canadians have used networks to connect collective action frames and disparate initiatives across various civil society organizations, food system regulatory sectors, and geo-political scales to develop a pan-Canadian food movement. Recognizing the dynamics of the growing movement and making these relationships visible, opens new theoretical and practical possibilities for food system transformation. We conclude with some general recommendations about future directions for research and action. While our case study is focused on the food movement in Canada, these perspectives on the growing food movement have implications for social movement scholars and global environmental efforts more broadly.

THE BROKEN FOOD SYSTEM

Understanding collective action frames necessitates an awareness of the "problematic" to which these frames respond. Critiques of the dominant food system provide important perspective in understanding the emergence and nature of a comprehensive food movement. Despite immense scientific and technological progress, the dominant food system does not work for the vast majority of the world's people. Friedmann and McMichael (1989, see also McMichael, 2009) have characterized the changes in food systems through the concept of food regimes – an analytical tool used to understand the stability and transitions in the food system over time and space. The food regimes analysis attempts to describe the broad configurations of industrial and social aspects of food production, distribution, and consumption by linking international patterns of agriculture with the development of the capitalist economy since the 1870s. Three distinct food regimes have been identified to date: the pre-industrial food regime (1870–1930s); the industrial food regime (1950–1970s) and; the current corporate food regime (1980s–). The corporate regime is distinguished as a hyper-industrial period characterized by neoliberalism, including free trade, privatization, the dispossession of peasant agriculture, flexible production, the reconstitution of food through industrial and biotech processes, and the power of corporate retailers consolidating differential supply chains.

Throughout the late 20th century, the corporate food regime fostered a profit-oriented market structure through horizontal and vertical integration within the food system (Lang & Heasman, 2004). Horizontal integration involves a trend towards increased concentration and control of capital, characterized by the dominance of corporate agribusiness, food processing, and retail entities in various sectors and the decline in small-scale, locally-owned autonomous production. Vertical integration entails consolidation across sectors into large corporate agri-food entities. Increasing market globalization initially led by political centralization and expansionism of Western states over the past three centuries encouraged trends towards vertical and horizontal integration through the development of transnational corporations (TNCs) that dominate commodity chains. Since food-related decision-making is relegated to market processes, market share has become increasingly concentrated in the hands of a few corporate interests (Albritton, 2009). For example, 78% of the retail market is captured by just three corporations (Oxfam, 2013), and most people remain outside or at the margins of the economic and political structures that control food. Operating within a "productionist" paradigm (Lang & Heasman, 2004), the corporate food regime facilitates the operation of a food system structure that creates ongoing challenges framed by concerns for social justice, ecological sustainability, and human health.

Social justice oriented critiques of the food system reflect on power and material (in)equity due to the differences in physical and economic accessibility to sufficient amounts of food along with the circumstances and conditions of those who produce food (Allen, 2008; Levkoe, 2011). Existing literature demonstrates that our food system fails to equitably meet people's needs in even the most basic sense, illustrated in part by increasing rates of hunger (Health Canada, 2007), and by imbalances in power and material resources (Patel, 2007). According to the UN Food and Agriculture Organization (FAO, 2012a), about 870 million people across the globe suffer from hunger and malnutrition, despite the fact that there is more than enough food being produced to feed the world's population (FAO, 2012b). Injustices within the food system are also evident in the difficult and dangerous conditions experienced by farmers (Qualman, 2011) and farm workers (Das, Steege, Baron, Beckman, & Harrison, 2001; UFCW & AWA, 2011), and in the unequal impact of these and other challenges on women (Allen & Sachs, 2007; Barndt, 1999, 2002; Shiva, 2009) and other marginalized groups throughout the food system (Alkon & Agyeman, 2011; Gottlieb & Joshi, 2010; Jayaraman, 2012; Sachs, Allen, Terman, Hayden, & Hatcher, 2014).

Environmentally harmful practices in all levels of the food systems have contributed to ecological challenges such as soil degradation and climate change. Concerns focus on unsustainable industrial farming and fishing practices, increasing agri-chemical use and technological dependencies, monoculture farming, intensive fish harvesting, and fish farming methods. However, issues related to water, soil, air pollution, and dwindling natural resources are also recognized as resulting from other problematic practices, including pre-production, processing, distribution, transportation, and food waste management (McMichael, 2007; Weis, 2007) and specifically the exploding consumption of meat (Weis, 2013). A recent study by GRAIN (2011) shows that between 44 and 57% of all greenhouse gas emissions can be attributed to food systems. This includes reliance on fossil fuels (e.g., for fertilizers, pesticides, and machinery), packaging, processing, refrigeration, transportation, deforestation for animal pastures, and industrial livestock operations contributing to water, soil, and air pollution.

Billions of people throughout the world are affected by health problems related to food. Undernourishment, malnourishment, diet-related illness, and food safety concerns are widespread (Patel, 2007; Schiff, 2007). Diet-related illnesses, such as cardiovascular disease, cancers, type-2 diabetes, and nutritional deficiencies have become increasingly prevalent among the overfed and underfed due to nutrient-poor diets (Halweil, 2002; Lang & Rayner, 2002). Despite technological advancements, food safety has also emerged in recent decades as a significant concern. Recently, severe outbreaks of foodborne illnesses such as salmonella enteritidis, bovine spongiform encephalopathy (BSE), and *Escherichia coli* (*E. coli*) have caused significant public concern over the safety of the food system (Schiff, 2007).

Food safety is also a concern in relation to the health effects potentially caused by certain forms of biotechnology used in agriculture and food-processing operations. Scientific communities and the general public have expressed significant concern over the health hazards that may result from the consumption of food produced with the use of genetically modified organisms (GMOs). While the health effects of genetically modified foods remain highly controversial and largely unknown, there has been significant documentation of unintended transgenic cross-contamination (Belcher, Nolan, & Phillips, 2005). There are also health-related concerns about other processes such as food irradiation.

These concerns are not new, as evidenced by the range of collective action frames developed to articulate and address food system challenges over the past century. In the following section, we describe the evolution of different collective action frames as well as master frames that have been developed with an aim to bring together disparate initiatives into a more comprehensive food movement.

THE EVOLUTION OF ISSUE-SPECIFIC FOOD FRAMES

In this section, we present a range of collective action frames that have been used to mobilize food-related actors in response to challenges in the dominant food system. Framing is a concept that has been used to understand how social movements function and gain support for their activities (McAdam, McCarthy, & Zald, 1996; Snow, Rochford, Worden, & Benford, 1986). Attempts to identify and label experiences result in collective action frames, which can be described as action-oriented sets of shared beliefs and meanings that inspire and legitimate the activities and campaigns of a social movement (Benford & Snow, 2000). Framing is based on the idea that actors are agents in the production and maintenance of ideas and meaning. Frames render events meaningful and guide action by articulating a problem, identifying causes, suggesting solutions, and issuing calls to action. Stevenson, Ruhf, Lezberg, and Clancy (2008) note that, "it is of strategic importance whether those seeking change in the modern food system can forge a master frame that both resonates and coalesces, or whether change initiatives proceed under diverse frames" (p. 37). We subsequently identify the development of food-related collective action frames beginning in the early 20th century to understand and articulate new issues of global production and distribution. These ideas were exasperated in the

1940s by the emergence of international concerns over hunger and workers' rights. However, it was not until the 1970s that collective action frames began to coalesce into a broader and more inclusive approach to address these issues from a systems perspective. As critiques of sector-specific and isolated frames emerged, a set of counter-hegemonic frames also evolved. These led to the development of master frames in the mid-1990s that have attempted to bring together disparate initiatives into networks that traverse sector, scale, and places. These master frames articulate more comprehensive understandings of, and responses to, the corporate food regime and have created a platform for the evolution of a global food movement.

Sustainable Agriculture

Calls for organic agriculture and other alternative farming methods, perhaps the earliest food-related frames, appeared in response to the rapid industrialization of agriculture during the first food regime, in the late 19th and early 20th century (Harwood, 1990). In the 1920s, Austrian philosopher Rudolf Steiner delivered a series of eight lectures in response to farmer's concerns over degraded soil conditions and the potential impacts of agricultural chemical use (Gold, 1999; Harwood, 1990). These lectures formed the conceptual basis for the development of biodynamic approaches to agriculture, or biodynamic farming, the first distinct and widely recognized "alternative" farming system. This approach to agriculture focused on minimal use of chemicals, closed-loop systems, and holistic management. Its tenets formed the value basis for subsequent alternative farming systems, and eventually for *sustainable agriculture* as a term encompassing a wide variety of farming approaches that understood agriculture as interdependent with natural systems.

Steiner's work began to influence a wide variety of scientists, philosophers, and farmers in the Global North. The following decades saw a growth of interest in biodynamic and other ideas of sustainable agriculture. Particularly, in the 1940s, the emerging organic agriculture movement began to challenge the use of synthetic fertilizers and pesticides in industrial agriculture. Pivotal works of Howard (1940), Balfour (1943), Faulkner (1943), Rodale (1945), and Bromfield (1950) signaled the emergence of the organic or "non-chemical" agriculture movement.

Efforts to promote sustainable farming and address other food-related issues grew steadily, although separately, between the 1940s and 1960s.

But it was Rachel Carson's *Silent Spring* (1962), a groundbreaking exposé of the dangers of synthetic pesticide use that laid the ground for a nascent environmental movement and breathed new life into organic agriculture and other progressive food efforts. The North American "back to the land" movement gained popularity in the late 1960s and 1970s, stressing production and consumption of locally produced food. Influenced by writers such as Schumacher (1973), back-to-the-landers and others began to consider the value of decentralized, localized economies. Further, Lappé's (1971) bestselling book *Diet for a Small Planet* was the first major book on the realities of industrial meat production and influenced a generation to think about the environmental impact of food production and consumption. Discourses around the environment, development, and the importance of local economies have persisted, having a significant impact on the development of *local food* as a popular frame in later decades.

The 1970s and 1980s also saw a renewal of critical perspectives on industrial farming and concepts of food security. Gold (1999) outlines the multitude of sustainable farming concepts and practices that emerged throughout this period, describing the different approaches of agroecology, biological or ecological farming, conservation tillage, farmland preservation, holistic management, integrated farming systems, integrated pest management, controlled grazing systems, low-input agriculture, precision farming, regenerative agriculture, and whole farm planning. In particular, the 1970s saw the creation of Masanobu Fukuoka's natural farming (Fukuoka, 1978) and Mollison and Holmgren's (1978) permaculture (see also Mollison & Slay, 1994). Government support for sustainable agriculture programs was also growing. For example, in the United States, the Sustainable Agriculture Research and Education Program (SARE) was founded in 1988 to promote and advance agricultural practices that balance environmental, economic, and social priorities. In the following years, sustainable agriculture programs emerged in countries around the world.

At the same time, industrial agriculture progressed into the initial success and eventual skepticism of genetic engineering and the Green Revolution (Patel, 2013; Schiff, 2007). *Sustainable agriculture* emerged as a frame to align farming aspirations with the broader goal of supporting environmental, economic, and social development for the entire planet, as discussed in the Brundtland Report (World Commission on Environment and Development, 1983). As Francis (1990) notes, "key individuals and organizations in agriculture have flocked to this term. After all, who would advocate a 'non-sustainable agriculture'?" (p. 97). Efforts to develop and promote sustainable agriculture in general, as well as those focused on specific manifestations such as organic, permaculture, and biodynamic, continued to evolve throughout the following decades.

In the 1990s, consumers also grew increasingly concerned about the potential threats to health from eating livestock treated with hormones and antibiotics, plant foods exposed to synthetic fertilizers, pesticides or irradiation, and genetically engineered crop varieties. These same industrial practices were also feared to damage the health of farm workers, pollute water and soil, cause fish stock depletion, and pose threats to wildlife and biodiversity (Gold, 1999; Weis, 2007).

Support for organic and sustainable farming continued to increase during 1990s and early decades of the 21st century. Standards for organic labeling were established in many countries, while non-profit organizations and producers' associations were created to promote organic farming methods. The organic market grew at unprecedented rates, moving from a niche market to a more mainstream and accepted component of the food industry. Large corporations such as Nestle, Cargill, and Walmart sought to profit from this growth, acquiring and creating their own brands of organic and "natural" foods.

Food Security

Alongside the calls for organic and biodynamic agriculture, he 1940's also saw the emergence of concerns over famines and widespread hunger, which had emerged in the aftermath of World Wars I and II. In the 20th century, nutrition, and consequently food, "moved from the sidelines of public health to being central ... to public health campaigns" (Lang & Heasman, 2004, p. 48). It was during this time that the FAO and the World Health Organization (WHO) were established under the newly created United Nations, in part to address these global concerns. Those organizations assumed the priority of preventing hunger by increasing food supply in every nation (Lang & Heasman, 2004).

In 1966, the General Assembly ratified the International Covenant on Economic, Social, and Cultural Rights (ICESCR), which recognized the universal right to adequate food, particularly as derived through systems that promote equitable distribution and appropriate use of natural resources. In accordance with the principle of interdependency of human rights, this chapter is taken in the context of all other chapters contained within the ICESCR and other covenants, including those reflecting on issues such as workers' and landowners' rights. However, it was not until 1999 that the UN would define a more holistic "right to food" concept and establish a Special Rapporteur to monitor states' progress towards the realization of that right.

The concept of *food security* continued to evolve through the mid to late 20th century, from the initial WHO and FAO definitions of the 1970s focusing on adequate supply, to eventually incorporating socially acceptable access as well as food safety. Food security, until the late 1970s and early 1980s, meant food sufficiency, generally concerning hunger on the macroeconomic scale of providing enough food for the growing world population. After being officially adopted by the UN's FAO at the World Food Summit in 1974, food security evolved into a multi-dimensional concept to include issues of nutrition and access (FAO, 2003).

Definitions of *food security* continued to evolve. By the 1990s, it was commonly perceived that appropriateness in quality and means of access, not just sufficiency, should be part of the conceptual framework of food security. By the time of the World Food Summit in 1996, food security for people was defined and widely accepted as "physical and economic access to sufficient, safe and nutritious food to meet their dietary needs and food preferences for an active and healthy life" (FAO, 1996). The updated definition, developed at the 1996 World Food Summit, signified a shift from the idea of meeting short-term nutritional sufficiency, as quantified by adequate caloric intake, towards a more subjectively qualified sustainable "livelihood" definition aimed at preferences and acceptability.

Fair Trade

The *fair trade* movement was also born in the aftermath of World War II, when a handful of religious groups and political NGOs became concerned about the working conditions and wages of goods producers in the Global South and particularly the poorest countries of Africa, Latin America, and Asia. Many credit the founding of the contemporary fair trade movement to the Central Mennonite Committee of Pennsylvania, and particularly to Edna Ruth Byler, who began by selling the needlework of Puerto Rican women from the trunk of her car (DeCarlo, 2007). In 1958, Byler and her colleagues opened a shop in Akron, Pennsylvania, that focused primarily on fair trade goods. But Byler was not the only one concerned with the conditions and wages of farm workers. By the late 1960s, fair trade shops and organizations were popping up across North America and Europe.

These efforts generally focused on handcrafts such as textiles, needlework, carvings in wood and stone, and ceramic products.

In the 1970s, fair trade products entered the mainstream food market when a few organizations began introducing goods such as coffee and tea. By the early 1980s, handcrafts were losing their novelty and appeal, and the explosion of demand for fair trade coffee led many organizations to explore the potential market for other foods (DeCarlo, 2007). Fruit, cocoa, sugar, and other food products were soon offered in fair trade shops and mail-order catalogues. In this same decade, interest grew in making fair trade products available through more accessible distribution channels, although some feared that the interests of large corporations would compromise the integrity of the fair trade concept. This concern gave rise to the labeling initiatives of the late 1980s. Like organic producers, fair trade proponents sought to assure consumers of the integrity of their products through labeling and certification schemes.

Through the mid to late 1990s, fair trade initiatives continued to grow alongside efforts aimed at sustainable agriculture. Fair trade certification organizations sprouted up across Europe, North America, Asia, and elsewhere. Fairtrade International launched an international certification regulation and label. Fair trade food products became widely distributed through mainstream vendors, and the market continued to grow exponentially.

Local Food

Following the Cornucopia Project in 1981, an investigation of the US national food system, regional food systems analyses focusing on local food activities expanded as an important tool for examining sustainability in food systems (Feenstra, 2002). In 1991, Getz (1991) popularized the term *foodshed* (adapted from the term watershed) to describe the geographic, as well as the socio-cultural and economic aspects of a sustainable food systems theory. Frames like *community food system* and *local food system* also surfaced to denote the social, economic, and geographic aspects of food systems.

Interest in *local food* exploded in the first decade of the 21st century, propelled by non-fiction bestsellers about local food such as *Fast Food Nation* (Schlosser, 2001), *The 100-Mile Diet* (Smith & Mackinnon, 2007), *What to Eat* (Nestle, 2006), and *The Omnivore's Dilemma* (Pollan, 2006). Many people, concerned with their personal health and the threat of global warming, saw buying locally as a way to address these issues. The local food movement introduced a variety of unique modes for the economic exchange of food – activities such as community-supported/shared agriculture (CSA), farmers' markets, farm-gate sales, food co-ops and buying clubs, community gardening, household gardening, and direct marketing to individual consumers and businesses such as restaurants and grocery stores.

Support for "local food" grew alongside increasingly widespread interest in organic agriculture and fair trade. Despite strong entry into the general marketplace and their continued growth, organics, fair trade products, and local food still represented only a small fraction of the mainstream food market. The classic criticism that "organic farming can't feed the world," was joined by other criticisms pointing to weaknesses in certification standards, elitism, definitions of "local," "fair," and other issues connected to food-related frames (e.g., Desrochers & Shimizu, 2012).

The Emergence of Master Frames

The various issue-specific frames discussed above, including organics, fair trade, food security, and local food, have played an important role in pushing forward debates about food and agriculture. These framings catalyzed many individuals to become interested and active in issues surrounding food systems. Nevertheless, these food-related frames have also become increasingly entrenched within dominant perspectives of the corporate food regime, leading many scholars and practitioners to become skeptical of their merit. Recognizing the proliferation of challenges in respect to social justice, ecological sustainability, and human health, there have been major critiques of issue-specific food frames. This has led to the emergence of a new set of master frames that have attempted to bring together the diversity of approaches and perspectives into a broader food movement.

Since its emergence in the early 1970s, *food security* had been adopted across the public and private sectors as the dominant frame for discussing the priorities, objectives, and challenges within food systems. Not discounting the many non-profit organizations and activist associations that were mobilized by its underlying ideals, food security frames have been critiqued for legitimizing the priorities of the corporate food regime (Koc, 2011). Central to these critiques is that the concept that food security frames the core problem of the global food system as scarcity and proposes technical solutions to feed increasing populations and solve existing problems (Allen, 1999). Those adopting food security as a frame have also been charged with focusing primarily on the needs of consumers and ignoring concerns of those who grow, catch, and forage for food. Specifically, food security has been said to come at the expense of small-scale peasant farmers across the globe that make up the majority of food producers (Desmarais, 2007). Overall, food security has been identified as an uncritical approach that has done little to address the problematic structures of the corporate food regime. It has also failed to shift government policy towards system-wide solutions. Instead, powerful governments and corporations have used the concept of food security to encourage a neoliberal, free trade agenda and to pursue a global, industrialized model for agriculture (Patel, 2009). This has resulted in efforts to maximize production and increase trade with little concern for how or by whom food is produced, distributed, and consumed.

Similar to these ideas of food security, critics have argued that the discourses and practices associated with organics, fair trade, and localism have been subsumed into neoliberal logics. Julie Guthman has written extensively about the paradoxes of organic farming in California and in the Global North more broadly (Guthman, 2004). Specifically, she has argued that while organics began with radical origins, organic farmers have been integrally connected to the capitalist transformation of the farming sector. For example, in 2002 the US Department of Agriculture institutionalized a federal organic regulation, which emphasized input-oriented standards as opposed to issues of social justice and broader ecological sustainability (a similar standard was developed in Canada in 2009). The new regulations resulted in price premiums being capitalized into land values. Guthman (2004) argues that the agrarian rhetoric used in organic discourse has ignored more structural issues of land ownership, labor practices, and producer–consumer relationships. She writes that the organic regulations have undermined alternatives: "organic production is not only bound by, but may also be contributing to, the trends in intensification and valorization that have characterized late twentieth-century agriculture" (p. 178). Through these shifts, it has been suggested that the organic sector has been complicit in reproducing neoliberal logics.

Fair trade initiatives have received similar criticism that, while rooted in progressive ideals, have had little transformative impact. While many fair trade initiatives have enabled acts of resistance that contest the social system and facilitate the redistribution of resources from the Global North to South, Shreck (2008) argues that they have had little impact on the unjust commodity system at the core of the corporate food regime. Similarly, Guthman (2007) suggests that fair trade (and other voluntary food labels) actively construct neoliberal subjects through conceding the market as the locus of regulation while creating capitalist markets where none previously existed.

As the locavore craze exploded over the past decade, there has been significant critique that ideas of localism have been uncritically associated with positive attributes such as quality and taste, ecological sustainability, better health, and producing higher economic gains within regional economies (Born & Purcell, 2006; DuPuis & Goodman, 2005; Levkoe, 2011; Winter, 2003). In other words, local becomes a problematic frame when it is seen to be a goal in and of itself and is assumed to be entirely desirable. Further, Allen, FitzSimmons, Goodman, and Warner (2003) note that ideas of social justice, ecological sustainability, and human health have different meanings, depending on the particular contexts in which they originate, and thus cannot be carried forward unexamined. In practice, essentializing the local scale can result in multiple tensions that increase the challenges to community-based food organizing. For example, the concept of *defensive localism* has been used to describe the construction of rigid barriers in the protection of local spaces while exhibiting reluctance towards difference (Hinrichs, 2003). These kinds of barriers have the potential to polarize and exclude particular cultural and social groups by portraying the perception of a homogeneous version of local that excludes the non-local other. Instead of relying on localism as the solution, it has been suggested that organizations should examine the difference between local places in respect to the particular social relations established where these initiatives arise. Further, ignoring the reality that food system challenges are often caused by factors beyond the immediate community may obscure other scalar options that might be more effective.

While these critiques have been an important part of understanding the limitations of issue-specific frames, they have also contributed to a set of master frames that aim to disrupt consent for the corporate food regime. The emergence of CFS in the mid-1990s as a master frame is emblematic of this shift, bringing together ecological and social justice issues as well as issues surrounding the quality and cultural appropriateness of food. Allen (2004) outlines the history of this master frame, tracing its origins to a UCLA research report on a low-income Los Angeles community (Ashman et al., 1993). The report, entitled *Seeds of Change: Strategies for Food Security in the Inner City*, defined the most pressing need of that "limited-resource" community as access to good-quality and affordable food. As a result, a group of organizations and individuals came together to share their interest in food-related issues. But rather than following the traditional route of adopting a set of isolated frames, they attempted to take a more comprehensive perspective that considered the multiple interconnected elements that make up food systems.

Central to CFS has been a call for *food justice*, an ideal that evolved from the environmental justice movement (Bullard, 1994). Focusing on the empowerment of racialized, gendered, and impoverished people, CFS advocates aim to assert their rights to ensure that the benefits and risks of the food system are shared equitably (Alkon & Agyeman, 2011; Allen, 2010; Gottlieb & Joshi, 2010). CFS also incorporated ideas of *food democracy* as both methodological and pragmatic processes for achieving its goals. Hassanein (2003) argued that achieving a just and sustainable food system required active participation among citizens: "food democracy is about citizens having the power to determine agro-food policies and practices locally, regionally, nationally, and globally" (p. 79). Food democracy points to active participation and political engagement as prerequisites for food system change. This requires that people develop the knowledge and skills necessary to actively participate. Thus, in its development, CFS was intended to be a comprehensive perspective that integrated ideas of social justice, ecological sustainability, and community health (Anderson & Cook, 1999; Clancy, 1994; Gottlieb & Fisher, 1996). While there is no one definition of CFS, Hamm and Bellows (2003) provide a description that reflects its comprehensive approach to addressing food system issues. They described CFS as "a situation in which all community residents obtain a safe, culturally acceptable, nutritionally adequate diet through a sustainable food system that maximizes community self-reliance and social justice" (p. 37).

The Community Food Security Coalition (CFSC), a non-profit organization incorporated in 1997, was a center point for the CFS movement in North America. The CFSC functioned as an umbrella for non-profits from across the United States working on CFS-related issues. Similar organizations exist in Canada, both at the federal level and in each of the Canadian provinces and territories (we will discuss these Canadian examples further in the case study below). Until its contentious collapse in 2013, the CFSC supported research, programs, and projects associated with the broad range of concerns from farmland protection and skill-building for farmers to urban agriculture, community food assessments, and food-related policy-making.

While CFS emerged primarily in North America, *food sovereignty* as a master frame evolved through collaborative dialogue among global peasant organizations to challenge political and economic power in the food system

(Pimbert, 2009; Wittman, Desmarais, & Wiebe, 2010). The global peasant movement, La Via Campesina first proposed the concept of food sovereignty at the World Food Summit in 1996 in Rome. Differing slightly from CFS, food sovereignty stands in direct opposition to the neoliberal market-driven agenda within the corporate food regime (Fairbairn, 2010; Patel, 2009). It evolved as a critical reaction to the experiences of peasant farmers around the world affected by shifts in national and international agricultural policy throughout the 1980s and 1990s. The argument, voiced by peasant farmers, suggested that until the people who are most affected by food- and agricultural-related policies have a democratic voice and a hand in implementing them, food security means very little. The concept became more formalized at the 2007 Forum for Food Sovereignty in Mali:

> Food sovereignty is the right of peoples to healthy and culturally appropriate food produced through ecologically sound and sustainable methods, and their right to define their own food and agriculture systems. (Nyéléni, 2007)

Food sovereignty as a master frame focuses on the rights of those who produce and consume food to reclaim control of the food system from corporate interests and global financial institutions. At its core, food sovereignty asserts the principle that food is a basic human right and that participatory democracy is fundamental to its realization. According to Windfuhr and Jonsen (2005), "While food security is more of technical concept, and the right to food a legal one, food sovereignty is essentially a political concept" (p. 15). Food sovereignty shifts the focus beyond the "market as movement" model to address core issues such as the redistribution and management of land and resources (Desmarais, 2007). Food sovereignty does not reject trade altogether, but is clear that it must be fair and equitable in feeding populations not corporate profits (Rosset, 2008). Incorporating critical perspectives of local food, food sovereignty focuses on the places of food production recognizing that food is different than other commodities and considers the relationships between nature, culture, and spirituality. Thus, food sovereignty is a call for democracy and a redistribution of power in the food system and entails people having a direct role in deciding how and where their food is produced and consumed. Over the past decade, the concept of food sovereignty has become more prominent in a wide range of literatures, among food and farm leaders and policy-makers. It has radicalized many grassroots initiatives (e.g., Desmarais, 2007; Holt-Giménez, 2011; Holt-Giménez & Patel, 2009; Wittman, Desmarais, & Wiebe, 2011) and contributed to the evolving conversation about the challenges and solutions to food systems issues.

Embracing the values of social justice, ecological sustainability and human health, CFS and food sovereignty have evolved as comprehensive food systems perspectives in response to criticisms towards isolated and issue-specific food-related frames and practices. As master frames, they have provided a platform for collective mobilization, moving beyond the limited scope of isolated initiatives, and providing protection against incorporation into the logic of the dominant system. The experience in Canada highlights the efforts of food initiatives to come together through collaborative networks and assert their place within a broader food movement. In Canada, there have been a number of attempts to build a food movement across sectors, scales, and places and to link local initiatives regionally, nationally, and globally. An exploration of this experience provides insight into the ways that networks can be used to connect different collective action frames and initiatives across various civil society organizations, food system regulatory sectors, and geo-political scales to develop a cohesive and resilient food movement. Recognizing the dynamics of the growing movement and making these relationships visible open new theoretical and practical possibilities for food system transformation.

THE CANADIAN FOOD MOVEMENT

The number and scope of AFIs in Canada has steadily expanded over the past several decades (e.g., CCA, 2009; Elton, 2010; Kirbyson, 2005; Levkoe & Wakefield, 2014; Palassio & Wilcox, 2009; Wittman et al., 2011). Activities in Canada represent the diverse range of isolated and master frames that aim to understand and address the challenges and solutions in the food system. They include networks of individuals and organizations focused on a range of issues with the corporate food regime. Levkoe (2014) writes,

> Canada has a rich tradition of collective action for social change. Prior to the 1970s, many of the food and agriculture related initiatives were focused on specific sectors and interests such as fishing, farming, health, poverty, labour, Indigenous peoples, and the environment. Due in part to Canada's large geographic size, different languages spoken, and a lack of communication technologies, unified action around the food system was complicated. (p. 5)

While most AFIs developed within a particular context and place, evolving networks at regional, provincial, and national levels have succeeded in drawing together disparate actors around master frames into an increasingly cohesive and resilient social movement. In this section, we highlight some of the key moments that have contributed to this collective mobilization.

The People's Food Commission

In Canada, the first large-scale, collective, and comprehensive mobilization focusing on food system issues coalesced around the People's Food Commission (PFC). Beginning in 1977, the PFC brought together thousands of diverse grassroots actors to discuss challenges and possible solutions to the increasing inequities and corporate concentration within the food system. Participants included a range of individuals and organizations such as the Canadian Labour Congress, the National Farmers Union, the Canadian Union of Students, the YWCA of Canada, and the National Indian Brotherhood, along with other national and provincial organizations. Through informal hearings and creative presentations, the objective of the PFC was to document a wide range of challenges and opportunities with the goal to transform the food system. The PFC's final publication, a booklet entitled *The Land of Milk and Money*, "began with the assumption that everyday experience is a valuable source of information, and that people's stories about how things work and fit together have an important validity" (People's Food Commission [PFC], 1980, p. 7). The PFC concluded, "behind the rise and fall of food prices, there were a handful of corporations who controlled and profited from the food system" (People's Food Commission, 1980, p. 81).

While participants used a multitude of issue-specific frames in their deputations, the PFP employed the concept of *food system* as an attempt to posit a master frame that would highlight the interrelationships within the range of sectors and processes that bring food to our plates. Over four years, the PFC mobilized thousands of people from across the country to articulate their concerns and suggestions around ways to transform the food system and generated significant energy and interest. It marked a turning point in Canada when people were first starting to think about collaborative and cross-sectoral approaches to food issues. It was also the first major mobilization in Canada that brought together diverse actors from different sectors and places to address the food system in a comprehensive way. Unfortunately, the lack of funds and energy limited the ability of the PFC to move forward and there was little political will to address the issues brought up in the final report. However, the ideas and inspiration that resulted from the PFC continued to permeate social mobilization for decades to follow.

Local and Regional Networks

Over the next several decades, a multitude of AFIs emerged. However, after the PFC concluded, there were few broad attempts to mobilize food issues beyond the local level and few attempts to bring the different sectors together. By the later part of the 20th century, a number of AFIs had been established that were working on broader food system issues. At a local level, mobilization included small businesses (e.g., small-scale and artisanal processors, grocers, cooperatives), rural farms, community gardens and urban farms, food hubs, regional health authorities, food policy councils, and other local food coalitions. At the same time, organizations were beginning to emerge with the aim of bringing together local and regional actors through regional networks (Levkoe, 2014). For example, The Nutrition and Food Security Network of Manitoba was established in 1992; the Food Security Working Group was established by the Ontario Public Health Association in 1993; and the Food Security Network of Newfoundland and Labrador in 1998. *Food security* was the most prominent frame adopted by the networks and, over time, many groups began to embrace the concept of *CFS* as a master frame. Today almost every province and territory in Canada has a provincial-level networking organization using a food system lens. These networks created the platform for individuals and groups to come together to dialogue about the food system beyond their locality and specific experience and were critical components in laying the groundwork for broader networks and actions in the following decades.

Various sector-specific organizations such as the National Farmers Union, the Canadian Organic Growers, and USC Canada also continued to work through regional and national chapters. Despite the growing level of activity, until the late 1990s, there was little means to bring these diverse initiatives together to address the food system in an integrated and comprehensive way. The one exception could be found in the emergence of food policy councils in North America and Australia. These organizations were critical actors in promoting integrated, systems approaches to food issues through education, program, and policy change at local and regional levels (Schiff, 2006, 2008). For example, the Toronto Food Policy Council (established in 1991) would come to have significant influence in the development of integrated, cross-sectoral approaches to food in Canada.

With the inauguration of the first World Food Summit in Rome in 1996, and the growing CFS and food sovereignty movements, Canadian activists and academics recognized a need to discuss food system issues in a more comprehensive way along with collaborative action. In 1999, a group of representatives from across Canada gathered in Toronto to discuss the possibility of forming a national food network that would consist of representatives from academia, community organizations, practitioners, and policy-makers. Their efforts were expedited by an invitation to prepare collective input for the Canadian government during the preparation of the official submission to the World Food Summit – five years later to be held in 2001 in Rome. In 2001, representatives from the public and private sectors were invited to attend a conference at Ryerson University in Toronto, and agreed to a resolution stating the need for a pan-Canadian food movement network (Koc & MacRae, 2001, p. 49).

National Networks

Following the Ryerson meeting, a National Food Assembly was held in 2004 in Manitoba with the goal of bringing together the broad array of food network representatives. At a second National Assembly in Waterloo the following year, participants agreed to three interlocking commitments that formed the vision for the new organization. These commitments were: (1) zero hunger; (2) a sustainable food system; and (3) healthy and safe food. Attempting to appeal to a wide range of food system actors, the organizations adopted CFS as a frame and took on the name Food Secure Canada-Réseau pour une alimentation durable (FSC-RAD). FSC-RAD formally incorporated as a non-profit organization in 2006 and adopted a collaborative approach and an inclusive politics. Its mission stated that FSC-RAD was to be a "Canada-wide alliance of civil society organizations and individuals collaborating to advance dialogue and cooperation for policies and programs that improve food security in Canada and globally."

At the same time, another organization was emerging with an interest in building partnerships among community and university food researchers. In April 2005, a meeting was held at Ryerson University that resulted in the decision to form an association that would bring together academics, community, and public sector researchers committed to generating new food-related knowledge in response to societal needs. Members emphasized a need for policy-relevant scholarship that could be used to build arguments aimed at decision-makers and support existing initiatives towards food system transformation. Taking the name the Canadian Association for Food Studies- l'Association canadienne des études sur l'alimentation (CAFS-ACÉA), founding President Koc (2006) wrote about the need for systemic perspectives in food research that supported existing network activities:

> Food is a process not a product. Looking at food as a process implies careful scrutiny of interlinkages and interconnections among these practices, processes, and structures. This awareness, in return implies the need for interdisciplinary collaboration so that we can benefit from different methodological and analytical strengths of diverse academic disciplines. (p. 1)

CAFS-ACÉA was born as a sister-organization to FSC-RAD to allow university and community-based researchers from diverse disciplines to meet regularly and share their work with each other and the broader public. The two organizations did, however, agree to keep an arms length relationship to allow for independent development. The development of both FSC-RAD and CAFS-ACÉA as national organizations was the result of more than two decades of mobilization around food issues. Many of the founding leaders of the organization were involved in the PFC and were directly connected to existing local, provincial, and global food networks. As the organizations developed, they embraced *CFS* as a master frame in their internal and external communications (and eventually sovereignty). As such, they were critical in bringing together disparate actors across sectors, scales, and places. Throughout the ensuing decade, their work would help to form and influence a cohesive, national, integrated approach to food issues.

The People's Food Policy

Building on the existing momentum, FSC-RAD proposed to renew the goals and processes of the PFC to create a collective, inclusive, comprehensive national food policy – the People's Food Policy (PFP). The idea emerged in 2007 when representatives from Canada attended the Nyéléni Forum for Food Sovereignty in Sélingué, West Africa. FSC-RAD partnered with a number of national-scale organizations (including the National Farmer's Union, USC Canada, and Inter Pares) to establish the PFP with the aim to develop Canada's first and only citizen-led comprehensive food policy. Over the course of two years, PFP volunteers facilitated hundreds of kitchen table talks, meetings, and events around the country. The PFP project engaged over 3,500 individuals and organizations from isolated frames and diverse sectors that collaboratively proposed and prioritized a food policy platform based on food sovereignty as a master frame. Writing teams gathered the recommendations and prepared several draft versions that were circulated publicly for comment and feedback and reviewed at FSC-RAD's 2010 National Assembly. In April 2011, the PFP was launched in Parliament Hill in Ottawa. The PFP consists of 10 detailed working-policy discussion papers, which include whole-of-government policy recommendations and guidelines for how the proposed changes could be put into action.

Reflecting on the PFP, FSC-RAD's national Chair, Cathleen Kneen (in Holt-Giménez et al., 2010) wrote,

> This project builds on the local organizing that is already going on in the multiplicity of food self-reliance projects in both rural and urban areas, and its method is to overcome the "individual" by starting with the personal. People are encouraged to examine the barriers to the food security projects they are engaged in, and to tease out the policies that support or have erected those barriers. (p. 234)

The PFP was developed as a living document, open for further revision and expansion and was adopted by FSC-RAD's as a set of guidelines for the organization's future work. The PFP came at a critical point and brought together diverse groups adopting a range of collective action frames. It catalyzed people by drawing on the emergent master frame of food sovereignty. The PFP attracted interest from within Canada and around the globe, resonating with others attempting to develop their own national food policy processes. It has also laid the groundwork to facilitate collective mobilization around other significant activities and initiatives in Canada, including the recent visit by the UN Special Rapporteur on the Right to Food.

The United Nations Special Rapporteur on the Right to Food

In November 2011, a number of individuals and organizations were contacted by staff of Olivier DeSchutter, the UN Special Rapporteur on the Right to Food. They were informed that the Special Rapporteur would be undertaking a formal country mission to Canada in May 2012, the first time a mission had been made to an OECD country. After discussions with the Rapporteur's staff, a civil society group was convened to initiate a collective process that would make suggestions supporting the upcoming visit. A Coordinating Committee facilitated by FSC-RAD and CAFS-ACÉA was formed and, together with input from stakeholders across Canada, created a 40-page document submitted to the Rapporteur staff, which detailed civil society assessment of priority issues regarding Canada's failure to respect, protect, and fulfill the right to food (Pascal, Murphy, Porter, Desmarais, & Schiff, 2011).

Based on this submission, civil society groups were invited to convene meetings on specific themes in Montreal, Ottawa, Toronto and Winnipeg. During the Special Rapporteur's mission to Canada on May 6–16, 2012, he and his team met with the Government of Canada and key stakeholders related to the right to food, including Indigenous Peoples' representatives, civil society organizations, and a wide spectrum of individuals and experts involved in food issues. The mission sparked a flurry of media attention across Canada's major news broadcasting channels. Nearing the end of the mission, the Special Rapporteur called for Canada to create a national food strategy, and acknowledged the PFP as an important step. The Special Rapporteur released a report, detailing his findings and policy recommendations to the UN Human Rights Council in 2013. He also indicated plans to return to Canada to track its progress.

The activity surrounding DeSchutter's mission to Canada was significant for national-level integrated food systems mobilization. The collective civil society process helped to cement the years of work of AFIs, local and regional networks, FSC-RAD, CAFS-ACÉA, and numerous other organizations working towards a systems approach to food issues, bridging gaps between issue-specific collective action frames and activities. It demonstrated the strength of collaboration and collective mobilization across sectors, scales, and places to create a cohesive and resilient food movement.

CONCLUSION

Over the past several decades, the food movement has grown dramatically across the globe. In this chapter, we have traced the way that disparate food-related initiatives scattered across sectors, scales, and places in Canada have employed master frames to encourage collective mobilization using a food systems lens. Many of the food initiatives we have discussed above developed within particular sectors (e.g., farming, fishing, consumption, etc.) and through specific collective action frames (e.g., fair trade, organics, food security, etc.). From the PFC in the late 1970s and the development of regional networks to the establishment of national networks like FSC-RAD and CAFS-ACÉA, Canadians who are concerned about the food system are finding strength and growth through social mobilization where the whole movement is much more than the aggregated actors. Collaborative approaches have provided a means for Canadians to work collectively across various AFIs, food system regulatory sectors, and geo-political scales to find possible solutions. Employing master frames like CFS and food sovereignty, the movement has begun to adopt a systems perspective that considers the interconnection of elements – from production to waste management – and elements of social justice, ecological sustainability, and human health throughout all aspects of the food system.

While the efforts discussed in this chapter are an important part of this process, there are significant areas that require further action and research. These include continuing to bring together disparate sectors, closing the gap between French- and English-speaking Canada, making AFIs and food movement organizations more inclusive and accessible to *all* populations, building connections with the global food movement, and developing a more critical analysis of political and economic structures that limit food system transformation. The adoption of master frames does not necessarily imply that the food movement is an entirely cohesive group. As others have demonstrated, there are a wide diversity of perspectives, approaches, objectives, and goals that constitute the movement in Canada (Levkoe & Wakefield, 2014) and globally (Holt-Gimenez & Shattuck, 2011). These differences may in fact be a strength of the food movement, and finding convergence in the diversity is the major task that lies ahead. While these recommendations are specific to Canada, the authors' experience suggests that they might resonate more broadly. As we have demonstrated, CFS and food sovereignty are not fixed, but constantly in process. As sociopolitical realities continue to shift, new research emerges, and new experiences occur, these master frames will, no doubt, continue to evolve.

Taking a social movement perspective offers new theoretical and practical possibilities for food system transformation. Instead of viewing initiatives in isolation, the adoption of master frames and the development of comprehensive food movement networks provide a platform to understand the connections between them. We encourage future research to take this perspective in studying resistance to the corporate food regime.

REFERENCES

Albritton, R. (2009). *Let them eat junk: How capitalism creates hunger and obesity*. London: Pluto Press.

Alkon, A., & Agyeman, J. (2011). *Cultivating food justice*. Cambridge, MA: MIT Press.

Allen, P. (1999). Reweaving the food security safety net: Mediating entitlement and entrepreneurship. *Agriculture and Human Values*, *16*(2), 117–129.

Allen, P. (2004). *Together at the table*. University Park, PA: Pennsylvania University Press.

Allen, P. (2008). Mining for justice in the food system: Perceptions, practices, and possibilities. *Agriculture and Human Values*, *25*(2), 157–161.

Allen, P. (2010). Realizing justice in local food systems. *Cambridge Journal of Regions, Economy and Sociology*, *3*(2), 295–308.

Allen, P., FitzSimmons, M., Goodman, M., & Warner, K. (2003). Shifting plates in the agrifood landscape: The tectonics of alternative agrifood initiatives in California. *Journal of Rural Studies*, *19*(1), 61–75.

Allen, P., & Sachs, C. (2007). Women and food chains: The gendered politics of food. *International Journal of Sociology of Food and Agriculture*, *15*(1), 1–23.

Anderson, M., & Cook, J. (1999). Community food security: Practice in need of theory. *Agriculture and Human Values*, *16*(2), 141–150.

Ashman, L., de la Vega, J., Dohan, M., Fisher, A., Hippler, R., & Romain, B. (1993). *Seeds of Change: Strategies for food security for the inner city, urban planning*. Los Angeles, CA: University of California.

Balfour, E. B. (1943). *The living soil*. London: Faber & Faber.

Barndt, D. (Ed.). (1999). *Women working the NAFTA food Chain: Women, food and globalization*. Toronto: Second Story Press.

Barndt, D. (2002). *Tangled routes: Women, work and globalization on the tomato trail*. Aurora: Garamond Press.

Belcher, K., Nolan, J., & Phillips, P. (2005). Genetically modified crops and agricultural landscapes: Spatial patterns of contamination. *Ecological Economics*, *53*(3), 387–401.

Benford, R., & Snow, D. (2000). Framing processes and social movements: An overview and assessment. *Annual Review of Sociology*, *26*, 611–639.

Born, B., & Purcell, M. (2006). Avoiding the local trap: Scale and food systems in planning research. *Journal of Planning Education and Research*, *26*(2), 195–207.

Bromfield, L. (1950). *Out of the earth*. New York, NY: Harper & Bros.

Bullard, R. (1994). *Unequal protection: Environmental justice and communities of color*. San Francisco, CA: Sierra Club Books.

Canadian Co-operative Association [CCA]. (2009). The Lay of the Land: Local Food Initiatives in Canada. Retrieved from www.coopscanada.coop. Accessed on April 20, 2014.

Carson, R. (1962). *Silent spring*. Boston: Houghton Mifflin.

Clancy, K. (1994). Social justice and sustainable agriculture: Moving beyond theory. *Agriculture and Human Values*, *11*(4), 77–83.

Das, R., Steege, A., Baron, S., Beckman, J., & Harrison, R. (2001). Pesticide-related illness among migrant farm workers in the United States. *International Journal of Environmental Health*, *7*(4), 303–312.

DeCarlo, J. (2007). *Fair trade: A beginner's guide*. Oxford: OneWorld Publications.

Desmarais, A. (2007). *La Via Campesina: Globalization and the power of peasants.* Halifax: Fernwood.

Desrochers, P., & Shimizu, H. (2012). *The locavore's Dilemma: In praise of the 10,000-mile diet.* New York, NY: Public Affairs.

DuPuis, E. M., & Goodman, D. (2005). Should we go "Home" to eat? toward a reflexive politics of localism. *Journal of Rural Studies, 21*(3), 359–371.

Elton, S. (2010). *Locavore: From farmers' fields to rooftop gardens – How Canadians are changing the way we eat.* Toronto: Harper Collins Publishers Ltd.

Fairbairn, M. (2010). Framing resistance: International food regimes and the roots of food sovereignty. In H. Wittman, A. Desmarais, & N. Wiebe (Eds.), *Food sovereignty: Reconnecting food, nature and community* (pp. 15–32). Halifax: Fernwood.

FAO (Food and Agriculture Organization). (1996). *Trade reforms and food security: Conceptualizing the linkages.* Rome: Food and Agriculture Organization of the United Nations.

FAO (Food and Agriculture Organization). (2003). *Rome declaration on world food security and world food summit plan of action.* Rome: World food Summit Food and Agriculture Organization of the United Nations.

FAO (Food and Agriculture Organization). (2012a). The State of Food Insecurity in the World. Retrieved from www.fao.org. Accessed on April 20, 2014.

FAO (Food and Agriculture Organization). (2012b). The State of Food and Agriculture 2012. Retrieved from www.fao.org. Accessed on April 20, 2014.

Faulkner, E. H. (1943). *Plowman's folly.* Norman, OK: Oklahoma University Press.

Feenstra, G. (2002). Creating space for sustainable food systems: Lessons from the field. *Agriculture and Human Values, 19*(2), 99–106.

Francis, C. A. (1990). Sustainable agriculture: Myths and realities. *Journal of Sustainable Agriculture, 1*(1), 97.

Friedmann, H., & McMichael, P. (1989). Agriculture and the state system: The rise and fall of national agricultures, 1870 to the present. *Sociologia Ruralis, 29*(2), 93–117.

Fukuoka, M. (1978). *The one-straw revolution: An introduction to natural farming.* Emmaus, PA: Rodale.

Getz, A. (1991). Urban foodsheds. *Permaculture Activist, 7*(3), 26–27.

Gold, M. V. (1999). *Sustainable agriculture: Definitions and terms* (Vol. 2004). SRB Series. Beltsville, MD: U.S. Department of Agriculture.

Gottlieb, R., & Fisher, A. (1996). Community food security and environmental justice: Searching for a common discourse. *Agriculture and Human Values, 3*(3), 23–32.

Gottlieb, R., & Joshi, A. (2010). *Food justice.* Cambridge, MA: MIT Press.

GRAIN. (2011). Food and climate change: The forgotten link. GRAIN. Retrieved from www.grain.org. Accessed on April 20, 2014.

Guthman, J. (2004). *Agrarian dreams.* Berkeley, CA: University of California Press.

Guthman, J. (2007). The polanyian way? Voluntary food labels as neoliberal governance. *Antipode, 39*(3), 456–476.

Halweil, B. (2002). *Home grown: The case for local food in a global market.* Washington, DC: The Worldwatch Institute.

Hamm, M., & Bellows, A. (2003). Community food security and nutrition educators. *Journal of Nutrition Education and Behavior, 35*(1), 37–43.

Harwood, M. (1990). A history of sustainable agriculture. In C. A. Edwards, R. Lal, P. Madden, R. H. Miller, & G. House (Eds.), *Sustainable agricultural systems.* Ankeny, Iowa: Soil and Water Conservation Society.

Hassanein, N. (2003). Practicing food democracy: A pragmatic politics of transformation. *Journal of Rural Studies*, *19*(1), 77–86.

Health Canada. (2007). *Canadian community health survey, cycle 2.2, nutrition (2004) – Income related household food security in Canada*. No. 4696. Ottawa, Ontario: Health Canada.

Hinrichs, C. (2003). The practice and politics of food system localization. *Journal of Rural Studies*, *19*(1), 33–45.

Holt-Giménez, E. (Ed.). (2011). *Food movements unite! Strategies to transform our food systems*. Oakland, CA: Food First Books.

Holt-Giménez, E., Bunch, R., Vasquez, J. I., Wilson, J., Pimbert, M. P., Boukary, B., & Kneen, C. (2010). Linking farmers' movements for advocacy and practice. *Journal of Peasant Studies*, *37*(1), 203–236.

Holt-Giménez, E., & Patel, R. (2009). *Food rebellions! Crisis and the hunger for justice*. Oakland, CA: Food First Books.

Holt-Giménez, E., & Shattuck, A. (2011). Food crisis, food regimes and food movements: Rumblings of reform or tides of transformation? *Journal of Peasant Studies*, *38*(1), 109–144.

Howard, A. (1940). *An agricultural testament*. London: Oxford University Press.

Jayaraman, S. (2012). *Behind the kitchen door*. Ithaca: Cornel University Press.

Kirbyson, A. (2005). *Recipes for success: A celebration of food security work in Canada*. Chicago, IL: Independent Publishers Group.

Koc, M. (2006). An exciting new organization. The Canadian Association for Food Studies. *Newsletter*, 1(1), p. 1.

Koc, M. (2011). Hyphenated alternatives: Discourses of food security. Paper presented at the Joint Annual Meetings of the Agriculture Food and Human Values Society, Association for the Study of Food and Society and Society for Anthropology of Food and Nutrition. The University of Montana, Missoula, Montana, USA. June 10, 2011.

Koc, M., & Macrae, R. (Eds.). (2001). *Working together: Civil Society working for food security in Canada*. Toronto: Media Studies Working Group.

Lang, T., & Heasman, M. (2004). *Food wars: The global battle for mouths, minds and markets*. Sterling, VA: Earthscan.

Lang, T., & Rayner, G. (Eds.). (2002). Why health is the key to the future of food and farming: A report on the future of farming and food. Policy Commission on the Future of Farming and Food. Retrieved from http://www.agobservatory.org/library.cfm?refID=30300. Accessed on April 20, 2014.

Lappé, F. M. (1971). *Diet for a small planet*. New York, NY: Random House.

Levkoe, C. (2011). Towards a transformative food politics. *Local Environment*, *16*(7), 687–705.

Levkoe, C. (2014). The food movement in Canada: A Social movement network perspective. *The Journal of Peasant Studies*, *41*(3), 1–19.

Levkoe, C., & Wakefield, S. (2014). Understanding contemporary networks of environmental and social change: Complex assemblages within Canada's "Food Movement". *Environmental Politics*, *23*(2), 302–320.

McAdam, D., McCarthy, J., & Zald, M. (1996). *Comparative perspectives on social movements: Political opportunities, mobilizing structures, and cultural framings*. Cambridge: University Press.

McMichael, P. (2007). Feeding the world: Agriculture, development and ecology. In L. Panich & C. Leys (Eds.), *The socialist register: Coming to terms with nature* (pp. 170–194). London: Merlin Press.

McMichael, P. (2009). A food regime genealogy. *Journal of Peasant Studies*, *36*(1), 139–169.

Mollison, B., & Slay, R. M. (1994). *Introduction to permaculture*. Tyalgum, New South Wales: Tagari Publications.

Mollison, B. D. H., & Holmgrin, D. (1978). *Permaculture 1: A perennial agricultural system for human settlement*. Melbourne: Corgi.

Nestle, M. (2006). *What to eat*. New York, NY: North Point Press.

Nyéleéni. (2007). Declaration of Nyeéleéni. Retrieved from http://www.nyeleni.org/spip.php?article290. Accessed on April 20, 2014.

Oxfam. (2013). Behind the brands. 166 Oxfam Briefing Paper. Retrieved from http://www.oxfam.org. Accessed on April 20, 2014.

Palassio, C., & Wilcox, A. (Eds.). (2009). *The edible city: Toronto's food from farm to fork*. Toronto: Coach House Books.

Pascal, A., Murphy, S., Porter, B., Desmarais, A., Schiff, R. et al. (2011). United Nations special rapporteur on the right to food mission to Canada: Joint civil society submission. Ottawa, Ontario: Report prepared by request of Olivier DeSchutter, United Nations Special Rapporteur on the Right to Food.

Patel, R. (2007). *Stuffed and starved: Markets, power and the hidden battle for the world's food system*. Toronto: Harper Collins.

Patel, R. (2009). Food sovereignty. *Journal of Peasant Studies*, *36*(3), 663–706.

Patel, R. (2013). The long green revolution. *Journal of Peasant Studies*, *40*(1), 1–63.

People's Food Commission. (1980). *The land of milk and money: The national report of the people's food commission*. Toronto: Between the Lines.

Pollan, M. (2006). *The omnivore's dilemma: A natural history of four meals*. New York, NY: Penguin Press.

Pimbert, M. (2009). *Towards food sovereignty: Reclaiming autonomous food systems*. London: CAFS, IIED and RCC.

Qualman, D. (2011). Advancing agriculture by destroying farms? The state of agriculture in Canada. In H. Wittman, A. Desmarais, & N. Wiebe (Eds.), *Food sovereignty in Canada: Creating just and sustainable food systems* (pp. 20–42). Halifax: Fernwood.

Rodale, J. I. (1945). *Paydirt: Farming & gardening with composts*. New York, NY: Devin-Adair Co.

Rosset, P. (2008). Food sovereignty and the contemporary food crisis. *Development*, *51*(4), 460–463.

Sachs, C., Allen, P., Terman, A. R., Hayden, J., & Hatcher, C. (2014). Front and back of the house: Socio-spatial inequalities in food work. *Agriculture and Human Values*, *31*(1), 3–17.

Schiff, R. (2006). Engaging cross-sectoral communication for development of sustainable food systems. In W. L. Filho (Ed.), *Innovation, education and communication for sustainable development*. Frankfurt: Peter Lang Scientific Publishers.

Schiff, R. (2007). *Food policy councils: An examination of organisational process, structure, and contribution to alternative food movements*. Ph.D. dissertation, Institute for Sustainability and Technology Policy, Murdoch University. Retrieved from http://researchrepository.murdoch.edu.au/293/

Schiff, R. (2008). The role of food policy councils in developing sustainable food systems. *Journal of Hunger and Environmental Nutrition*, *3*(2–3), 206–228.

Schlosser, E. (2001). *Fast food nation: The dark side of the all American meal*. New York, NY: Houghton Mifflin Company.

Schumacher, E. F. (1973). *Small is beautiful*. London: Blond & Briggs.

Shiva, V. (2009). Women and the gendered politics of food. *Philosophical Topics*, *37*(2), 1–32.

Shreck, A. (2008). Resistance, redistribution and power in the fair trade banana industry. In W. Wright & G. Middendorf (Eds.), *The fight over food: Producers, consumers and activists challenge the global food system* (pp. 121–144). University Park, PA: Pennsylvania State University Press.

Smith, A., & Mackinnon, J. B. (2007). *The 100-mile diet: A year of local eating*. New York, NY: Random House.

Snow, D., Rochford, E. B., Jr., Worden, S. K., & Benford, R. D. (1986). Frame alignment processes, micromobilization, and movement participation. *American Sociological Review*, *51*, 464–481.

Stevenson, G. W., Ruhf, K., Lezberg, S., & Clancy, K. (2008). Warrior, builder, and weaver work: Strategies for changing the food system. In C. Hinriches & T. Lyson (Eds.), *Remaking the north American food system* (pp. 33–62). Lincoln, NE: University of Nebraska Press.

UFCW and AWA (United Food and Commercial Workers and Agricultural Workers Alliance). (2011). Status of Migrant Farm Workers in Canada. Retrieved from www.ufcw.ca. Accessed on April 20, 2014.

Weis, T. (2007). *The global food economy: The battle for the future of farming*. Halifax: Fernwood.

Weis, T. (2013). *The ecological hoofprint: The global burdon of industrial livestock*. New York, NY: Zed Books.

Windfuhr, M., & Jonsen, J. (2005). *Food sovereignty: Towards democracy in localized food systems*. Rugby: ITDG Publishing.

Winter, M. (2003). Embeddedness, The new food economy and defensive localism. *Journal of Rural Studies*, *19*(1), 23–32.

Wittman, H., Desmarais, A., & Wiebe, N. (Eds.). (2010). *Food sovereignty: Reconnecting food, nature and community*. Halifax: Fernwood.

Wittman, H., Desmarais, A., & Wiebe, N. (Eds.). (2011). *Food sovereignty in Canada: Creating just and sustainable food systems*. Halifax: Fernwood.

World Commission on Environment and Development. (1983). *Our common future (the Brundtland report), in world commission on environment and development*. Oxford: Oxford University Press.

IMPLICATIONS OF THE ENVIRONMENTAL JUSTICE MOVEMENT ON REDISTRIBUTIVE URBAN POLITICS: AN EXAMPLE FROM MEGACITY DHAKA, BANGLADESH

Saleh Ahmed

ABSTRACT

Many megacities in the Global South have developed without any systematically regularized modernization processes. Therefore, urbanization is accompanied by the inappropriate level of economic and social opportunities. This macro-structural constraint of urbanization contributes to frame the patterns and features of many location-based poverty and inequality.

Dhaka followed the similar trajectory of development without creating substantial opportunities for its underclass, poor, and marginalized citizens. This unequal pattern of development influences the disproportionate environmental burdens of the marginalized people, who have little or no

Occupy the Earth: Global Environmental Movements
Advances in Sustainability and Environmental Justice, Volume 15, 255–274

ISSN: 2051-5030/doi:10.1108/S2051-503020140000015010

voice in the urban decision-making process. People experience unequal environmental inequality through their differential exposures to poor housing and/or living conditions, along with a lack of access to safe drinking water, sewage, adequate waste disposal systems, and so on.

In this context of urban injustice, this environmental justice movement (EJM) is an evidence of demanding basic rights among the unprivileged citizens. To some extent, this is a response to the state's structural failure to provide urban environmental services (and social justice) to majority of its poor citizens.

Evidence suggests that in context of fragile democracy, an EJM can generate opportunities for progressive urban politics as well as contribute to redistributive development opportunities for its poor and marginalized citizens. Most megacities in the Global South inherit urban governance from their colonial heritages, which were often criticized for its citizen-detached and top-down nature.

Dhaka is not an exception in this context. In the present socio-political condition, even though the environmental movement is not very welcomed by the government – more particularly from the ruling political party – it can have tremendous implications on redistributive urban politics. This chapter focuses on Dhaka, which is currently one of the world's largest megacities in the Global South.

This chapter highlights the local patterns and process of the unequal environmental burdens and the subsequent mobilization, demanding environmental justice in that process, as well as how that movement contributes to redistributive urban politics. The discussions of this chapter also have the implications on public policy discourse focusing on environmental rights, movements, and citizens-centered politics in the megacities of the Global South, with renewed relationships between the state and its citizens.

Keywords: Dhaka; Environmental Justice Movement; Global South; Megacity; Redistributive Urban Politics

INTRODUCTION

The current political and economic model in many countries of the Global South is centered on neoliberal policy. Apart from offering several social and economic opportunities, this policy is often criticized for making democratic spaces smaller, with unsuccessful attempts at cultural hegemony being under the façade of "modernization" or "globalization." Now unequal distributions of wealth and opportunity are not uncommon in many societies. Citizens of the Global South are now experiencing a "race to the bottom" by generating mass poverty, inequality, and marginalization. This indicates that the citizens are poorer today than before (Red de Solidaridad con Mexico, 2012).

With some mixed economic, social, and political experiences of neoliberalism, Bangladesh received moderate exposure to democratic transitions, particularly since 1991, after the fall of a decade-long military regime (Ahmed, 2012). Since its birth as a sovereign country in 1971, the country has experienced different forms of governance, ranging from military authoritative to one-party dominated democracy (Sobhan, 2007).

Even though with this transition, the citizens of Bangladesh are still largely poor, marginalized, and disadvantaged, and deprived from the meaningful contributions of democracy. Citizens usually get rare chances of having a voice in decision-making processes (Ahmed, 2012). To many citizens, the only impact of democracy is to give them the opportunities to vote in every five years. Although democracy should create opportunities for its citizens, it has virtually failed to ensure a fair chance for all people in Bangladesh to develop a thriving voice as well as livelihood opportunities. However, it is not the issue with the core principles of democracy; it is rather more how democracy is perceived and being translated into local contexts. This unequal access to power and distribution of resources for the poor and marginalized causes unequal political representation (Ahmed, 2012). In this context, democracy has very limited capacity to fulfill people's desires for better lives.

Currently, 28% of Dhaka city-dwellers live below the poverty line, and among them, approximately 47.5% of those are living in slums or in poor living conditions (The World Bank, 2014). Many of these citizens have minimal to no political representation in the national or local decision-making process due to their marginalized situation in society. However, this does not necessarily indicate that the poor are apathetic to political issues or not capable of political engagement. Evidence shows that poor and marginalized citizens are motivated to engage in urban politics with a desire to have direct access to decision-making processes, particularly with the decisions that might affect their life and livelihood opportunities (Hossain, 2005). In many parts of the Global South, demands for social, economic, and political rights among poor and marginalized citizens are increasing, mostly due to the structural failure of the states such as how the citizens are being governed or whom the state includes in the decision-making process. This rising voice for rights often translates to citizens' movements for environmental justice.

This chapter is based on a field research in megacity Dhaka during the summer of 2011 A.D. The findings highlight issues related to poor and marginalized citizens' access to urban open space as means for recreation and shelter local urban crises, such as urban floods. This chapter also illustrates the process of how marginalized citizens organize against the state to ensure their environmental rights and justice, and the process can have implications on public policy outcomes. Even though this chapter focuses on Dhaka, the learning is expected to transferable to other cities in the Global South with similar social, economic, and political situations.

URBAN CHALLENGES IN DHAKA

Dhaka has emerged as one of the fastest growing megacities in the world in recent years (United Nations Population Fund, 2007). It had a manageable population of 2.2 million in 1975 and that became 12.3 million by 2000. The growth rate of this urbanizing population during 1974–2000 was, on average, 6.9% annually (United Nations, 1998). There are very few cities in the world that have ever experienced such phenomenal population growth during this period. It is predicted that the growth rate of urbanizing people in Dhaka will also continue to remain high even in the coming years. Even though the recent population growth in Dhaka slightly slowed, it has increased by 2.53%, and will reach a total population of 16.6 million by 2015 (United Nations Human Settlements Programme, 2013). Currently, Dhaka is one of the top 10 most populous cities of the world (United Nations Human Settlements Programme, 2013; United Nations Population Fund, 2007).

In Bangladesh, Dhaka has always been a city of attraction for those seeking a better life, particularly among the unprivileged citizens, since there are few comparable social, economic, and political services in other parts of the country. It is not only the capital of Bangladesh, but is also the most dominant cultural, political, and economic center of the country. Apart from this national importance, Dhaka suffers from its poor "governance" performance, in that the citizens often have little or no access to the decision-making process (Ahmed, 2012). Currently, it is a megacity that can be criticized for top-down planning interventions with little to no citizen engagement. Now it is divided into two legislative parts: Dhaka North and Dhaka South. The argument of division of city (and city services) was that the size and extent of urbanization are now in unmanageable shape due to its large population and area. Therefore, to increase urban services, it was decided to divide the city. Without any political or citizen consensus, the Awami League, the present ruling party, signed into law the separation of Dhaka in 2011. Citizens were surprised with the decision, since such a big decision was taken without any public discussion or consultation. It shows one of the examples that how citizens, particularly the poor in Dhaka, are not part of decision-making, which can likely affect their lives. World Governance Indicators tell the similar story (Table 1).

Apart from governance challenges, like many other megacities in the Global South, urbanization in Dhaka comes with dual consequences: (a) it provides opportunities for millions of people and (b) it puts pressure on local and regional resource bases, which might end up with further human suffering and losses (United Nations Department of Economic and Social Affairs, 2013).

The scale and velocity of urbanization in megacities like Dhaka completely dwarf those of Victorian Europe. London in 1910 was seven times larger than it had been in 1800. On the contrary, cities like Dhaka, Kinshasa, or Lagos are today approximately 40 times larger than they were in the 1950s (Davis, 2006). The most striking result will be the increase of new megacities with populations in excess of 8 million, and even more spectacularly, hyper cities experiencing hyperurbanization with more than 20 million inhabitants, which was the estimated urban population of the world at the time of the French Revolution (Davis, 2006). In 1995, only Tokyo had incontestably reached that threshold population. By 2025, Asia alone could have 10 or 11 megacities that large (Davis, 2004).

Table 1. World Governance Indicators – Bangladesh.

Indicator	Year	Governance Score (−2.5 to +2.5)	Percentile Rank (0–100)	Standard Error
Voice and accountability	2002	−0.52	33.65	0.17
	2007	−0.54	31.25	0.12
	2012	−0.42	34.12	0.11
Government effectiveness	2002	−0.70	27.32	0.16
	2007	−0.68	27.67	0.18
	2012	−0.83	22.49	0.19

Source: World Governance Indicators (2014).

The major challenge will be how these hyper-megacities will react to the changing socio-economic and political scenarios within their massive populations. The present trend of urban services in megacities like Dhaka has already demonstrated critical structural failure, ends up often with urban chaos, and social and political unrest (Ahmed, 2012). The emerging cities of the Global South will likely follow in the industrializing footsteps of Manchester, Berlin, and Chicago, if there is no major social, technological, or political change. These megacities are more like Victorian Dublin, which suffered more from the problems of de-industrialization than industrialization between 1800 and 1850 (Prunty, 1998). Here, urban population growth with stagnant or negative urban economic growth is the extreme face of "over-urbanization."

THEORIZING ENVIRONMENTAL JUSTICE MOVEMENT

Environmental justice research tends to focus on the distribution of hazardous facilities in poor, vulnerable communities (Pellow et al., 2001), and on providing the outcomes of unequal environmental exposures without having an in-depth understanding on the process of how it occurred (Weinberg, 1998).

Wilson (1996) argues, in a study focusing on US context, environmental injustice has a racial dimension, and usually with some exceptions, the people of color tend to have limited economic and political power to influence the decision-making process that potentially impacts their lives. Therefore, poor marginalized people of color are most likely vulnerable to environmental injustice or inequalities. Many people often confuse environmentalism or EJM to be about a "Green Party." However, it is more about the green ideas with their evolution, meaning, and importance to the society and citizens (Paehlke, 1989). It covers a range of issues involving the political and social dimensions of daily life.

EJM comes up with environmentalism, which is a distinct ideology in its unwillingness to maximize economic benefits for the dominant systems or structures (Paehlke, 1989). It has minimal appeal by the way of any personal economic gains; it concerns more for common goals, and goods with some lasting benefits. The success of EJM largely depends on a reasonable level of democracy and democratic rights for the majority (Paehlke, 1989). A society with a fragile democracy provides limited public space for any type of movement, and that's why the success of EJM is also limited. But one can argue that even though EJM might not make any large structural change or shift in fragile democracy, it can at least offer citizens with an opportunity to raise their needs, values, or desires and can make the first step toward redistributive urban politics.

UNEQUAL ENVIRONMENTAL BURDEN IN DHAKA

The spatial concentration of pollution is often determined by those who hold social, economic, and political power in the society. Research consistently indicates that those from the lower social classes will be more exposed to environmental risks or burdens (Ringquist, 2000). This environmental injustice can be defined as any kind of environmental policy and/or practice that negatively affects those of a lower social, economic, and political status (Bullard, 1994). Environmental injustice has been expanded to include members of other disadvantaged communities with heightened environmental risk based on race, class, gender, education, and political power (Leon-Guerrero, 2014). The fragmented nature of Dhaka's social and economic opportunities usually contributes to frame the patterns and features of location-based poverty and environmental injustice (Ahmed, 2013a).

This unequal pattern of development gradually worsens the disproportionate environmental burdens of the marginalized citizens, who have little to no voice in the political decision-making process (Ahmed, 2013a). Citizens experience this through their different (unequal) exposures to poor housing and/or living conditions along with a lack of access to safe drinking water, adequate waste disposal systems, and so on. In most societies, these environmental conditions are generally more prevalent in the areas where the poor, marginalized citizens live (Pellow, 2007), and therefore, it is often argued that poverty is highly related to unequal environmental exposures. Environmental injustice is not just about a correlation between hazards and populations; it is also about the power of citizens to challenge and reverse this injustice. Class dynamics and political power play decisive roles in shaping the context of local patterns of environmental injustice.

Unequal environmental burdens directly correlate with stratified social classes and unequal political relationships in the megacities of the Global South. Impoverished citizens and/or minorities are the major victims of environmental burdens, and in most cases, they have fewer economic opportunities along with limited political voice or participation in the decision-making process.

In Dhaka, poor, marginalized citizens are involved in different types of informal business activities and usually live in unhealthy conditions. Poor environmental conditions are a predominant feature in these types of neighborhoods (Chowdhury & Amin, 2006). Therefore, with some exceptions, poverty is ultimately blamed for these unequal environmental burdens (United Nations Human Settlements Program, 2003).

THE OLD DHAKA CASE STUDY

Dhaka originated from a few small settlements in the 7th century CE. Initially, it was a small local trading center on the Buriganga river banks (Wikipedia, 2014). During that time, it was ruled by the Buddhist kingdom of Kamarupa before passing to the control of the Sena dynasty in the 9th century CE (Dhaka City Corporation, 2006). Modern Dhaka, which is often termed as New Dhaka, originated from Old Dhaka, which still belongs to city's glorious past.

Even with numerous social, economic, and political transformations, Old Dhaka retained its history and traditional way of life and business. In different times, it refused embracing the urban modernity like New Dhaka did. However, with the dominance of modernity and present day urbanization, Old Dhaka lost its dominance over arts, culture, literature, and economics. Now Old Dhaka is perceived to be the decaying and poor section of the city. Nevertheless, there are still some areas with locational or family dominance over cultural activities and the economy. These areas are reminiscent of its glorious past. Overall, it is clearly differentiable the cultural and economic differences between Old and New Dhaka.

Due to its deteriorating environment and poor living conditions in Old Dhaka, many of the local elites have moved their places of residence to New Dhaka, particularly after 1971, when Dhaka became the capital of a sovereign country for the first time in its history (Afsar, 2003). Since Dhaka is divided now into two legislative parts, Old Dhaka falls within Dhaka South (Ahmed, 2013a). But like any other larger community, the city boundary is not very clear among the citizens. Community is often fluid and continuous, so for people it is very difficult to identify where Old Dhaka or New Dhaka starts or ends. However, the city authority has some boundaries (on maps), but the major criticisms could be that there is still confusion based on the issues related to urban services, such as how to tackle urban service delivery if a citizen's residence falls in Dhaka South, but their children's schools are in Dhaka North.

The majority of city's poor working class resides in Old Dhaka due to lower living expenses (Ahmed, 2013a), who actively engage in formal and informal occupations near their residences or in other places in New Dhaka. In this case, access to work is highly important. This works is usually located near their place of residence, and informal in nature.

However, if Dhaka is one of the densest megacities in the world, then Old Dhaka may have the densest population in contemporary human history (Cox, 2012), which is approximately 150,000 people/square kilometer (Standing, 2009). Even though there are controversies about the exact number, it is widely accepted that the population density in Old Dhaka has already exceeded the tipping point for human risk and vulnerability (Alam & Rabbani, 2007).

Context

As already mentioned, the discussion of this chapter is based on an empirical evidence from Old Dhaka, particularly a small neighborhood named Kayethtuli. During the field research, there were attempts to find the exact size and population of Kayethtuli from some reliable sources. Unfortunately, none of the previous research focused on the local context has accurately reported the population of Kayethtuli. Without any doubt, Kayethtuli is highly densed with limited to no provision of urban services. These service deficiencies (e.g., lack of pure drinking water, medical services, and waste management facilities) have adverse impacts on local environment and livability. In Kayethtuli, there is only one public open space, locally known as Kayethtuli Water Pump Playground (KWPP). Apart from the essential water pump, the open space provides a play area for children. There are also some trees, which contribute to improving the poor air quality in Kayethtuli. In addition, this open space provides a secure location in cases of emergency. In recent years, urban disasters (such as fires, waterlogging, or urban floods during the rainy season) have increased in frequency (Ahmed, 2013a). Furthermore, as Dhaka falls in an earthquake-prone zone, there is a higher perceived risk for earthquakes. Citizens in Kayethtuli know that in an emergency situation, this open space is meant to be a shelter and meeting place. Therefore, the KWPP not only has recreational value, but can be viewed as an evacuation zone during any natural or man-made crises.

The KWPP is itself owned by the government, but the local citizens have been using and maintaining it over the last several years. In that way, these citizens developed an informal communal ownership of the KWPP.

Increased Tension

Water Supply & Sewerage Authority (WASA), which is one of the major developmental representations of the Government of Bangladesh in Dhaka city, initiated to build high-rise buildings in the KWPP for its middle income public officials in 2011. Since it would further marginalize the local people by decreasing local open space, people protested. As mentioned earlier, to the local people, KWPP was not only an open space, but also a place for shelter during any emergency. Local businessmen also mentioned that this park has positive impacts on their business, since the visitors often spend some money for purchasing snacks, and so on while visiting the park. This is one of the very few open public spaces in Old Dhaka. Many local people also mentioned that this is a major meeting place for the community members, which helps them to have social interactions. Therefore, it is clearly understandable that the KWPP is significant from economic, social, and environmental perspectives (Ahmed, 2013a).

However, on the other hand, it is really true that Dhaka suffers from an acute shortage of low-cost housing. WASA's proposed housing project by no means contributed to lessen the urban housing problem. Rather, it had potentials to generate many urban issues, such as the losing public space, including the children's playground, and decreasing opportunities for social interactions.

There was also some context for this initiative. Water shortage, particularly during the summer months, is a major issue among the Old Dhaka residents. To solve this problem, WASA built a water pump at the corner of this open space. People did not show any disagreement with this, since it aimed to solve one of the major local problems. However, even for doing this, WASA or any other government agency did not consult with the local people, whether or how much the local people wanted to have a water pump in this playground. WASA initiated this solution based on their assumed intensity of the problem. Even though it contributed positively to the community's needs, no participatory mechanism was visible in the process.

For building the housing complex in the KWPP, WASA did the same. They did not consult with the local people, whether they really wanted this or not. Local people were, therefore in the dark and had no idea how much this housing complex could positively contribute to their community welfare. Apart from this, they had some fears that they would completely lose the KWPP with increased social problems, such as increased inequality, lack of opportunities for social interactions, and urban environmental risks. There was also another concern: if the housing complex was built then whether the water pump that was then located in the KWPP would perform at full capacity and force the citizens to accept sub-standard services.

When the local community discovered the government's initiative regarding the construction of the high-rise building in KWPP, they organized themselves against further construction in this open space. Officially, WASA initiated this high-rise building project on April 26, 2010, but they were forced to stop the project due to local protests. After that, the government used unfair tactics to stop this protest by doing different kinds of local political manipulation, oppresion, and fear. In the local political context, any form of social (or environmental) movement is treated by the ruling party/parties as a potential threat to their power. The government tried to start the project again on April 18, 2011. However, this time they also faced similar protests and were not successful. They had to postpone the project indefinitely due to local protests and disagreements with the community (Ahmed, 2013a).

Street Politics to Organized Demonstration

The people realized they needed to organize themselves, not only for the good of their own community, but also that their voice could be seriously considered by the government, media, and the people on the outside. On May 14, 2011, the *Kayethtuli Maat o Panir Pump Rokkha Committee* (Kayethtuli Field and Water Pump Protection Committee) organized a large-scale demonstration, initiated by the local people and was endorsed by social and political activists in the area. Even though only a fraction of local citizens (almost 500 people) actively participated in the demonstration, it was widely supported by the other residents of Kayethtuli. National environmental activist group *Poribesh Bachao Andolon* (POBA, translated, Save the Environment Movement) was also actively engaged in this three-hour long demonstration. Throughout the protest, citizens demonstrated their voice regarding their demands for environmental justice (Ahmed, 2013a). They forcefully stated their community's demands to the government to recognize their needs and respond accordingly.

Members of the local community participated in or supported the demonstration despite the fact that they could be directly or indirectly affected by this high-rise building project. This demonstration got substantial media attention, since locals successfully connected their voices and demands with the national environmental movement organization POBA as well as the local public authorities, who represent the government in the community. Local intellectuals and social elites were actively engaged in this demonstration. Altogether, the community successfully expressed its concerns for this locally unwanted and unwelcomed development initiative.

Rising tensions between the community and the state were clearly visible, while at the same time, collective participation of local marginalized citizens on a common issue indicated how a crisis could generate the notion of community and the sense of belonging (Fig. 1). This notion of community and the sense of belonging not only helped the local community to confront any form of environmental injustice, but also helped to develop resiliency against any other man-made or natural disasters.

After the demonstration on May 14, 2011, the local community organized an informal meeting with all the involved interest groups, for example, local activists, POBA, and local social, economic, and political elites. They discussed their future agenda on this local environmental issue and government intiative and how they could interact or negotiate with the government if the same situation were to occur in the future. They agreed that

Fig. 1. Demonstration by the Kayethtuli Mat o Panir Pump Rokkha Committee.

if they wanted to prevent future government initiatives on building this high-rise complex, they needed to continue these mobilizations and demonstrations as part of applying social pressures on the government (Ahmed, 2013a).

They also tried to give a formal shape of their movement by forming a committee with 11 executive members and more than 100 general members so that they could properly incorporate community views, values, and needs. In that meeting, they decided that they would plant greenery surrounding the open space and would organize different types of community activities and festivals. Through their initatives, the local community tried to give the space more social and environmental value and meaning than before (Ahmed, 2013a).

PROCESS WITH REDISTRIBUTIVE URBAN POLITICS

Although marginalized citizens often live in poor conditions, scholars of urban politics have found that many underprivileged classes have developed effective methods to fight against environmental injustice (Harper & Rajan, 2004). However, the citizens, particularly the poor, are not consistently involved in the decision-making process. The state's occasional failure to ensure justice, rights, and/or inclusion of citizens in the development process, plays a decisive role in enhancing social capital, which can contribute to address and improve local environmental challenges.

Evidence across the world shows that the continuous denial of citizens' incorporation in the decision-making process actually helps the local citizens to organize themselves with the possible support of locally available resources (Ahmed, 2013a; Bayat, 1997). The media also contributes broader societal attention, which often makes further stimulus and/or positive spillover impacts in the society. Usually, movements help citizens raise their values and demands with a possible impact on public policy.

In the recent past, citizens in Bangladesh have greatly mobilized by a series of poor governmental decisions, such as building multistoried markets or housing complexes in urban open spaces or parks, building some large structures by destroying agricultural lands, and pursuing environmentally hazardous resource extraction policies. All these initiatives contributed to increasing communal tensions. Dhaka's recent EJMs were largely initiated by the poor and middle-class, with some support from environmental activist groups and media (Ahmed, 2013a). Often, these movements are very much location-, time-, or issue-specific. Protests often focus on a single neighborhood issue or cascading, domino-like problems. These movements can be perceived as reactive, short-sighted, and short-lived without having any lasting societal impacts or having impacts on national public policy. However, despite these criticisms, EJMs can generate opportunities for the redistribution of social and environmental opportunities, goals, and goods in the local context and can improve community well-being, as well as with an increased sense of community and social belonging.

It is important to remember that environment and environmental conditions are closely tied to the lives and livelihoods of millions of poor citizens in Dhaka. Even though in many cases, governmental agencies and individual communities surmise that food and shelter are the most important aspects of daily life, they neglect to take account of poor environmental quality and unequal environmental burdens as the major factors in a positive living environment. Environmental quality has a large impact on human health, income, and other associated aspects of life. Many poor neighborhoods in Bangladesh are aware of their environmental choices as they are exposed to the challenges without having alternative choices due to their poor economic and political capacities.

The concept of social power is linked to the way one person can influence the actions of others. A classical definition of such power was given by Max Weber, and where he mentioned that power is the opportunity of a person or a group of persons to realize their own interest in a communal action, even against the resistance of others who are participating in some degree in the action (Gerth & Mills, 1946). Through the lens of Weberian argument, we can see that the people in Kayethtuli are in the lowest strata in this power dynamics, who are poor and at the same time have limited control over the decision-making process that can impact their lives.

The environmental movement in Kayethtuli has gone through a gradual transformation from spontaneous street politics to more organized civic engagement on the issue of local environmental justice. As a consequence, the local community experienced the emergence of (social) resources and innovations, such as an enhanced sense of community, along with social belonging, which ultimately contributed to the renewed urban politics and more redistributive opportunities for the local citizens. In the Kayethtuli environmental movement, local marginalized people were the key agents of protests and demonstrations, with support from the national environmental organization, POBA. These social agents were largely from the disparate clusters of the urban unemployed, underemployed, or other marginalized groups. The process demonstrated new forms of civic innovations, where citizens organized themselves, opted for common goals, demanded their rights, and opposed environmental injustice collectively.

The Kayethtuli environmental movement frames a new dimension of urban politics among the marginalized citizens, who usually have minimal or no voice in the urban decision-making process. The nature of urban politics in the developed society is substantially different than the nature of urban politics in the Global South, such as in places like Kayethtuli in megacity Dhaka. The marginalized people neither have access to the decision-making process, nor have any voice that can represent their desires or demands. The environmental movement was an option in Kayethtuli to enhance citizens' engagements with urban decision making and to make the process more inclusive. More specifically, it contributed certain aspects of urban life.

Political Powers

Due to the fragile nature of the local democracy and political opportunities, poor, marginalized citizens get minimal or no representation in the urban decision-making process (Ahmed, 2012). The entire process of policy and planning has little to offer to the Kayethtuli people, majority of who represent the lower socioeconomic strata. In this context, the redistribution of politics and power is minimal.

Kayethtuli experience suggests that the failure of bureaucracies and state institutions to deliver services for the poor marginalized citizens contributed to initiate an EJM (Ahmed, 2013a). Initially, local citizens demanded for their basic rights, and government disagreed, mostly for their "pro-market" but citizen-detached attitude. As a government, it is easier to confront the poor citizens as disadvantaged individuals who demand better (often very basic) social services, than as determined revolutionaries who demand social justice (Warren, 1973). However, local citizens transitioned from spontaneous street politics to organized environmental movements, which helped them to be strong enough to negotiate with the government for their rights, in a way that the government could not just refuse.

Despite the structural constraints or challenges of local politics and urbanization, the EJM in Kayethtuli offered a new avenue for redistribution of public goods and power. This experience showed how the local EJM could contribute to redistributive urban politics in areas hugely constrained by social, political, and economic limitations. This renewed political culture also helped to develop local resources, such as social capital and social innovations among the marginalized citizens. Initially, the local community was not organized to generate social movements for their environmental justice. But through community participation, they organized themselves, developed their sense of community and social belonging, which ultimately contributed to raise their voices and demands for environmental rights, and increased political voices and power along with redistributive urban opportunities.

Social Opportunities

Renewed political power and opportunities also allow for enhancement of social opportunities (Ahmed, 2011). In Kayethtuli, through the EJM, citizens developed their sense of community and social belonging. Local citizens were largely poor and suffered from many urban ills. Increased interaction helped them to build social networks, which contributed positively to their increased social security in times of any form of crisis. The local community gained better access to education and health facilities as well as their access to public authorities for their community needs.

Enhanced Transparency

Lack of transparency is one of the major developmental challenges in the Global South, since in most cases, they either do not practice proper democratic norms, even though they are labeled as a democratic society, or they do not have a democracy at all (Ahmed, 2011). The experiences from Kayethtuli give an impression that the local marginalized citizens, by demanding their environmental rights, formed a new socio-political entity in the larger framework of urban politics. Equalized power relationships between citizens and government also contributed to enhanced social opportunities as well as increased transparencies in the process. Transparency actually helps people to get informed about the process that might impact their livelihoods in the future. In a fragile society, like in Dhaka, transparency needs to be guaranteed if the people care for lasting development in the local community (Ahmed, 2011). Increased but fair demands for environmental rights in Kayethtuli contribute to enhanced transparency in the governance mechanism, which can influence positively the redistributive urban politics and associated opportunities.

Political powers, social opportunities, and enhanced transparency are closely interdependent and critical in a society for lasting development. An absence of any one of these makes it difficult to attain meaningful development and can jeopardize the entire framework of redistribution and justice. The Kayethtuli example shows that renewed political powers, social opportunities, and enhanced transparency not only contribute to environmental justice, but also eventually to the local sustainability, which is the most crucial component of development in any locality.

SUMMARY

An environmental movement can create meaningful changes and social opportunities in a society that suffers from a fragile democracy. The Kayethtuli example shows that it helps both the citizens and the state to negotiate locally desired choices for their livelihoods and sustainability to prevent any urban tyranny. It also helps the state to adopt more inclusive and distributive governance and the citizens to gain more access to the decision-making process. When the citizens have the access to the decision-making process, they can influence the public policy toward their desired trajectory of life, opportunity, and growth. In a way, the Kayethtuli environmental movement contributed to both the state and the citizens to improve their capabilities for local sustainability.

The future of Dhaka involves complex interdependencies among multiple factors, including increased population pressure, stress from climate factors, and shortages of basic urban service delivery, along with possible limitations due to poor government capabilities (Ahmed, 2013b). Without the citizens' involvement in the decision-making process, the scenario can get even worse. Achieving local sustainability, which is crucial for millions of people in the megacity, can only be a feasible urban development target if the city and city's government can meaningfully incorporate the citizens in planning, management, and implementation phases of development initiatives.

The present trend of urban scenarios in Dhaka does not indicate a promising path or future. The city experiences colonial influence in urban governance, which is incompatible for the citizens in this 21st century, post-colonial capitalist era. This can potentially hamper citizens' lives, economy, and society. In this regard, the key is to let citizens make their decisions about their own lives, so that they can choose the kind of life they value.

The Kayethtuli environmental movement shows how citizens have different views of their communities and lifestyles than the government has. In this context, public policy failed to address citizens' needs, desires, or values, which caused a confrontation between the citizens and the state. In the end, however, the government was forced, to some extent, to address citizens' demands in Kayethtuli. This local experience demonstrates social innovation, such as an increased sense of community and community rights through citizens' mobilization.

Generally, an EJM can generate opportunities for redistributive urban politics or experiences. Sometimes, environmental movements in a fragile society like Dhaka usually experience obstacles or oppressions from the government or elite class. The poor are marginalized; the current urban challenges, such as increased population, climate impacts, or governance challenges will make them even more marginalized (Ahmed, 2012, 2013b). Therefore, if there is no continuous social innovation or pressure on the traditional system, the citizens from the lower strata will have little to receive from or offer to the society. No city or society can attain meaningful development if the majority of its citizens are not part of the development process. Non-inclusive, partial development efforts can only initiate further poverty, inequality, social injustice, and/or crime.

Right now, it is not uncommon to experience citizens' initiatives in pursuit of producing common interests through the redistribution of goods, opportunities, power, and so on. The Kayethtuli environmental movement illustrates similar outcomes, where power and opportunities are elitist and the majority of citizens are deprived of democratic dividends. Traditionally, post-colonial urban development initiatives often neglect the voice of the marginalized, and/or un-represented citizens. On the contrary, neoliberal forces help to form the citizens' movement, which is instrumental for increasing social innovations and social capital.

In this context, with the designated title of world's most densely populated city, if Dhaka cannot attain a renewed and transformative governance framework, then it is likely that the megacity will have no other choice but to be entrapped with poverty, tyranny, and poor economic opportunities. In the end, citizens with systematic social deprivation, intolerance, or decreasing democratic space by the repressive states can cause disaster through drastic human and/or economic loss.

REFERENCES

Afsar, R. A. (2003). Internal migration and the development nexus: The case of Bangladesh. Paper presented at the Regional Conference on Migration, Development and Pro-Poor Policy Choices in Asia, 2003. Retrieved from http://www.eldis.org/vfile/upload/1/document/0903/Dhaka_CP_6.pdf. Accessed on October 3, 2012.

Ahmed, S. (2011). Transparency and redistributive urban politics in Dhaka. Re-public: Re-imagining democracy. Retrieved from http://www.re-public.gr/en/?p=4263. Accessed on January 12, 2010.

Ahmed, S. (2012). The redress of urban governance and marginality in Dhaka, Bangladesh. In C. C. Yeakey (Ed.), *Living on the Boundaries: Urban Marginality in National and International Contexts* (Vol. 8, pp. 511–533). Advances in Education in Diverse Communities: Research, Policy and Praxis. Bingley, UK: Emerald Group Publishing Limited.

Ahmed, S. (2013a). Civic environmentalism and the politics of marginalized people: A case study from megacity Dhaka, Bangladesh. *Environmental Justice*, *6*(2), 56–61.

Ahmed, S. (2013b). Resiliency and adaptation in Dhaka, Bangladesh. In ISSC/UNESCO (Ed.), *World social science report 2013: Changing global environments* (pp. 246–249). Paris: OECD Publishing and UNESCO Publishing.

Alam, M., & Rabbani, M. D. G. (2007). Vulnerabilities and responses to climate change for Dhaka. *Environment & Urbanization, 19*(1), 81–97.

Bayat, A. (1997). Un-civil society: The politics of the "informal people". *Third World Quarterly*, *18*(1), 53–72.

Bullard, R. (1994). The legacy of American apartheid and environmental racism. *St. John's Journal of Legal Commentary*, *9*, 445–474.

Chowdhury, F. J., & Amin, A. T. M. N. (2006). Environmental assessment in slum improvement programs: Some evidence from a study on infrastructure projects in two Dhaka slums. *Environmental Impact Assessment Review*, *26*, 530–552.

Cox, W. (2012). World Urban Areas Population and Density: A 2012 Update. *New Geography*, May 3. Retrieved from http://www.newgeography.com/content/002808-world-urban-areas-population-and-density-a-2012-update. Accessed on June 2.

Davis, M. (2004). Planet of slums: Urban involution and the informal proletariat. *New Left Review*, *24*, 5–34.

Davis, M. (2006). *Planet of slums*. London: Verso.

Dhaka City Corporation. (2006). *Pre-Mughal History of Dhaka*. Retrieved from http://www.dhakacity.org/Page/SiteMap/Link_1/2/List_id_1/22/Subid_1/34/Pre_Mughal_Dhaka. Accessed on March 23, 2014.

Gerth, H. H., & Mills, C. W. (Eds. and trans.). (1946). *From Max Weber: Essays in sociology*. New York, NY: Oxford University Press.

Harper, K., & Rajan, S. R. (2004). *International environmental justice: Building the natural assets of the world's poor*. Working Paper Series 87. Political Economy Research Institute, University of Massachusetts Amherst.

Hossain, S. (2005). The socio-spatial differentials of political behavior of the urban poor in Bangladesh. *Bangladesh e-Journal of Sociology*, *2*, 74–89. Retrieved from http://www.bangladeshsociology.org/BEJS%20-%20Issue%202.2ShahadatHosain.pdf. Accessed on May 12, 2012.

Leon-Guerrero, A. (2014). *Social problems: Community, policy, and social action*. Los Angeles, CA: Sage.

Paehlke, R. C. (1989). *Environmentalism and the future of progressive politics*. New Haven, CT: Yale University Press.

Pellow, D. N. (2007). *Garbage wars: The struggle for environmental justice in Chicago*. Cambridge, MA: The MIT Press.

Pellow, D. N., Weinberg, A., & Schnaiberg, A. (2001). The environmental justice movement: Equitable allocation of the costs and benefits of environmental management outcomes. *Social Justice Research*, *14*(4), 423–439.

Prunty, J. (1998). *Dublin slums, 1800–1925: A study in urban geography*. Dublin: Irish Academic Press.

Red de Solidaridad con Mexico. (2012). *Social and Political Analysis*. Retrieved from http://www.mexicosolidarity.org/about/analysis/en. Accessed on February 22.

Ringquist, E. J. (2000). Environmental justice: Normative concerns and empirical evidence. In N. Vig & M. Kraft (Eds.), *Environmental policy in the 1990s* (4th ed.). Washington, DC: Congressional Quarterly.

Sobhan, R. (2007). *The political economy of malgovernance in Bangladesh* (Vol. 3). Dhaka, Bangladesh: Centre for Policy Dialogue.

Standing, H. (2009). *Dhaka, The Environmental Danger Zone*. Retrieved from http://www.ids.ac.uk/go/news/dhaka-the-environmental-danger-zone. Accessed on February 2, 2014.

The World Bank. (2014). *South Asia population-urban growth: A challenge and an opportunity*. Washington, DC: The World Bank. Retrieved from http://web.worldbank.org/WBSITE/EXTERNAL/COUNTRIES/SOUTHASIAEXT/0,,contentMDK:21393869~pagePK:146736~piPK:146830~theSitePK:223547,00.html#add-resources. Accessed on June 3, 2012.

The Worldwide Governance Indicators. (2014). *The Worldwide Governance Indicators*. Retrieved from www.govindicators.org. Accessed on May 22.

United Nations. (1998). *World urbanization prospects: The 1996 revision*. New York, NY: United Nations.

United Nations Department of Economic and Social Affairs. (2013). *World economic and social survey 2013: Sustainable development challenges*. New York, NY: United Nations.

United Nations Human Settlements Program. (2003). *The challenges of slums: Global report on human settlements*. London: Earthscan Publications Ltd.

United Nations Human Settlements Programme. (2013). *State of cities 2012/2013: Prosperity of cities*. New York, NY: Routledge.

United Nations Population Fund. (2007). *State of world population: Unleashing the potential of urban growth*. New York, NY: UNFPA.

Warren, R. L. (1973). Comprehensive planning and coordination: Some functional aspects. *Social Problems*, *20*(3), 355–364.

Weinberg, A. (1998). The environmental justice debate: A commentary on methodological issues and practical concerns. *Sociological Forum*, *13*, 25–32.

Wikipedia. (2014). *Dhaka City Corporation*. Retrieved from http://en.wikipedia.org/wiki/Dhaka_City_Corporation. Accessed on March 11.

Wilson, W. J. (1996). *When work disappears: The world of the new urban poor*. New York, NY: Vintage Books.

SPIRITUAL ECOLOGY AS AN INTERNATIONAL ENVIRONMENTAL MOVEMENT

Leslie E. Sponsel

ABSTRACT

Spiritual ecology explores the interface of religions and spiritualities on the one hand, and environments, ecologies, and environmentalisms on the other. As an international environmental movement, spiritual ecology involves a multitude of diverse leaders, organizations, and initiatives. They share a common concern and commitment to pursuing the vital role of religion and spirituality in environmentalism to complement secular approaches to environmental problems and issues from the local to the global levels. Here, after some background, spiritual ecology as a component of the phenomena of international environmental movements is exemplified through three cases: the Green Belt Movement beginning in Kenya, the Green Patriarch of the Eastern Orthodox Church, and the Alliance of Religions and Conservation affiliate of the World Wide Fund for Nature (formerly World Wildlife Fund).

Keywords: Alliance of Religions and Conservation; biodiversity conservation; Green Belt Movement; religious environmentalism; tree planting; waterway pollution

Occupy the Earth: Global Environmental Movements
Advances in Sustainability and Environmental Justice, Volume 15, 275–293

ISSN: 2051-5030/doi:10.1108/S2051-503020140000015011

INTRODUCTION

Spiritual ecology as a generic category encompasses a vast, diverse, complex, and dynamic arena of intellectual and practical activities at the interfaces of religions and spiritualities on the one hand, and ecologies, environments, and environmentalisms on the other. The plural terms reflect the breadth and multiplicity of this arena. Other labels are far narrower, referring only to a particular aspect of spiritual ecology, such as ecomysticism, ecotheology, or religious environmentalism. Spiritual ecology is growing exponentially as a response to the dire and urgent need for actions to promote the survival, welfare, and flourishing of Earth including humankind. As an international environmental movement, spiritual ecology's strength is that it is a diffuse worldwide initiative involving a multitude of diverse individuals and organizations, with no single leader or institution at the helm over all (Sponsel, 2012, 2014a). Here as in evolution, diversity is crucial for future adaptability. Next some background is provided followed by three exemplary cases.

BACKGROUND

Environmentalism has grown beyond its roots in North America and Europe to become a global force. For example, Hawken (2007, p. 2) estimates that more than one million non-governmental organizations (NGOs) associated with over a billion people are engaged with environmental and related matters such as environmental justice. Although usually overtly secular, many if not most of the individuals involved in environmentalism are motivated by some kind of extraordinary experience(s) in nature that many would identify as mystical or spiritual, even if they are not necessarily religious (cf. Crosby, 2013). However, seldom is this explicit in their statements and actions.

Many biologists, for example, although operating mainly, if not exclusively, within the framework of Western science, variously recognize the mutual relevance of biodiversity and religion or spirituality. Takacs (1996, pp. 254–270) identifies spiritual value as among the several different kinds of values of biodiversity, based on interviews with numerous prominent biologists. As scientists many, if not most of them, would not admit to being religious in the sense of adhering to any particular religious faith or institution. However, many of them admit to experiencing extraordinary emotions during their field research in nature that they variously identify as a sense of wonder, awe, joy, exhilaration, tranquility, reverence, mystery, or spirituality. Most of them distinguish between religion as a social institution and spirituality as an individual experience. Takacs (1996, p. 270) concludes: "Some biologists have found their own brand of religion, and it is based on biodiversity. The biologists portrayed here attach the label spiritual to deep, driving feelings they can't understand, but that give their lives meaning, impel their professional activities, and make them ardent conservationists." Closely related to such feelings is the phenomenon of biophilia, a deep affective bond with nature (Kellert & Wilson, 1993).

A rare example of someone who freely displayed his spiritual attachment to nature is wilderness prophet and conservationist John Muir (1838–1914) (Lacy, 2011; Sierra Club, 2014; Worster, 2008). Moreover, he was the single most important individual in the development of the national park system in the United States of America, beginning with Yellowstone in 1872 (National Park Service, 2014). By now, there are 58 national parks in the United States and its territories. In turn, today worldwide there are more than 6,000 national parks in some 100 countries, albeit many are not adequately administered and inventoried for biodiversity (National Parks Worldwide, 2014). The practical significance of Muir's spiritual ecology cannot be overestimated (Burns, 2009; Stevens, 2014).

As another case in point, yet relatively little recognized, let alone appreciated, is the connection between sustainability and religion, as demonstrated in detail by Johnston (2013, p. 55): "Even ostensibly secular international organizations utilize the religious dimensions of sustainability (in the form of spiritualized language, religious metaphors, or discourses of awe and reverence) for their own, often different ends." He asserts that sustainability serves as a political religion to the degree that its underlying core values function as a vehicle for connecting affective or emotionally evocative states with political issues regarding resource use. Furthermore, Johnston (2013, p. 70) asserts that: "… the key motivation for participating in the quest for a better, more sustainable world was often religious … the emergence of sustainability and the resurgence of religion may be viewed as complementary trends." Thus, Johnston argues that sustainability cannot be fully understood without considering its religious dimension (see also Kitchell, 2012; Taylor, 2010, p. 195).

Increasingly in recent years, environmental organizations, such as the Worldwatch Institute, have tried to leverage some of the power of religions to advance their concerns and initiatives. Worldwatch Institute researcher Gardner (2002, p. 48, 2006) points out that religious assets include environmentally relevant aspects of sacred texts, worldview, and ethics; moral authority and persuasion; communication venues such as bulletins, newsletters, and websites as a platform; leveraging material resources, including funds, investments, physical facilities, and land; educating and engaging members in sustainability initiatives as well as encouraging environmental protests, boycotts, and letter-writing campaigns to exert their political weight; and building community relations and providing services that are environmentally sound.

The Alliance of Religions and Conservation (ARC), an affiliate of the World Wide Fund for Nature, formerly World Wildlife Fund (WWF), also recognizes and effectively pursues the tremendous potential of religion. ARC (2014) identifies additional points of the relevance of religion for environmentalism: worldwide religions are the largest organized sector of civil society; 85% of humans self-identify with some religion; religions are trusted as an important source of authority, information, and wisdom; worldwide up to 8% of the habitable surface of the planet is owned by religions as are 5% of the forests; religions are among the largest investment blocks in the global stock market; they are involved in their community for the long-term extending over generations; more than 50% of schools are founded, funded, or run by religious organizations; and more than 50% of the world's tourist destinations are religious, including pilgrimage sites and routes.

In general, however, world religions have lagged behind secular initiatives on behalf of the environment. This is a curious discrepancy, since the first Earth Day in the United States on April 22, 1970, has continued annually. The first Earth Day was a nationwide grassroots demonstration to place the environment on the American political agenda as well as a teach-in about worsening environmental problems and issues. It was observed by hundreds of communities across the United States, with about 20 million people participating. It marked a historic turning point, reflecting and then perpetuating a greatly heightened level of environmental information, awareness, concern, and action. Today, Earth Day is observed around the world by more than 500 million people in some 175 countries (Earth Day Network, 2014).

Although commonly viewed as a secular event, in fact, the first Earth Day was influenced to some degree by religious concerns. For example, it was officially supported and endorsed by the National Council of Churches, among other religious organizations (Christofferson, 2004). Yet, many religious individuals and organizations, and especially some Protestants, remained resistant to environmentalism until more information accumulated over successive decades about the gravity of accumulating environmental problems and issues (Shabecoff, 1993, pp. 125–128).

Tucker (2010) provides one of the most clear, concise, and forceful rationales for the relevance of religion to environmentalism. For instance, Tucker (2010, p. 3) writes: "The common values that most of the world's religions hold in relation to the natural world might be summarized as reverence, respect, restraint, redistribution, responsibility and renewal. While there are clearly variations of interpretation within and between religions regarding principles, it may be said that religions are moving toward an expanded understanding of their cosmological orientations and ethical obligations. Although these principles have been previously understood primarily with regard to relations toward other humans, the challenge now is to extend them to the natural world."

Yet another relatively neglected component of environmentalism is the many rural communities that are actually closest to nature, especially indigenous societies. While the 350 million indigenous persons in the world comprise only 5% of the human population, they inhabit 20% of the world's landmass, living in 70 countries (Maybury-Lewis, 2002, pp. 7–8). Their intimate daily environmental experience, accumulated over millennia, generates profound knowledge and wisdom for their survival and adaptation that can also provide invaluable lessons for the rest of humanity (Anderson, 2014). At the same time, indigenous societies do have a cumulative environmental impact that can be significant over the long term, but it is far less than that of industrial societies (Sponsel, 2014c). Often indigenes are in some ways and degrees ecologists, environmentalists, and conservationists with attributes of these phenomena embedded in their larger cultural worldviews, values, attitudes, and practices. Indeed, indigenes are the descendants of the original spiritual ecologists (Sponsel, 2012, pp. 13–30). Indigenous sacred places are mostly in nature, and often they protect higher concentrations of biological diversity than their secular counterparts (McLeod, 2013). Indigenous societies are a growing global force for environmentalism, including spiritual ecology. Indigenes are becoming more political, vocal, organized, and influential as particular societies and together collectively, this in the face of increasing encroachment from external agencies such as natural resource extraction by industrial societies (Bodley, 2014; Coates, 2004; Corry, 2011; Niezen, 2003; Sponsel, 2014b).

Behind much of spiritual ecology, as a component of the international environmental movement in general, is the growing realization that although purely secular approaches to environmental matters are absolutely necessary and have achieved much progress, they have proven insufficient. Environmental problems persist, new ones are being discovered each year, and the global situation is becoming worse rather than better. For instance, the UN Millennium Ecosystem Assessment was researched by 1,360 experts from 2001 to 2005. Also, there have been five reports since 1988 of the UN Intergovernmental Panel on Climate Change involving thousands of scientific and other authorities. Even the possibility of the extinction of the human species is increasingly being taken seriously (Guterl, 2012; Hartman, 1999; Leslie, 1996; McKibben, 2010). Such well-documented scientific alerts have generated relatively little concern and action globally, compared to what is really needed.

Most secular approaches treat only the superficial symptoms of the environmental crisis, not its underlying causes (Kitchell, 2012). Many well-informed thinkers consider the root causes to be essentially moral and ethical (Moore & Nelson, 2010). They believe that no less than a revolution of consciousness is required to finally turn things around for the better. Only a most profound rethinking and transformation of culture – encompassing worldview, values, attitudes, behavior, and institutions – holds any real promise for creating a more sustainable, green, just, and harmonious relationship between the humankind and nature. This may be the last hope for turning things around for the better, thereby saving humanity as well as nature. While it may be decades hence before it is known whether or not spiritual ecology has really helped in this respect, there are already a multitude of diverse achievements as well as great potential for the future. (For complementary overviews of spiritual ecology see Bauman, Bohannon II, & O'Brien, 2011a, 2011b; Gottlieb, 2006a; Grim & Tucker, 2014; Sponsel, 2012; Taylor, 2010. Gottlieb, 2006b, contains chapters on different religions, particular issues, and regions including Africa and Latin America. See Sponsel, 2014a, for a bibliographic essay of more than 100 books on spiritual ecology.)

To adequately survey this subject of spiritual ecology would require an encyclopedia of several volumes (see, e.g., FORE, 2014; Taylor, 2005). Here because of the limited space, just three cases must suffice to illustrate spiritual ecology as a growing component of the international environmental movement in general: the planting of billions of trees, initiated by the members of the Green Belt Movement (GBM) developed by Wangari Maathai, which started in Kenya and spread into many parts of Africa and then beyond worldwide; the conferences and practical actions on pollution in major waterways led by the Green Patriarch Bartholomew I with some 300 million adherents of the Eastern Orthodox Church; and many thousands of field projects for biodiversity conservation, facilitated by Martin

Palmer as Secretary General of the Alliance of Religions and Conservation in association with the WWF with a membership of 1.2 million people in 100 countries (WWF, 2014).

GREEN BELT MOVEMENT

Biologist Wangari Muta Maathai (1940–2011), a Kikuyu from Kenya, was awarded the Nobel Peace Prize in 2004, the first African woman to get this award so far. The award recognized her contribution to sustainable development, democracy, and peace. It was one of her triumphs on a long, difficult, and sometimes hazardous path. That path started with the planting of trees in rural Kenya for the practical reason of advancing the welfare of women who struggled with increasing scarcity of fuelwood and water along with other symptoms of environmental deterioration, poverty, and hardship. At the time, only 3% of the country's original forest remained after clearing for timber and plantations of coffee and tea crops for export under British colonialism (Breton, 1998, p. 11).

Maathai's environmental concerns emerged in the mid-1970s as she observed the environmental deterioration of Kenya with deforestation, watersheds and streams declining, and even desertification (MacDonald, 2006, p. 87). She believed that these problems could be relieved through tree planting which would promote soil conservation and fertility plus produce fuelwood, water, and also food for improving the nutrition of the rural poor.

The GBM was launched on the World Environment Day in June 1977. By now, there are more than 6,000 tree nurseries and 150,000 women volunteers in Kenya. The planting of more than 51 million trees of indigenous species on private farms and community lands, like schools and church yards, transformed the social as well as the biophysical landscape of Kenya. Furthermore, the United Nations followed Maathai's example with the Billion Tree Campaign. In 1986, funds from the UN Environmental Programme (UNEP) facilitated the movement's spread to 36 organizations in 15 countries in Africa, with the planting of more than 35 million trees and the establishment of the Pan-African Green Belt Network (Mburu, 2005, pp. 960–961). By now, this has become a global initiative with more than 12 billion trees planted in 170 countries (UNEP Billion Tree Campaign, 2014). In 1987, the UNEP recognized the GBM for its success in ecological restoration with the Global 500 Award.

The community-based GBM mobilizes and empowers women through promoting community tree planting, sustainable organic farming of indigenous crops, and environmental conservation including of biodiversity and local species. Women are paid a modest stipend to search nearby forests for seeds to grow native trees and then establish nurseries. Through civic and environmental education seminars and practical activities such as tree planting the GBM improves the livelihood of rural communities, household food security, and women's rights. It tries to develop a value-driven society emphasizing personal and communal improvement through advancing social and environmental justice. While the driving force in the movement is women, men and children have become involved as well (Mburu, 2005, pp. 958–959). (The GBM is described in detail by Maathai, 2004, 2010, and in documentary film by Merton & Dater, 2008.)

The GBM developed from the roots of traditional African culture and religion wherein trees and nature were usually valued and respected, even though this was weakened by colonialism and Christian missions. Yet, eventually, the GBM also emphasized aspects of Christian stewardship for God's creation. About 90% of Kenyans are Christians, and the percentage is even higher among the GBM members. Church leaders and members became increasingly concerned and active in healing their natural environment, even though most evangelicals were resistant until recently (Mburu, 2005, p. 958).

A diversity of landscape features and natural phenomena are considered sacred in many African traditions, and taboos often restricted resource extraction (Warner & Hoskins, 2005, pp. 26–28). African traditions must have had some influence on Maathai and the GBM. As Warner and Hoskins (2005, p. 26) observe: "Traditionally, the gods are all around: present in the nature in which people are a part. Therefore to respect the spiritual is to respect nature. The converse is equally true. The rape of the land is traditionally seen as a violation of the spirit world, and only to be undertaken at peril." Furthermore, "… spiritual well-being depends on a harmonic relationship with the gods and with the environment. Life is a unity, a web of relationships extending outwards with all that is around: humans, flora, fauna, and gods. Indeed, even to separate those terms, as if distinct categories, is to miss the point that in African religious life everything is interconnected. As such, if one part suffers, all do" (Warner & Hoskins, 2005, p. 27).

Maathai (2010, p. 27) writes: "It gradually became clear that the Green Belt Movement's work with communities to repair the degraded environment could not be done effectively without participants embracing a set of core spiritual values." The spiritual ecology behind Maathi and the GBM is detailed in Maathai's (2010) most recent book, *Replenishing the Earth: Spiritual Values for Healing Ourselves and the World.*

Maathai was often asked why she was undertaking tree planting and related initiatives. Gradually, she began to realize that beyond practical needs, passion, and vision, underlying the GBM was a set of four basic core values, and that ultimately they were spiritual and shared by many religions. Maathai (2010) identifies these values as: 1. love for the environment; 2. gratitude and respect for Earth's resources; 3. self-empowerment and self-betterment; and 4. the spirit of service and volunteerism. She adds: "These values encapsulate the intangible, subtle, nonmaterial aspects of the GMB as an organization .... Such values are not unique to the Green Belt Movement. They are universal, but they can't be touched or seen. We cannot place a monetary value on them; in effect, they are priceless. They define humanity" (Maathai, 2010, p. 16, cf. p. 35).

Interestingly, Maathai (2010, p. 19) identifies the existence of "our inner ecology, our soul and sense of being human." Throughout her 2010 book, she uses the word "God" not only in connection with the monotheistic Abrahamic traditions of Christianity, Islam, and Judaism, but also the Kikuyu religious worldview. However, she also applies the term "the Source" as "the place of all knowledge and awareness" and the "repository of all that we cannot explain" (Maathai, 2010, p. 21). She writes: "I came to identify trees as the answer to the environmental problems the Kenyan women faced, partly because I grew up in the countryside surrounded by trees and green vegetation .... I can also point to the Source as the wellspring for all of the ideas that came" (Maathai, 2010, p. 31). (Also, see Daneel, 2001, 2006; Maathai, 2010, pp. 90, 154–155, 193.)

One source that Maathai discusses frequently in the book is the traditions of her own Kikuyu culture, such as sacred trees inhabited by spirits (e.g., Maathai, 2010, pp. 50–53, 78, 85, 120–122). Another is the Holy Bible of Christians. which she quotes. However, her attention to ecologically relevant aspects of religion is much broader and more diverse, including references to Judaism, Buddhism, Hinduism, and Shintoism, among other traditions (e.g., Chapter 4: *The Power of the Tree*, pp. 77–91). It is also noteworthy that although a biological scientist, Maathai (2010, p. 59) finds no incompatibility with spirituality: "... [one] can be committed to the scientific method and still experience ecstasy at the great mystery of the cosmos." Later she writes: "We need to remind ourselves, however, that for all of its extraordinary sophistication, there are many issues that modern science can't explain" (Maathai, 2010, p. 74).

Finally, Maathai (2010) points to several different sources of inspiration for her life and work with the GBM: her family (p. 199); Catholic schooling (pp. 173–174); other spiritually motivated heroes, including those in the civil rights movement in the United States and Kenya (pp. 159–160); liberation theology in Latin America and beyond (pp. 160–164); the Source (pp. 171–183); leaders in the spiritual ecology revolution like Mary Evelyn Tucker, John Grim, Al Gore, and Terry Tempest Williams (pp. 174–176, 197); the Earth Charter initiative (pp. 176–179); and members of the GBM (p. 198). (See Maathai, 2006, for more information on her life, work, and the GBM.)

EASTERN ORTHODOX CHRISTIANITY

His All Holiness Bartholomew I is Archbishop of Constantinople, New Rome, and Ecumenical Patriarch. His website states: "The Ecumenical Patriarch has the historical and theological responsibility to initiate and coordinate actions among the Churches of Alexandria, Antioch, Jerusalem, Russia, Serbia, Romania, Bulgaria, Georgia, Cyprus, Greece, Poland, Albania, The Czech Land and Slovakia, Finland, Estonia, and numerous archdioceses in the old and new worlds. This includes the convening of councils or meetings, facilitating inter-church and inter-faith dialogues and serving as the primary expresser of Church unity as a whole. As Ecumenical Patriarch he transcends every national and ethnic group on a global level and today is the spiritual leader of approximately 300 million faithful world-wide."

Popularly known as the Green Patriarch, for Bartholomew I, the environment is primarily a religious and spiritual issue, although scientific, technological, economic, and political matters are also clearly important for him. A month after his election in 1991 as Ecumenical Patriarch, he convened an environmental conference in Crete called "Living in the Creation of the Lord." In the years since, he co-sponsored with His Royal Highness (HRH) Prince Philip, International Chairman of WWF, international scientific environmental symposia in Patmos (1995), Black Sea (1997), Danube River (1999), Adriatic Sea (2002), Baltic Sea (2003), Amazon River (2006), Arctic (2007), and Mississippi River (2009). The Amazon and Arctic symposia were under the patronage of Kofi Annan, Secretary General of the United Nations, as well as Bartholomew I. These international, interdisciplinary, and interfaith conferences are intended to generate reflection on the fate of the rivers and seas as well as interfaith debate on the environment by visiting and discussing some of the most ecologically threatened areas on the planet. The venue for the symposia is usually a voyage aboard a ship that follows the route of pollution from its source to its points of impact, thereby vividly demonstrating the interconnectedness of the world's waters and associated ecosystems. Incidentally, these conferences also pursue carbon offsets and other measures to reduce their ecological footprint. (The Patriarch's website has a wealth of information on these symposia and other matters as well as documentary films on The Green Patriarch, Arctic, and Amazon.)

These symposia are not limited to informing the participants, but also discuss practical solutions and lead to their implementation. Among these are the development of the WWF Wetlands Conservation Programme in Greece; a Pew Fellowship in Marine Conservation awarded to Professor Laurence Mee to develop public information materials on the Black Sea and to organize a two-year program for training a group of teachers from all six associated countries; representatives of the Lower Danube countries of Bulgaria, Romania, Moldova, and Ukraine agreed to sign a declaration for the creation of a Lower Danube Green Corridor for restoration and conservation of an interconnected network of floodplains along the Lower Danube; pressure from journalists on the Albanian government to clean up a toxic waste dump from abandoned chemical factories in Porto Romano; and a Bread and Fish project to bring together farmers and fishermen to consider environmental issues that connect these industries in the Baltic Sea region.

In June 1999, the Halki Ecological Institute was organized to advance broader regional collaboration and education among 75 clergy, theologians, educators, students, scientists, policy-makers, and journalists. It developed a new direction in interdisciplinary vision and dialogue about environment matters (Chryssavgis, 2003, pp. 8, 11, 2007).

In the ecological symposia and in the Halki Ecological Institute, the Green Patriarch functions as a facilitator for participants to transcend their differences by realizing their commonalities and complementarities for their common good in the context of facing the serious challenges of the continuing global environmental crisis. Most initiatives involve interfaith dialogue and collaboration among leaders of different religions and likewise between religious and scientific leaders. This is a very constructive achievement, considering the tensions and conflicts that have often emerged over the centuries within and among these different interest groups. As John (Zizioulas) Metropolitan of Pergamon writes: "… the Orthodox Church realizes that without a profound spiritual transformation of the human being there is no hope for salvation of the natural environment from the human interventions that threaten it with destruction, and that in order to achieve such a spiritual transformation all religious communities should cooperate, together with scientists, politicians, ethicists, etc." (Chryssavgis, 2003, p. viii).

For more than two decades, the Ecumenical Patriarch has given over 136 environmental addresses and issued an annual encyclical on the environment on the first day of each September, which is designated as a day of prayer for the protection of creation for the entire Orthodox Church (see Chryssavgis, 2003).

In conclusion, a statement by Chryssavgis (2003, p. 23) merits quoting at length because it reflects much of the heart of spiritual ecology as an international environmental movement in a very clear, concise, and forceful way:

> We tend to call this an 'ecological' crisis, and this is a fair description insofar as its results are manifested in the ecological sphere. The message is quite clear: our way of life is humanly and environmentally suicidal. Unless we change it radically, we cannot hope to avoid or reverse cosmic catastrophe.
>
> Yet, the crisis is not first of all ecological. It is a crisis concerning the way we perceive reality, the way we imagine or image our world. We are treating our environment, our planet, in an inhuman, god-forsaken manner-precisely because we see it in this way, precisely because we see ourselves in this way.
>
> Therefore, before we can effectively deal with problems of our environment, we must change our world-image, which in turn means that we must change our self-image. Otherwise, we are simply dealing with symptoms, not with their causes.

[Chryssavgis (2003) reprints many of the statements by the Green Patriarch and provides a very useful overview as background. However, statements by Bartholomew I, and a great deal of additional interesting information can be readily found on his website as well. Also see Chryssavgis (2006, 2007), Krueger (2012), and the film The Green Patriarch, 2009.]

ALLIANCE OF RELIGIONS AND CONSERVATION

The WWF celebrated its 25th anniversary on September 29, 1986, at the Basilica of Saint Francis in Assisi, Italy. HRH Prince Phillip, the then WWF President, welcomed the participants with this directive: "Come, proud of your own tradition, but humble enough to learn from others."

His statement referred to all participants in a meeting that engaged scientists, environmentalists, conservationists, and government officials as well as religious leaders and adherents. One concrete result was the publication of a 33-page booklet called *The Assisi Declarations* in which leaders from Buddhism, Christianity, Hinduism, Islam, and Judaism each offered a brief statement on environmental ethics for their own followers (WWF, 1986).

The format of the Assisi meeting was a model emulated by successive conferences sponsored by the WWF and later independently by other organizations. In 1995, another WWF conference at Windsor Castle in England yielded additional declarations on environmental ethics from Baha'i, Jainism, Sikhism, and Taoism. A parallel WWF conference was held that same year in Ohito, Japan. It produced the interfaith "Declaration on Religion, Land and Conservation" and added statements by representatives of Shintoism and Zoroastrianism. All focused on exploring the relationship between religion and ecology through interfaith dialogue with meaningful interaction among religious and secular leaders, the latter including scientists, scholars, environmentalists, conservationists, journalists, and others (Edwards & Palmer, 1997).

These WWF conferences generated another important development, the ARC, initially based at Manchester Metropolitan University under the leadership of Martin Palmer. Currently, he is Secretary General of ARC, which he co-founded with HRH Prince Philip Duke of Edinburgh in 1995. In addition, he is a regular contributor to BBC World Service and BBC Radios 3 & 4 as well as television programs for the BBC, including the series Exploring Planet Earth. Furthermore, Palmer is the author or the editor of over 50 books as well as a popular lecturer in over 50 countries, including the United States, Turkey, India, and Mongolia (Palmer, 2011).

ARC outreach also generated a pioneering inexpensive series of six paperback books by Cassell Publishers Limited for the general public, each on a different religion in relation to ecology. These foreshadowed the later more ambitious academic series from the Forum on Religion and Ecology, published by Harvard University Press (FORE, 2014). Since then, there has been an explosion of literature on this subject (Sponsel, 2014a).

The vision of ARC is to reduce human environmental impact through greater adherence to ecologically relevant aspects of religious beliefs and values. Their strategy is to help faiths realize their potential to be proactive on environmental matters and also to assist secular groups in becoming active partners in collaborating with religious organizations. ARC encourages and assists religious communities in the application of their powerful influences and resources for the widest possible environmental benefits through a focus on these six key areas: stewardship of land and assets; education at the community level and beyond; information campaigns in the print, radio, television, and internet media; relating human and environmental health; promoting lifestyle changes that take advantage of consumer power; and advocacy for environmental practices and policies.

Among the multitude of diverse projects implemented by ARC are Sacred Land, Living Churchyards, Sacred Gardens, Pilgrimage Trails and Shrines, Sacred Gifts for a Living Planet, International Interfaith Investment Group, and the ARC website and Bookshop. For instance, the Churchyards Project encompasses 6,000 sacred ecosystems in Britain alone without the use of pesticides and mowing grass only once a year in order to allow birds, bats, reptiles, and insects to flourish. In addition, 6,000 saplings from 2000-year-old churchyard trees were planted in churchyards throughout Britain. Many of them will still be alive in another 1,000 years hence.

During November 2–4, 2009, the conference "Many Heavens, One Earth" was held at Windsor Castle in England with 200 religious and secular leaders attending. It was co-sponsored by ARC and the UN Development Programme (UNDP) with host H.R.H. Prince Phillip and the U.N. Secretary General Ban Ki-moon participating. A major focus was global climate change. There were 31 commitments made by religious leaders for their organizations to pursue appropriate environmental action and an additional 18 commitments by the leaders of secular organizations to collaborate with religious groups. In an ARC (2010) report a year later, Palmer notes that: "The great strength of the faiths is that they are the most sustainable of all human institutions having outlived empires, dynasties and ideologies." Importantly, many of the commitments are long-term and are being discussed, developed, and pursued.

The Green Pilgrimage Network was launched by ARC in late 2011 with 10 religions. Pilgrimage is the largest movement of people worldwide. The largest is the Hajj when around two million Muslim pilgrims visit Mecca in Saudi Arabia each year. The primary goal is for pilgrims on route and at sacred sites to be more environmentally friendly, reducing their ecological footprint as much as possible (ARC, 2014). (For more on ARC, see its website plus Dudley, Higgins-Zogib, & Mansourian, 2005; Edwards & Palmer, 1997; Jensen, 1999; Johnston, 2013, pp. 136–151. For related resources from other international environmental organizations, see Posey, 1999; Pungetti, Oviedo, & Hooke, 2012; Verschuuren, Wild, McNeely, & Oviedo, 2010.)

CONCLUSION

Our contemporary environmental problems and crises, from the local to the global levels, ultimately result from the worldview and associated values, attitudes, practices, and impacts of industrialism, materialism, consumerism, and capitalism, especially when they are fueled by rapacious greed. In particular, they operate on the fallacy that unlimited growth is possible on a limited base. That base is not only non-renewable natural resources, but also the capacity of Earth's biogeochemical systems to absorb pollution and other stresses. Spiritual ecology tries to help awaken people to such issues and help them find their own path toward a more sustainable, green, just, and harmonious future in nature.

The leadership and respective organizations of Wangari Maathai, Ecumenical Patriarch Bartholomew I, and Martin Palmer are just three specific examples among a multitude of diverse cases that could be cited of spiritual ecology in action and coalescing as an international environmental movement of growing significance. They illustrate the power of leveraging religions on behalf of the survival and welfare of nature and humanity. As the largest organized sector of civil society, religions can reach ordinary people on a massive scale, a dire necessity given the scale, complexity, difficulty, gravity, and urgency of many environmental problems. While the positive effectiveness of such initiatives may not always be readily demonstrated, the sheer numbers involved as indicated above are surely suggestive of substantial impact and progress. Yet there are obstacles, such as opposition from conservative sectors of religions, science (especially scientism), academia, governments, politicians, the economy, and society.

Spiritual ecology is a quiet revolution in the sense that it is nonviolent, gradual, without any single leader or organization, and relatively little known. It is a revolution in the sense that it calls for no less than a fundamental re-thinking, re-feeling, and re-visioning of personal lifestyles and societies. Ultimately, the most important choice we face today is between ecocide and ecosanity. Individuals and organizations engaged in spiritual ecology hope that this growing international environmental movement can help turn things around for the better before some critical threshold or tipping point is reached, leading to worldwide catastrophe. This may well be the last hope for humanity and the planet. Yet, even though the momentum of spiritual ecology is accelerating rapidly, it may be decades into the future before the practical results of this revolution become readily apparent on a global scale.

REFERENCES

Alliance of Religions and Conservation (ARC). (2010). *A year after Windsor*. Bath, UK: Alliance of Religions and Conservation. Retrieved from http://www.arcworld.org/downloads/One%20year%20after%20Windsor%20report%20Nov%202010.pdf. Accessed on May 1, 2014.

Alliance of Religions and Conservation (ARC). (2014). Retrieved from http://www.arcworld.org. Accessed on May 1, 2014.

Anderson, E. N. (2014). *Caring for place: Ecology, ideology, and emotion in traditional landscape management*. Walnut Creek, CA: Left Coast Press.

Bartholomew I. (2014). His All Holiness, Archbishop of Constantinople, New Rome and Ecumenical Patriarch. Retrieved from http://www.patriarchate.org. Accessed on May 1, 2014.

Bauman, W. A., Bohannon, R. R., II, & O'Brien, K. J. (Eds.). (2011a). *Inherited land: The changing grounds of religion and ecology*. Eugene, OR: Pickwick Publications.

Bauman, W. A., Bohannon, R. R., II, & O'Brien, K. J. (Eds.). (2011b). *Grounding religion: A field guide to the study of religion and ecology*. New York, NY: Routledge.

Bodley, J. H. (2014). *Victims of progress* (6th ed.). Lanham, MD: Rowman and Littlefield Publishers.

Breton, M. J. (1998). Wangari Maathai (1940–): Recruits army of green belt foresters. In *Women pioneers for the environment* (pp. 11–18). Boston, MA: Northeastern University Press.

Burns, K. (2009). *The national parks: America's best idea* (720 mi.). Arlington, VA: Public Broadcasting Service.

Christofferson, B. (2004). *The man from clear lake: Earth day founder senator Gaylord Nelson*. Madison, WI: University of Wisconsin Press.

Chryssavgis, J. (Ed.). (2003). *Cosmic grace humble prayer: The ecological vision of the Green Patriarch Bartholomew I*. Grand Rapids, MI: William B. Eerdmans Publishing Company.

Chryssavgis, J. (2006). The earth as sacrament: Insights from Orthodox Christian theology and spirituality. In R. S. Gottlieb (Ed.), *The Oxford handbook of religion and ecology* (pp. 92–114). New York, NY: Oxford University Press.

Chryssavgis, J. (Ed.). (2007). Ecumenical Patriarch Bartholomew: Insights into an Orthodox Christian worldview. *The International Journal of Environmental Studies*, 64(1), 9–18.

Coates, K. S. (2004). *A global history of indigenous peoples: Struggle and survival*. Hampshire, UK: Palgrave Macmillan.

Corry, S. (2011). *Tribal peoples for tomorrow's world*. London: Survival International. Retrieved from http://www.survivalinternational.org/. Accessed on May 1, 2014.

Crosby, D. A. (2013). *The thou of nature: Religious naturalism and reverence for sentient life*. Albany, NY: State University of New York Press.

Daneel, I. (2001). *African earthkeepers – Wholistic interfaith mission*. New York, NY: Orbis Books.

Daneel, I. (2006). African initiated churches as vehicles of earth-care in Africa. In R. S. Gottlieb (Ed.), *The Oxford handbook of religion and ecology* (pp. 535–567). New York, NY: Oxford University Press.

Dudley, N., Higgins-Zogib, L., & Mansourian, S. (2005). *Beyond belief: Linking faiths and protected areas to support biodiversity conservation*. Gland, Switzerland: World Wildlife

Fund International/Manchester, UK: Alliance of Religions and Conservation. Retrieved from http://www.arcworld.org/projects.asp?projectID=294, http://www.arcworld.org/downloads/WWF%20Beyond%20Belief.pdf. Accessed on May 1, 2014.

Earth Day Network. (2014). Retrieved from http://www.earthday.net. Accessed on May 1, 2014.

Edwards, J., & Palmer, M. (1997). *Holy ground: The guide to faith and ecology*. Northamptonshire, UK: Pilkington Press.

Forum on Religion and Ecology (FORE). (2014). Retrieved from http://fore.research.yale.edu/. Accessed on May 1, 2014.

Gardner, G. T. (2002). *Invoking the spirit: Religion and spirituality in the quest for a sustainable world*. Washington, DC: Worldwatch Paper 164. Retrieved from http://www.worldwatch.org/system/files/EWP164.pdf. Accessed on May 1, 2014.

Gardner, G. T. (2006). *Inspiring progress: Religions' contribution to sustainable development*. Washington, DC: W.W. Norton & Company/Worldwatch Institute.

Gottlieb, R. S. (2006a). *A greener faith: Religious environmentalism and our planet's future*. New York, NY: Oxford University Press.

Gottlieb, R. S. (2006b). In R. S. Gottlieb (Ed.), *The Oxford handbook of religion and ecology*. New York, NY: Oxford University Press.

Green Patriarch. (2009). *The Green Patriarch* (42 min.). Ojai, CA: Becket Films.

Green Patriarch, His All Holiness the Ecumenical Patriarch Bartholomew. (n.d.). The Green Patriarch.

Grim, J. A., & Tucker, M. E. (2014). *Ecology and religion*. Washington, DC: Island Press.

Guterl, F. (2012). *The fate of the human species: Why the human race may cause its own extinction and how we can stop it*. New York, NY: Bloomsbury USA.

Hartman, T. (1999). *The last hours of ancient sunlight: Waking up to personal and global transformation*. New York, NY: Three Rivers Press.

Hawken, P. (2007). *The blessed unrest: How the largest movement in the world came into being and why no one saw it coming*. New York, NY: Viking.

Jensen, T. (1999). Forming 'the alliance of religions and conservation'. In D. A. Posey (Ed.), *Cultural and spiritual values of biodiversity* (pp. 492–499). London: Intermediate Technology Publications/Nairobi, Kenya: United Nations Environmental Programme (UNEP).

Johnston, L. F. (2013). *Religion and sustainability: Social movements and the politics of the environment*. New York, NY: Equinox Publishing Ltd.

Kellert, S. R., & Wilson, E. O. (Eds.). (1993). *The biophilia hypothesis*. Washington, DC: Island Press.

Kitchell, M. (2012). *A fierce green fire: The battle for a living planet* (101 min.). First Run Features. Retrieved from http://www.afiercegreenfire.com/index.html. Accessed on May 1, 2014.

Lacy, S. (2011). *John Muir: In the new world* (85 min.). West Long Branch, NJ: Kultur.

Leslie, J. (1996). *The end of the world: The science and ethics of human extinction*. New York, NY: Routledge.

Maathai, W. (2004). *The green belt movement: Sharing the approach and the experience*. New York, NY: Lantern Books.

Maathai, W. (2006). *Unbowed: A memoir*. New York, NY: Alfred A. Knopf.

Maathai, W. (2010). *Replenishing the earth: Spiritual values for healing ourselves and the world*. New York, NY: Doubleday.

MacDonald, M. (2006). Wangari Maathai: Mother of the trees. In S. Kumar & F. Whitefield (Eds.), *Visionaries: The 20th century's 100 most important inspirational leaders* (pp. 86–87). White River Junction, VT: Chelsea Green Publishing.
Maybury-Lewis, D. (2002). *Indigenous peoples, ethnic groups, and the state* (2nd ed.). New York, NY: Pearson. Retrieved from http://www.culturalsurvival.org/. Accessed on May 1, 2014.
Mburu, G. (2005). Kenya green belt movement. In B. R. Taylor (Ed.), *The encyclopedia of religion and nature* (Vol. 2, pp. 957–961). New York, NY: Thoemmes Continuum.
McKibben, B. (2010). *Eaarth: Making life on a tough new planet.* New York, NY: Henry Holt and Company, LLC.
McLeod, C. (2013). *Standing on sacred ground* (228 min.). Bullfrog Films, PA: Bullfrog Films. Retrieved from http://www.sacredland.org. Accessed on May 1, 2014.
Merton, L., & Dater, A. (2008). *Taking root: The vision of Wangari Maathi* (80 min.). Marlboro, VT: Marlboro Productions.
Moore, K. D., & Nelson, M. P. (Eds.). (2010). *Moral ground: Ethical action for a planet in Peril.* San Antonio, TX: Trinity University Press.
National Park Service. (2014). Retrieved from http://www.nps.gov. Accessed on May 1, 2014.
National Parks Worldwide. (2014). Retrieved from http://nationalparksworldwide.com/. Accessed on May 1, 2014.
Niezen, R. (2003). *The origins of indigenism: Human rights and the politics of identity.* Berkeley, CA: University of California Press.
Palmer, M. (2011, December 6). *Touching hearts and minds: The experience of the faiths in changing lives and the planet* (23 min.). Worldwide Fund for Nature (WWF) YouTube. Retrieved from http://www.youtube.com/watch?v=rPWx8DtioRE. Accessed on May 1, 2014.
Posey, D. A. (Ed.). (1999). *Cultural and spiritual values of biodiversity.* London: Intermediate Technology Publications/Nairobi, Kenya: United Nations Environmental Programme (UNEP). Retrieved from http://www.unep.org/pdf/Cultural_Spiritual_thebible.pdf. Accessed on May 1, 2014.
Pungetti, G., Oviedo, G., & Hooke, D. (Eds.). (2012). *Sacred species and sites: Advances in biocultural conservation.* Cambridge: Cambridge University Press.
Shabecoff, P. (1993). *A fierce green fire: The American environmental movement.* New York, NY: Hill and Wang.
Sierra Club. (2014). Retrieved from http://www.sierraclub.org. Accessed on May 1, 2014.
Sponsel, L. E. (2012). *Spiritual ecology: A quiet revolution.* Santa Barbara, CA: Praeger. Retrieved from http://spiritualecology.info. Accessed on May 1, 2014.
Sponsel, L. E. (2014a). Spiritual ecology: Is it the ultimate solution for the environmental crisis. *Choice, 51*(8), 1339–1342, 1344–1348.
Sponsel, L. E. (2014b). Indigenous sacred places: Threats and responses (review of standing on sacred ground) American anthropological association anthropology news/society for the anthropology of religion section. *Anthropology News, 54*(3). Retrieved from http://www.anthropology-news.org/index.php/2014/04/11/indigenous-sacred-places/. Accessed on May 1, 2014.
Sponsel, L. E. (2014c). Human impact on biodiversity: Overview. In S. A. Levin (Ed.), *The encyclopedia of biodiversity* (2nd ed.). New York, NY: Elsevier. (in press).
Stevens, S. (Ed.). (2014). *Indigenous peoples, national parks, and protected areas: A new paradigm linking conservation, culture, and rights.* Tucson, AZ: University of Arizona Press.

Takacs, D. (1996). *The idea of biodiversity: Philosophies of paradise*. Baltimore, MD: Johns Hopkins University Press.

Taylor, B. (Ed.). (2005). *The encyclopedia of religion and nature*. (Vols. 1–2). New York, NY: Thoemmes Continuum Press. Retrieved from http://www.religionandnature.com. Accessed on May 1, 2014

Taylor, B. (Ed.). (2010). *Dark green religion: Nature spirituality and the planetary future*. Berkeley, CA: University of California Press. Retrieved from http://brontaylor.com. Accessed on May 1, 2014.

Tucker, M. E. (2010). The alliance of world religions and ecology. *SGI Quarterly*, (61), 1–4. Retrieved from http://www.sgiquarterly.org/feature2010jly-2.html. Accessed on May 1, 2014.

United Nations Environmental Programme (UNEP) Billion Tree Campaign. (2014). Retrieved from http://www.unep.org/pdf/BTC_Booklet.pdf. Accessed on May 1, 2014.

United Nations Intergovernmental Panel on Climate Change. (2014). Retrieved from http://www.ipcc.ch/. Accessed on May 1, 2014.

United Nations Millennium Ecosystem Assessment. (2014). Retrieved from http://www.maweb.org/en/index.aspx. Accessed on May 1, 2014.

Verschuuren, B., Wild, R., McNeely, J., & Oviedo, G. (Eds.). (2010). *Sacred natural sites: Conserving nature and culture*. London: Earthscan Ltd.

Warner, F., & Hoskins, R. (2005). African religions and nature conservation. In B. R. Taylor (Ed.), *The encyclopedia of religion and nature* (Vol. 1, pp. 26–29). New York, NY: Thoemmes Continuum.

World Wide Fund for Nature (WWF). (2014). Retrieved from http://wwf.panda.org/. Accessed on May 1, 2014.

World Wildlife Fund. (1986). *The Assisi declarations: Messages on man and nature from Buddhism, Christianity, Hinduism, Islam, and Judaism*. London: Waterloo Printing Company Ltd. Retrieved from http://www.arcworld.org/downloads/THE%20ASSISI%20DECLARATIONS.pdf. Accessed on May 1, 2014.

Worster, D. (2008). *A passion for nature: The life of John Muir*. New York, NY: Oxford University Press.

BIRD HUNTING IN EUROPEAN MALTA: A CASE OF ENGO EMPOWERMENT?

Michael Briguglio

ABSTRACT

This chapter analyses the politics of bird hunting in relation to the empowerment of environmental non-governmental organizations (ENGOs) in the European Union (EU), with specific reference to Malta's first years of EU accession.

In particular, the analysis focuses on the activism of Maltese and International ENGOs – with special focus on Birdlife Malta and Birdlife International – on this issue, which is characterized by extensive EU legislation and by constant lobbying.

This chapter argues that ENGOs, both Maltese and European, were influential on State power in Malta, especially by resorting to the EU, and also being given prominence by the media. Yet the hunting lobby was influential too, and its influence on Malta's main political parties is an overdetermining factor, which remained in place even after EU accession.

Occupy the Earth: Global Environmental Movements
Advances in Sustainability and Environmental Justice, Volume 15, 295–324

ISSN: 2051-5030/doi:10.1108/S2051-503020140000015012

This chapter concludes that despite Malta's EU accession, national political factors remain highly influential in the Maltese hunting issue, and that one can expect more antagonism in the years to come.

Keywords: Malta; environmental NGOs; bird hunting; EU; social movements; discourse theory

INTRODUCTION

This chapter analyses the politics of bird hunting in relation to the empowerment of environmental non-governmental organizations (ENGOs) in the European Union (EU), with specific reference to Malta's first years of EU accession.

In particular, analysis will be made of the activism of Maltese and international ENGOs – with special focus on Birdlife Malta and Birdlife International – on this issue, which is characterized by extensive EU legislation and by constant lobbying.

For this purpose, this chapter employs discourse theory (Howarth, 2005; Howarth & Stavrakakis, 2000; Laclau & Mouffe, 1985), thus enabling an analysis of the articulation of hunting as an environmental issue; the construction of chains of equivalence and antagonism in the politicization of hunting; and the empowerment of ENGOs.

This chapter argues that ENGOs, both Maltese and European, were influential on State power in Malta, especially by resorting to the EU, and also being given prominence by the media. Yet the hunting lobby was influential too, and its influence on Malta's main political parties is an overdetermining factor, which remained in place even after EU accession.

This claim is substantiated through the analysis of impacts of social movements (Carter, 2001; Giugni, 1995; Rootes, 2007a; van der Heijden, 1999) and the conceptualization of the EU as a gateway for empowerment (Börzel, 2008; Dimitrakopoulos & Richardson, 2001; Johnston, 2011).

This chapter concludes that despite Malta's EU accession, national political factors remain highly influential in the Maltese hunting issue, and that one can expect more antagonism in the years to come.

DISCOURSE THEORY

This chapter analyzes the empowerment of ENGOs by focusing on discursive constructions and hegemonic struggles. In this respect, the social agents in question are seen as constructing nodal points through articulatory practice (Laclau & Mouffe, 1985), where meaning is partially fixed through empty signifiers (Howarth & Stavrakakis, 2000; Laclau, 1996, 2005).

In this respect, a "logic of equivalence" (Laclau & Mouffe, 1985) takes place, wherein discursive coalitions form boundaries between insiders and outsiders, yet also covering over the differences that exist between the former. Internal differences are thus minimized in opposition to the exterior enemy (Griggs, 2005, p. 116).

Chains of equivalence may be formed, where the particularities of specific socio-political demands are partially surrendered to the common aspects of such particular demands. Different antagonisms and conflicts may result in different chains of equivalence, and this process might affect the identity of the subjects involved in it (Laclau & Mouffe, 1985).

The formation of empty signifiers, chains of equivalence, and nodal points gives rise to the possibility of partial social order, and, possibly, of hegemonic formations. In this respect, hegemonic formations may be formed through the construction and stabilization of systems of meaning, which are articulated through nodal points that organize social orders by means of hegemonic practices. In Howarth's (2000) words, "these privileged condensations of meaning confer partially fixed meaning on a particular set of signifiers" (p. 110).

An example in this regard was the 1980s New Right Thatcherite hegemony, which was organized around the nodal points of "free economy" and "strong State" (Howarth, 2004, p. 259). Therefore, it can be said that "every conflict implies a conflict between *convictions, beliefs* and *representations of the world*" (Althusser, 1990, p. 36).

Hegemonic formations are, therefore, characterized by signifiers of a demand that is more universal than the particular demands that constitute them. Yet, every hegemonic formation must have some form of exclusion, which in turn can result in the creation of counter-hegemonic practices (Mouffe, 2005). This means that hegemonic formations are contingent and partial. When there is a universal representation of a demand, the hegemonic stability it represents is only relative.

Hegemonic discourse can also face dislocation. This refers to extra-discursive disruption to social formations, when new events result in a structural or organic crisis represented by hegemonic struggles, which attempt to fix floating signifiers within new nodal points (Torfing, 2005, p. 16). Laclau (cited in Howarth, 2000, p. 111) argues that in late modern societies such experiences are becoming more common due to processes such as globalization and commodification, in a context of "accelerated tempo" and "combined and uneven development." EU membership can be discursively situated accordingly.

EXTERNAL IMPACTS OF SOCIAL MOVEMENTS

In order to substantiate the conceptualization of ENGO empowerment and possible hegemonic formations, this chapter makes use of the conceptualization of external impacts of social movements as proposed by Giugni (1995), Van der Heijden (1999), Carter (2001), and Rootes (2007a). External impacts in this regard are referred to as sensitizing impacts, procedural impacts, structural impacts, and substantive impacts.

Sensitizing impacts have to do with influence in the "public arena, which goes in the direction of the goals of the movement" (Giugni, 1995, p. 211). Examples may include public opinion and political party agendas.

Procedural impacts refer to increased participation of social movements in decision-making structures such as those within the State, for example, through "consultation procedures, by undertaking negotiations, through formal recognition, or through general inclusion of challengers" (Giugni, 1995, p. 210). This may be both occasional and permanent (*ibid.*), and could refer to both an entire movement or to specific organizations within it (Van der Heijden, 1999, p. 202)

Structural impacts deal with changes in State structures, political institutions, and alliances, such as the setting up of government agencies, and substantive impacts refers to the acquisition of policy-making power (Giugni, 1995, p. 210), examples of which being the setting up of legislation and enforcement measures.

INSTITUTIONALIZATION OF SOCIAL MOVEMENTS

Concern has been shown that just as how social movements may have social and political impacts, they may also be co-opted within existing structures to the detriment of their transformative potential. For example, whilst asserting that civil society and the State are different institutions, Edwards (2009) notes the perils of having both too much disconnection between the State and civil society, resulting in the negation of positive effects on each other, and of becoming too close. In the latter case, governments might end up being caught up within particular interests from within civil society, and civil society might lose its critical edge.

Social movements may also be professionalized and made more moderate through institutionalization (Johnston, 2011, pp. 72–73), and this might be at the expense of radical organizations and demands. As Touraine (1981) puts it in this regard, "once relations become institutionalized, social movements degenerate into political pressure" (100).

However, a movement may maintain various aspects of non-institutionalization whilst enjoying benefits through a certain degree of institutionalization. According to Rootes (2007a), institutionalization is not a barrier to the mobilization of protest. He notes how the environmental movement is diverse, with various organizations being institutionalized whilst others "remain vital and resist the complete co-optation that would render them toothless" (*ibid.*, pp. 625–633).

SOCIAL MOVEMENTS IN AN EU CONTEXT, WITH A FOCUS ON ENVIRONMENTAL NGOS

This section discusses the activism of social movements on a European level. In the first instance, however, a brief conceptualization of the political dimension of the EU will be put forward.

The EU can be considered as an example of politicization at an international level. In this regard, there are mixed views on the influence of the EU on national politics, or what some refer to as "Europeanization," or the "the domestic adaptation to European regional integration" (Vink & Graziano, 2008, p. 7). In this chapter, the EU is considered to be characterized by political process at multi-directional levels and at different speeds (Delanty & Rumford, 2005, p. 49), in which the national State keeps having considerable political influence (Lenschow, 2006, p. 67).

At the same time, the "Europe has widened the political and discursive context of national policy-makers and societal actors" (*ibid.*, p. 68), and new "institutional opportunity structures and also the normative consequences in terms of substantial political issues" (Vink & Graziano, 2008, p. 8) are in place. Hence the EU can be seen as a "social complex" in terms of structure and change, characterized by commonalities and differences (Roche, 2010).

As is the case with the national State, the EU is characterized by hegemonic struggles (Jessop, 2004), and its processes cannot be entirely reduced to a single determining essence. For example, EU accession could have a dislocatory effect within new member states, yet, in turn each new member State has its own overdetermining characteristics and hegemonic formations.

National states transpose EU law by using the power assigned to them by their national parliament and in having various means at their disposal in this regard. They may take unilateral decisions to opt out from certain EU policies, or may choose to implement specific directives in particular ways, which may differ from the policy process of other member states. Thus, EU policy processes might have differentiated outcomes (Steunenberg, 2006; Wallace, Wallace & Pollack, 2005). Processes involving governmental and civil society stakeholders in specific sectors, rather than generic structures, tend to influence the transposition of directives (Steunenberg & Rhinard, 2010).

At the same time, EU courts have relative influence as regards jurisprudence on the obligations and rights of governments in the implementation of EU policy (Dimitrakopoulos & Richardson, 2001).

EU policy-making and governance should, therefore, be seen as a complex and fluid process, involving actors that are relatively autonomous though related to each other, such as the European Commission, the European Parliament, the Council of Ministers, the European Court of Justice, national states, and transnational institutions. None of these actors has exclusive power, and none can be defined to be the ultimate form of State. Hence, there is no such thing as a European super-State, but at the same time, the national State does not have a monopoly of power in its territory. State power is multi-leveled, as are popular hegemonic conflicts across local, national, European, and global borders.

In relation to social movements, the EU has been described as being "incredibly open and permeable to interest group lobbying compared even with the more pluralistic member states" (Mazey & Richardson, 2006, p. 251). Hence, social movements may lobby different institutions including the European Commission, the European Parliament, the European Court of Justice, and the Council of Ministers, through different mechanisms and strategies. National governments still play an important role, especially in relation to the Council of Ministers, but their importance varies in terms of issues, policies, interest groups, and type of national governments themselves.

In every case, specific hegemonic formations are in place. In this regard, not all civil society antagonisms and conflicts within the EU member-states have shifted to an EU-level. Relations of power are articulated in various specific configurations, including the local, the national, the regional, and the global. Local and national political alliances are still of great importance. Hence, "The multiplicity of nodal points calls for a variety of strategies, and the conflict cannot simply be envisaged at the global level" (Mouffe, 2005, p. 114).

Social movements may have an important role in the implementation of EU policy. Apart from benefiting from EU funding for this purpose, such actors may experience empowerment, for example, through EU legislation, whereby they can take public authorities to court for failing to implement EU policy. Besides, civil society actors use links with the European Commission to promote their own interests when they feel that they reached have a dead end with national authorities (Dimitrakopoulos & Richardson, 2001, pp. 344–345). The "boomerang effect" (Johnston, 2011, pp. 186–187) may take place, through which EU accession is used as a gateway for empowerment.

If one focuses on ENGOs as examples of the environmental movement, one can observe that they are active in a context of competing environmental and economic pressures in issues such as climate change. The European Commission itself is not a homogenous bloc – as regards the environment, one can witness clashes between "competitiveness-first" commissioners and the commissioner responsible for the environment (Euractiv.com, 2005). Yet, such a tension is not necessarily present in all environmental struggles. For example, the issue being discussed in this chapter – bird hunting – is usually alien to discourses on economic factors.

Börzel and Buzogany (2010) argue that ENGOs strive to "'pull down' [...] EU policy to the domestic level by pressuring public administration to legally incorporate, practically apply and enforce it" (712). In this regard, together with the media, they can shame national governments by using the EU as a watchdog. *Natura 2000*, which is based on the Habitats Directive and on the Wild Birds Directive, exemplifies EU policy, which enables ENGOs to make use of such strategies (*ibid.*).

ENGOs may, therefore, use the EU as a gateway for empowerment through the boomerang effect referred to above. The dislocatory effect of EU accession could provide new discursive opportunities for ENGOs in the formation of nodal points through hegemonic struggles, which in turn can be analyzed through the external impacts referred to earlier in this chapter.

Börzel argues that ENGOs coming from Southern European countries, which are most likely to benefit from opportunities provided by the EU, which can be used to put pressure on their respective governments, are at the same time most likely to take advantage of them, given their lack of resources (Börzel, 2008, pp. 232–233). Indeed, she concludes that actors that do not exploit resources to their advantage on the international level are likely to be weakened, due to the fact that a substantial degree of environmental policy is being carried out beyond the nation State. The same applies also to institutions such as national parliaments.

Another important point concerning ENGO activism in the EU is whether they are forming alliances – or chains of equivalence – on a European level. In this regard, whilst environmental activism across the EU has increased, also due to increasing connectedness of environmentalists (Rootes, 2007a, p. 622), Chris Rootes argues that it is still the national and local levels which are mostly characterized by such activism.

Indeed, Rootes (2007b) says that there is "at best a very modest trend towards the Europeanization of the issues raised by environmental protests and the forms they have taken" (255). Most protests are focused on national and sub-national levels, and the mobilization of protesters takes place accordingly. Hence, the localism of southern European political cultures is considered to be influential on environmental activism in countries such as Spain, Greece, and Italy (Rootes, 2007a, p. 622).

Webster (2000) analyzes the collaboration between European ENGOs and reaches similar conclusions, stating that collaboration between them is not so common, and when they do collaborate, they tend to prefer informal coalitions. The differences among such ENGOs, as well as their constraints are reasons for this. Examples include competition overfunding, contrasting strategies, and aims to achieve success on an individual basis.

THE MALTESE CONTEXT

Given that the issue under investigation relates to Malta in a European context, it is important to discuss the political context of this country.

Malta is a small island state in southern Europe, with a land area of about 320 square kilometers, making it the smallest member of the EU. It is situated south of Sicily and north of Libya in the center of the Mediterranean Sea. Malta has a population of just over 400,000 and has the highest population density in the EU, by far exceeding other member states (Eurostat, 2010, pp. 163–164).

The Maltese political context has traditionally been characterized by strong party loyalty towards the Nationalist and Labour parties, which is very much rooted in social and family background (Hirczy, 1995).[1] Voting turnout in general elections in Malta always exceeded 90 percent of the electorate (Lane, 2000, p. 212). According to Baldacchino (2002), the dominant Nationalist and Labour parties are "catch-all" parties with strong social networks and modern media apparatuses, acting like total institutions. As he puts it:

> Only the members of the troika – the two main political parties and the Catholic Church – loom large as anchors of identity. The "national interest" has been sabotaged, imploded into the frenzied partisanship internally, replaced by integrationism externally. (p. 199)

Patronage plays a strong role in Maltese politics. According to Boissevain (1993), "politics has become a corrosive zero-sum contest characterized by factional loyalty that reaches a veritable frenzy just before elections" (150). In this context, government ministers have become "the new super 'saints'" (*ibid.*, p. 154), and members of parliament have become "brokers, at best intermediaries between ministerial patrons and their own constituents" (*ibid.*). In this context, "Malta is a paradise for political saints" (*ibid.*, p. 156).

The Nationalist Party was in power between 1987 and 1996, and between 1998 and 2013. Subsequently, Labour won the general elections in 2013 through a landslide victory, winning 54.83 percent of votes. The Nationalist Party had won the previous general elections in 2008 by a razor-thin minority of 1,580 votes.

Malta joined the EU in 2004. Between 1998 and 2003, the Malta Labour party consistently opposed Malta's EU membership bid that was championed by the Nationalist government and other civil society organizations, including the non-parliamentary Green Party.

On the environmental front, EU membership was discursively constructed as being beneficial to Malta's environment, which possibly meant that ENGOs would be empowered in the process. Indeed, following EU accession, Maltese civil society activism on an EU level was witnessed, such as the conflict between environmentalists and hunters on the hunting issue, whereby the former pressed for Malta's conformity to the EU birds' directive, whereas the latter argued that Malta should have a right to derogate (Camilleri, 2010).

Birdlife Malta, one of Malta's oldest ENGOs, was actively involved on a Maltese and European level. Previously known as the Malta Ornithological Society, it was founded in the 1960s (Briguglio & Brown, 2009). Following Malta's EU accession, its demands were supported by

Birdlife International, the International Committee Against Bird Slaughter (CABS), other Maltese ENGOs, Malta's Green Party, and the Greens in the European Parliament. Indeed, the interplay between national and international factors across civil society and State institutions will be analyzed below. Through discourse theory, the chapter will proceed by focusing on how hunting was articulated as an environmental issue in the first five years following Malta's EU accession, and whether ENGOs were empowered in the process.

THE ARTICULATION OF HUNTING AS AN ENVIRONMENTAL ISSUE

Malta's first legislation aimed at preserving bird species was published in 1911 (Camilleri, 2010, p. 38), and in the last decades of the twentieth century, subsequent Labour and Nationalist administrations carried out various changes in legislation (*ibid.*, p. 31), yet Malta was not in line with EU rules. The latter

> developed largely out of public concern about the annual killing of migratory birds, which was widespread in Southern Europe, as well as a result of pressure from citizens' organizations and the European Parliament. (Falter & Scheuer, 2005, p. 37)

For example, Maltese legislation published in 1980 left out various species that were protected under the Berne Convention, apart from the fact that it was hardly enforced (Camilleri, 2010, p. 39). In 1993, and in reply to Malta's 1990 application to join the EU, the European Commission expressed its position that traditional practices would have to conform to the European wildlife preservation conventions, and during the same year, the Nationalist government published new regulations (Pollacco, 2004, p. 131).

These introduced various restrictions on hunting, and were very controversial, to such an extent that in 1994, the Labour party – which won the subsequent general election in 1996 – entered into an agreement with hunters' and trappers' organizations to tone down such regulations, to the delight of the latter (Briguglio, 2001, p. 205). The Federation of Hunters, Trappers and Conservationists (FKNK)[2] – the largest hunting organization in Malta – directed its members not to vote Nationalist in these elections (Pollacco, 2004, p. 135) and also used the slogan "10,000 members = 10,000 votes" (Debono, 2009). Hence, the agreement had an impact on Labour's victory (Boissevain, 2000, p. 7). Subsequently, the Nationalist Party – which was re-elected to power in 1998 – made an agreement with hunters and trappers, promising that hunting legislation would not be affected by EU legislation (Camilleri, 2010, p. 32). The Nationalists remained in power until 2013, when it was replaced by the Labour party.

Malta's 2004 EU accession obliged conformity with the Birds Directive, which prohibits hunting in spring. When the Nationalist Government negotiated Malta's EU membership, it claimed to have won a concession which enabled hunting for quail and turtle dove in spring, subject to strict monitoring and enforcement (Government of Malta, 2003, p. 106). A new committee, known as "Ornis," was set up to regulate hunting and trapping.[3] It comprised representatives appointed by the government, hunters, and environmentalists.

Hunting was amongst the most controversial environmental issues in Malta during and following EU accession. Even though Lino Farrugia – the General Secretary of the Hunters' Federation – did not do well in the 2004 European Parliament elections in 2004 (winning 1.3 percent of votes as an independent candidate), hunters still had a "very significant bargaining counter ... played to their advantage" (Falzon, 2008, p. 17). By 2007, there were 11,929 licensed hunters (The Times, February 8, 2007).

According to Falzon (2008), hunters still retain a sense of "control" over the countryside (p. 17). Lack of State resources, particularly through poor enforcement on illegal hunting practices, does not help to stop illegal hunting, even though, for example, in 2006, 419 persons were arraigned in this regard (The Times, 2007b). Falzon (2008) notes that, as is the case with other Mediterranean hunters, hunters in Malta have produced a discourse that is at odds with that of sustainability, which, in turn, is produced by the "northern European conception of hunting" through the EU Birds Directive and ENGOs (p. 20). Yet, hunting has been subject to significant opposition both in Malta and within the EU. Indeed, 64 percent of Maltese people believed that hunting is detrimental to other recreational activities in the countryside (MEPA, 2010, p. 31).

Whilst both the Nationalist and Labour parties have attempted to strike what they deem to be a balance between the hunting and environmentalist lobbies, the Green Party has been consistently speaking for Malta's complete adherence to the Bern Convention and to the EU's Birds Directive. In this regard, the European Greens lobbied within the European Parliament and pressurized the European Commission to investigate the hunting issue in Malta (Harwood, 2009, p. 126). As elaborated upon below, the European Parliament issued a resolution against the state of affairs in

Malta, also due to the pressure from petitions received from ENGOs in the EU and from the general public (Camilleri, 2010, p. 44). An ongoing infringement procedure against Malta characterized this issue, where the EU and the media have been "used" by those actively opposing Malta's hunting situation (Harwood, 2009, p. 186).

A few days after Malta's adoption of the Birds Directive on March 29, 2006 (Mallia, 2006), Birdlife International appealed to the European Commission to ensure that Malta bans hunting in spring. Birdlife described Malta as "Europe's worst offender on illegal bird hunting" (Camilleri, 2006a). The Government of Malta had allowed spring hunting in both springs of 2005 and 2006 (Camilleri, 2009a). Yet, Malta's change in relations through EU membership meant that a new legal and political context was now in place.

Maltese ENGOs, with the support of their European counterparts, the European Commission, the European Parliament (EP), and the Greens, have articulated an empty signifier (Laclau, 2005), which can be called "Stop spring hunting in line with the Birds Directive." The signifier has been filled with various conservationist claims, for example, by linking hunting in Malta to the destruction of migratory species, which are a common European heritage, by using scientific and technical language to justify claims for sustainability, by lobbying for Malta's full conformity with the EU legislation and, failing this, on the financial sanctions imposed on Malta. Meaning was partially fixed in this respect, creating a nodal point. Yet, the Birds Directive forms part of the *EU Natura 2000 Network*, which, though "intended to be a unifying discourse ... it often has to reckon with an unruly political environment" (Van der Heijden, 2010, p. 98), particularly as it is in competition with other policies at various levels (*ibid.*).

According to Van der Heijden (2010), ENGOs have created "new green public spheres" around *Natura 2000*, through reports, usage of the media, and activism at EU, national, and local levels. In turn, their opponents include landowners, hunters, and farmers amongst others (p. 121).

As stated above, Birdlife took a leading role on the hunting issue in Malta. It was not only equipped with its affiliation with Birdlife International, and it not only enjoyed direct or indirect support of other Maltese and European ENGOs and the Green Party (at European and Maltese levels), but, more significantly, it formed part of a chain of equivalence (Laclau & Mouffe, 1985), which also comprised the European Commission and the European Parliament. On the other hand, the Nationalist and Labour parties attempted to neutralize conflict with both sides, by aiming for consensus. At the same time, both supported the hunting lobby by trying to defend their aspirations, within European and national politics.

Once Malta had formally adopted the Birds Directive in 2006, the European Commission said that it would be taking legal action against Malta on spring hunting (Camilleri, 2006b), and the Petitions Committee of the European Parliament recommended the non-renewal of the derogation on spring hunting in 2008 and the abolition of trapping after 2008.

ANTAGONISMS IN THE IMPLEMENTATION OF THE BIRDS DIRECTIVE

For the ENGOs campaigning for the implementation of the Birds Directive, the "other" was clearly Malta's hunting lobby, which, in turn returned the compliment with confrontational discourse. A few days after the European Commission announced that it would be taking legal action against Malta on hunting, Lino Farrugia, speaking on behalf of the Federation of Hunters, Trappers and Conservationists, said that Birdlife would have to step over the formers' "dead bodies" before abolishing "traditions" like hunting, and that the Maltese government was determined to allow spring hunting to take place through a derogation.

> Birdlife and co. will feel the full weight of the wrath of FKNK and Maltese hunters and trappers ...[We would] gladly accept the easy road of confrontation with BirdLife. (The Sunday Times, 2006)

Both organizations accused the government of giving in to the demands of their adversaries. Both of them were later on invited for a meeting organized by the European Commission on January 17, 2007, to discuss spring hunting, trapping, and illegal hunting. Other invitees included the Federation of EU Hunting Associations, Birdlife International, MEPA, and Malta's permanent representation in Brussels (Fenech, 2007a; The Times, 2007a). When the meeting took place at the DG Environment building in Brussels, a protest against spring hunting was held by the Gaia for Animal Rights Organisation. During the meeting, the European Commission reportedly reiterated that Malta was not justified in asking for derogation for spring hunting from the Birds Directive, and that trapping must stop after 2008. Yet the Maltese government declared that it intended to challenge the Commission's stand. Lino Farrugia, on behalf of the Hunting Federation, said, "we are very satisfied with the position being adopted by the government" (Camilleri, 2007a). Later on, following an ultimatum to the government, which was subsequently withdrawn, Farrugia said that he was cautiously optimistic on the likelihood of opening the spring hunting season, adding that his Federation provided "enough scientific and socio-cultural information to yield fair regulations, including the permission for spring hunting" (Grech, 2007b).

On the other hand, after the meeting, Birdlife said that it expected Malta's Ornis committee to be in line with EU directives. In the meantime, the European Commission released an information note which stated that

> following an assessment of the information provided (by the government), the Commission has concluded that a satisfactory solution to spring hunting existing during the autumn hunting season and, therefore, no derogation from the Birds Directive was justifiable. (Camilleri, 2007b)

Birdlife International was on the lookout for Malta's "delaying tactics" (Grech, 2007a) on spring hunting. The ENGO's EU policy manager Konstantic Kreiser said Malta was the only country, which was permitting spring hunting for two bird species and that derogation on this would only be applicable "if Malta proves that there is no other solution to spring hunting – and the EU is clearly providing autumn as an alternative" (*ibid.*). The Government of Malta did not take heed of the European Commission, the European Parliament, and the ENGOs, and once again gave its approval for spring hunting for 2007, through an Ornis Committee meeting chaired by Louis Cilia, following the resignation of Tonio Azzopardi a few days before the meeting. Birdlife officials walked out of the meeting in protest, and called the decision "scandalous and farcical" and said that it would keep the European Commission informed of the sanctioning of illegal hunting. In contrast, the Hunters' Federation said the decision was "unsurprising and too long in coming." The government also introduced harsher sanctions for illegal hunting (Grech, 2007c). The European Commission immediately condemned the government's decision and said that this will lead Malta to the European Court of Justice, following an infringement procedure that had commenced in June 2006 (Camilleri, 2007c).

Environment Minister George Pullicino defended his government's position in an article in *The Times*, entitled "Striking a balance," which attempted to produce a "rational arbiter" discourse. In his words,

> It is easy to be in favour or against. But it is not easy to seek a balanced approach that seeks to strike a balance between nature protection and respect for tradition ... Beyond emotions, one must underscore a fact. This government has a

> political mandate, twice confirmed by universal suffrage, to safeguard traditional hunting , including in spring. People voting for the government's programme, following the lengthy EU accession negotiations, consented to the conditions negotiated by the Maltese government. Hunting in spring was part of the package ... The Maltese government considers its application of the derogation to be well-founded and within the legal framework of the EU Birds Directive and it is prepared to defend its position even in front of the European Court of Justice ... The message cannot be clearer. We will be tough on abusive hunting practices. I cannot hide my surprise at the fact that Birdlife went as far as criticizing even their own proposals!... Even though practised by a minority, one should give due consideration to the fact that hunting are deeply rooted among our society, It is highly illogical to think that this tradition can be taken away by one single decision. Of course, those who oppose hunting cannot be satisfied with this week's announcements. I appreciate that. But likewise, I appeal to them to appreciate that we have a political mandate for sustainable hunting and that is exactly what we are trying to implement. (Pullicino, 2007)

The European Commission remained consistent in its position on hunting in Malta, and in a meeting between Environment Minister George Pullicino and Environment Commissioner Stavros Dimas in 2007, the latter insisted that Malta should not allow spring hunting any longer (Camilleri, 2007d). In an address to the European Parliament, Dimas said that

> the government's decision to keep spring hunting this year is a perfect example of misinterpretation of EU law. I do have a glimmer of hope that following our second warning the Maltese government will comply with community law. (Camilleri, 2007f)

Dimas added that Malta's opening of spring hunting seasons in 2005, 2006, and 2007 was illegal, adding that an additional warning letter would be sent to the Government of Malta by the Commission (Vella, 2007). Subsequently, a resolution by the Greens and the United for Europe of the Nations group (*ibid.*), which urged the European Commission "to redouble its efforts to persuade the Maltese authorities to comply fully with Community law" on hunting, was approved by the European Parliament. However, amendments by Maltese MEPs resulted in the removal of text, which urged the government to reconsider its recent position on hunting in spring. Malta's Nationalist MEPs voted against the resolution, whilst the Labour ones abstained. On the other hand, Alternattiva Demokratika and Birdlife Malta welcomed the vote (The Times, 2007c).

As promised, the European Commission did write to Malta and gave it a month, as against the usual two months, to submit a formal reply. Should a reply not have been submitted by then, or were it not satisfactory, this would have led the Commission to proceed to the second stage of its legal proceedings, namely, the reasoned opinion stage. Commissioner

Dimas said that when other countries with similar problems, namely, Italy, France, and Finland, were taken to court, they had to stop hunting in spring (Camilleri, 2007g). When Malta duly replied, it downplayed the quantities of quail and turtle dove species that migrate over Malta but added that a scientific study will be carried out on actual figures. The government also said that given the Maltese context, it was conforming to the accepted term of "no other satisfactory solution" given that spring is Malta's main hunting season (Camilleri, 2007e). Malta's official reply was later on harshly criticized by Birdlife, which replied through a technical report compiled by wildlife biologist Andre' Raine. He basically argued that Malta's migratory route between Europe and Africa is important on an international level, and that killing of birds was very high (Grech, 2007d).

It is interesting to note that in the same year, Cyprus was officially reprimanded by the European Commission for allowing two days of hunting during Spring 2007 (Camilleri, 2007h). In the following year, Cyprus permitted bird hunting for six days in May, despite the EU Birds Directive (Camilleri, 2008e). Birdlife Cyprus estimated that around 500,000 migrating songbirds, especially robins and warblers, were trapped and killed in the island, mainly to be cooked (Reuters, 2007). Italy, Spain, and Austria had also been subject to legal action by the European Commission for breaking Bird Directive laws (Camilleri, 2006c).

In the meantime, the CABS announced the organization of a bird protection camp in Malta in mid-September 2007, comprising 22 campaigners from the EU who would be surveilling the activities of hunters (Fenech, 2007b). In addition, Birdlife Malta announced a similar camp in September, comprising 53 activists (also including scientists and conservationists similarly to CABS) from different countries – the biggest to-date since they were first organized in the late 1990s. The main objective of Birdlife's "raptor camp. [was]... to collect data on migratory species," though "the second aim. [was]. to act as a deterrent for hunters who break the law" (The Times, 2007d).

Once the monitoring camps commenced, activists reported various cases of indiscriminate hunting activities. Commenting at a press conference, one such activist said, "I knew about the illegal hunting here, I had read about it but I wasn't prepared for the scale of it. There are hunters' hides everywhere" (Micallef, 2007a). In the meantime, various CABS activists said that they were being harassed by some hunters (Micallef, 2007b), amidst a tense atmosphere, which was also characterized by rhetoric from the Hunters' Federation against "German extremists" and "foreign interference," adding further "foreign leadership" to that of the "Turkish"[4] director of Birdlife Malta (Busuttil, 2007).

By the time the monitoring camps came to an end, both CABS and Birdlife reported various instances of illegal bird hunting. Birdlife Malta alone reported over 209 incidents of illegal hunting in the few days covered by its camp (Grech, 2007e). The Federation for Hunting and Conservation announced that hunters who were convicted of hunting illegally would lose their membership with the Federation. This decision was of particular significance as according to Maltese legislation, endorsement from a recognized hunting association is required to obtain a hunting license (The Times, 2007e).

The European Commission was closely following developments, and it decided to issue a reasoned opinion as part of its infringement procedures against Malta. Commissioner Divas referred to Malta's spring hunting as "madness," and said that this issue was one of the most controversial that he had faced in recent years, with some other countries having similar cases as Malta's (Camilleri, 2007i). As spring was approaching, Birdlife Malta criticized Malta's delaying tactics as regards responses to warnings from the European Commission, and it also criticized the Nationalist and Labour parties for trying to appease the hunting lobby in order to win votes, and in disregard of EU legislation and Malta's "natural heritage" (Grech, 2008). The ENGO also wrote to the Prime Minister in an appeal not to open the upcoming hunting season. In reply the Federation for Hunting and Conservation and its EU counterpart, the Federation for Associations for Hunting and Conservation in the EU (FACE), said that Malta can be granted derogation under the EU Birds Directive,

> which would make it possible for hunters to catch quail and turtle dove in small numbers and under strictly supervised conditions in spring by taking into account the particular bio-geographical and socio-cultural situation of Malta. (Fenech, 2008)

In this regard, FACE and the FKNK issued a document with arguments for a derogation that permits a short spring hunting season. It also said that the latter would commit itself to strict observance of legislation and that it supported the Birds Directive (The Sunday Times, 2008). This did not impress the European Commission, which stuck to its position on the issue (Camilleri, 2008a). Indeed, on January 30, 2008, the European Commission decided to take legal action at the European Court of Justice. Alternattiva Demokratika immediately appealed to the government to ban spring hunting (Camilleri, 2008b). Birdlife International and Birdlife Malta joined suit, whilst the Government of Malta stated that the European Court of Justice was indeed the best venue for a final decision to be taken on the issue, which it would respect, though it would be defending its position in favor of spring hunting (The Times, 2008). The Labour party, in the meantime, said that it will stand by the decision of the European Court of Justice (Massa, 2008).

HUNTING AND THE EUROPEAN COURT OF JUSTICE

With Malta's upcoming general election just a few days away, the hunting issue remained a major controversy. On February 26, 2008, at the European Court of Justice, the European Commission formally presented its case against Malta regarding spring hunting. It called for an urgent decision on the matter, and failing this, for interim measures so as to disallow any spring hunting in Malta till the matter is decided upon (Camilleri, 2008c). In the meantime, the Nationalist Party was returned to power on March 9, 2008.

On April 25, 2008, after three weeks of deliberations, the European Court of Justice decided to issue interim measures through which spring hunting would be banned during 2008 (Micallef, 2008). The government duly announced that it would be standing by this decision, and, therefore, would not be opening the spring hunting season for the current year. Birdlife Malta and Birdlife International welcomed the decision, and interpreted the interim measure as an

> urgent need to prevent irreversible damage to these migratory bird species while a final ruling on this case is pending and not expected before 2009. (Xuereb, 2008)

The European Commission said that it expected the government to permanently stop allowing spring hunting. Yet as the European Court of Justice put it,

> it is apparent that the Commission's application for interim measures cannot be dismissed for lack of urgency as far as concerns the current year. With regard to 2009, on the other hand, urgency is not made out in the present application for interim measures. (Camilleri, 2008d)

In spring 2009, the hunting season was not opened, even though the Hunters' Federation urged the government to do so (The Times, 2009a). In a letter to Malta's members of parliament, also including Prime Minister Gonzi, the Federation said that the latter

> has once again on this issue failed to exercise the powers of decision that as a determined leader of a member state, equal to all other member states in the European Union, he is expected to exercise. (The Times, 2009b)

The Federation went as far as offering "to give-up most of the autumn season in return for the socio-cultural tradition of spring hunting" (*ibid.*). The ban of spring hunting was effective to an extent that CABS observed that the majority of hunters were respecting it (The Times, 2009c).

Following the 2009 European Parliamentary elections, which resulted in a majority of seats for Labour and a minority of seats for the Nationalist Party, the Hunters' Federation once again boycotted the Ornis Committee, as the latter decided to ban autumn trapping of finches – which were the most sought after birds – whilst permitting trapping for four other species. Birdlife – clinging on Malta's EU Accession Treaty – welcomed the decision but appealed to the prime minister to explain the conditions that would be in place and how many trappers would be permitted to trap. The Hunters' Federation, on the other hand, said that Maltese trappers should not be treated as "second-class European citizen(s)," adding that trapping was a "socio-cultural practice that dates back centuries" (Peregin, 2009).

On September 10, 2009, the European Court of Justice was due to decide on Malta's spring hunting issue. Before the verdict, Birdlife International and Federation of Associations for Hunting and Conservation of the EU issued a joint statement urging respect of the Court's decision in respect of biodiversity (Camilleri, 2010, pp. 46–47). In their words,

> We remain convinced that the EU Nature Directives provide sufficient possibilities to reconcile nature protection, sustainable hunting, bird-watching and other recreational activities and socio-cultural traditions, in line with the Guidance Document on Sustainable Hunting. (The Times, 2009d)

Previously, both organizations had reached an agreement in 2004 for the continuation of hunting in the EU provided that it conforms to the Bird Directive (Camilleri, 2010, p, 18). An example of "conflictual consensus" (Mouffe, 2000, p. 103) was in place. The European Court of Justice ruled that "by authorising spring hunting of quails and turtle doves from 2004 to 2007, Malta has failed to comply with the Wild Birds directive" (European Commission, 2009). At the same time, however, the ruling asserted that hunting for quails and turtle doves in autumn were not a satisfactory solution, even though, at the same time, the amounts of killings of the species in spring was not in line with the Birds Directive. The Maltese government, which made use of the assistance from FACE in its legal team, said that it would respect the court's decision (Camilleri, 2009b).

Birdlife Malta said that the judgment confirmed that spring hunting should be banned permanently; adding that hunting in autumn could take place on turtle dove, quail, and another 30 species in accordance with the Birds Directive. Birdlife International also welcomed the ruling, adding that this ruling was not only positive for millions of birds, which migrate back from Africa to Europe, but also showed how the Birds Directive could help "our common heritage" (The Times, 2009e). Yet, the other camp claimed victory too. The Federation of Hunters, Trappers and Conservationists did not give up:

> The decision does not mean that spring hunting is banned for ever. The court could not remove Malta's right to apply for a derogation in the future, and in view of new data which we already have, and with political will, we can get a derogation, although it will be more difficult. (The Times, 2009f)

During the 2009 FACE general assembly, the Federation of Hunters, Trappers and Conservationists said it welcomed a statement by the government for the studying of a possibility to have a limited hunting season involving quail and turtle dove. The Federation said that it expected BirdLife Malta to be constructive with respect to the Court's decision, and added that it would not tolerate neither illegal hunting nor attempts to stop "the traditional Maltese socio-cultural passion of spring hunting" (The Sunday Times, 2009).

During the same assembly, international lawyer Jan Bouckaert, who had been appointed by FACE to assist Malta's Attorney General Silvio Camilleri and the Hunters' Federation lawyer Alex Perici Calascione at the European Court of Justice, said that the Court's decision was a case of losing a battle but winning the war. In particular, Bouckaert said that the Court looked at specific circumstances rather than adopting a generic approach. He believed that this "victory" applied to all EU member states. FACE President Gilbert de Turckheim congratulates the Maltese Hunters' Federation for speaking the truth in a determined way, thus leading to a "milestone judgment" (*ibid.*).

To the hunters' delight, Prime Minister Lawrence Gonzi was reported by the *Malta Today* as telling their representatives that the European Court's decision made possible a future derogation on spring hunting, though he did not make any commitments regarding the upcoming spring season. Gonzi reportedly said this in a meeting with the Federation of Hunters, Trappers and Conservationists and the St Hubert Hunters' Association (Vassallo, 2009).

Subsequently, in 2010, the government announced that there would be a limited hunting season in spring, between 24 and 30 April (excluding Sunday), which was subject to quotas on numbers of birds allowed to be shot and a limited number of licenses, amongst other conditions (Camilleri, 2010, p. 63). It was more than evident that the hunting issue would remain characterized by antagonism in the realm of Maltese environmental politics.

WERE ENGOs EMPOWERED IN RELATION TO HUNTING IN MALTA WITHIN AN EU CONTEXT?

The above narrative shows that as regards hunting, ENGOs, both Maltese and European, were influential on State power in Malta, especially by resorting to the EU, and also being given prominence by the media. Yet, the hunting lobby was influential too, and its influence on Malta's main political parties is an overdetermining factor, which remained in place even after EU accession. As Camilleri (2010) declares,

> Bird hunting has metamorphosed into a political issue, which led the Maltese Government to a situation where it was walking on a tight-rope involving the respective competing interests. Through consultation and frequent discussions on hunting, Government strives to strike a balance between maintaining a long-practiced tradition alive within the limits of EU hunting regulations. The Maltese system follows the pluralist model in politics, where interest groups (albeit their differences, including degree of influence) can lobby Government and provide checks and balances against state institutions and officials. In turn the State mediates between competing interests. However, this process is peculiar as it is characterized by the dominance of two political parties, namely the Nationalist and Labour parties respectively. Judging by the recent history of hunting in Malta, it seems that the issue will remain politically controversial at least for the immediate future. (p. 65)

The environment and the hunting lobbies were articulating the hunting issues in different ways, yet they both managed to create nodal points. In itself, this had an important role in the politicization of hunting. As far as environmentalist demands go, chains of equivalence were formed around the "Stop spring hunting as in Birds Directive" signifier, to the benefit of conservationist ENGO Birdlife Malta. Indeed, Birdlife's demands were not only supported and/or shared by a myriad of ENGOs (Maltese and European) and other NGOs, which, at times, issued joint statements, but also by the Green Party and its European Counterparts, and, significantly, of the European Commission and the European Parliament. As regards the latter, however, certain hurdles were in place, especially from the European Peoples' Party and also from Maltese MEPs. Personalities also supported Birdlife Malta, and the ENGO gained widespread sympathy when members or property were harassed or attacked.

The European Commission was clearly on the side of the ENGOs, whilst the European Court of Justice gave a ruling which had different interpretations, though being ultimately restrictive on hunting, when the latter was compared to its status prior to Malta's EU accession. In the meantime, the Government of Malta and the major political parties tried to present a position of arbiters, though it is evident that they did much to try and accommodate the hunters' lobby notwithstanding the EU legislation.

The hunting lobby, affiliated to FACE and armed with nationalistic, confrontational, and traditionalistic discourse, tried to play the Nationalist and Labour parties against each other, especially during election time. This won them various concessions from the two parties. On the other hand, they now knew they were being watched and monitored. Indeed, despite the lack of widespread enforcement on hunting infringements in Malta, the presence of activists in monitoring camps acted as a detriment to hunters.

Significantly, even though the decision of the European Court of Justice was given different interpretations by the environmentalist and hunting lobbies, it ultimately enhanced empowerment of Maltese ENGOs. An anti-hunting hegemonic formation did not exist in Malta in the first years after EU accession, but notwithstanding this, and despite the attempts of the Maltese government to gain an exceptional status for Malta, hunters are in a much more restricted position than they were prior to Malta's EU membership. The dislocatory effect of EU accession resulted in discursive opportunities for ENGOs in Southern European Malta to resort to the EU (Börzel, 2008, p. 232) through the "boomerang effect" (Johnston, 2011, pp. 186–187).

Therefore, "Stop spring hunting as in Birds Directive" has the status of a nodal point, despite the political antagonisms in place (Howarth & Stavrakakis, 2000, p. 8; Laclau & Mouffe, 1985, p. 112). It is also interesting to note that both the pro-hunting and anti-hunting lobbies gave legitimacy to EU legislation. Both Birdlife International and FACE said that

> EU Nature Directives provide sufficient possibilities to reconcile nature protection, sustainable hunting, bird-watching and other recreational activities and socio-cultural traditions, in line with the Guidance Document on Sustainable Hunting. (The Times September 9, 2009a)

Yet, as shown previously, this is accompanied by antagonistic relations and different interpretations of EU policy and court decisions.

An application of the conceptualization of external impacts (Carter, 2001; Giugni, 1995; Rootes, 2007a; van der Heijden, 1999) of the environmental movement, namely, sensitizing, procedural, structural, and substantive would conclude as follows as regards empowerment of Maltese ENGOs with respect to hunting.

The "Stop spring hunting in line with the Birds Directive" sensitized the political agenda of both the EU and Malta. The signifier was used by ENGOs – both Maltese and European – to sensitize EU institutions. In Malta, a political discourse was clearly in place, not only within the realm of party and ENGO politics, but, also within the mass media, where the issue was frequently reported and commented upon. This discourse was in antagonism with the discourse propagated by the hunting lobby.

Unlike Malta, within the EU, a hegemonic formation around the Birds Directive was already in place prior to Malta's EU accession. This was institutionalized in EU legislation, particularly through the Birds Directive. Malta's EU accession thus had a dislocatory effect, which disrupted existing national legislation, thus having a substantive impact.

Parallel to the structural set-up of the EU as regards hunting – for example, through its Environment Directorate within the European Commission – structural changes took place in Malta due to hunting. The Maltese government also created the Ornis Committee featuring representation from the different camps, which, in itself meant that Birdlife Malta was institutionalized within the official consultation structure. The constant lobbying and the creation of monitoring camps by Maltese and European ENGOs also meant that there was a strong presence from civil society. On the other hand, enforcement against irregularities was rather weak, thus revealing a shortcoming in Maltese domestic structures. Along the same lines, the government was criticized by environmentalists for underreporting when presenting official data to the EU.

Linked to structural changes, there were also procedural impacts which enhanced empowerment of ENGOs. The Ornis Committee itself was a case in point. But so was the fact that nature conservation sites – often forming part of EU programs – were being managed by ENGOs. The European Commission gave legitimacy to the ENGOs through consultation, and the European Parliament took heed of ENGO lobbying by endorsing calls for Malta's conformity to the EU legislation. The European Greens, in particular, were very vocal on this.

EU ENGOs, in general, and Maltese ENGOs, in particular, used the EU as a gateway for empowerment, through which their claims could be articulated (Börzel, 2008, p. 232; Dimitrakopoulos & Richardson, 2001, pp. 344–345; Johnston, 2011, pp. 186–187). Birdlife Malta – the most prominent ENGO in this regard – clearly relied on its European counterpart, and it also benefitted from the lobbying of other ENGOs within the EU. The dislocatory effect of EU accession, therefore, also affected the discursive coalitions between Maltese and EU ENGOs. Other ENGOs in Malta which spoke up on the issue also added pressure.

The conservationist ideology related to the preservation of species, through which the "Stop spring hunting in line with the Birds Directive" was articulated, enabled the creation of a nodal point in relation to the adaptation of EU policy in Malta. Apart from Birdlife Malta, other ENGOs used such discourse.

A quick-fix solution on hunting issue, therefore, does not seem to exist, meaning that one can expect more antagonism on this issue in southern European Malta. The hunting issue exemplifies environmental politics in an EU context, which interplays with the national dimension.

HUNTING GOES NATIONAL AGAIN: THE REFERENDUM COALITION

Fast forward three years of wrangling between environmentalists and hunters, and in 2013, Malta elected a new government, The Labour party achieved a massive majority, ending almost 25 years of Nationalist Party rule, the latter being interrupted only by a 22-month interval of Labour rule between 1996 and 1998. Labour's campaign attempted to create a "politics without adversaries" (Briguglio, 2010), which, amongst other strategies, aimed to reconcile hunters with environmentalists. It was, however, evident that hunters had much to gain through a Labour victory. Indeed it was observed that Labour "woos bird hunters' votes by sympathizing with their claims ... But at the same time the LP declares that it will stick by the EU birds directive" (Briguglio & Pace, 2013, p. 281).

A few months after Labour's victory, a coalition of Maltese ENGOs and the Green Party formed a campaign with the intention of collecting the amount of signatures necessary – 10 percent of the electorate – for an abrogative referendum, which is permitted by Maltese legislation to remove existing legislation. The specific aim of the referendum was to remove legislation in Malta which permits hunting in Spring.

The required number of signatures were collected after a few months (Mangion, 2014), which in itself represented a first in Malta. In reaction, the hunters' lobby collected signatures to safeguard pastimes and minority rights, amidst speculation that the numbers collected greatly exceeded those collected by environmentalists (Martin, 2014). However, this has no legal validity, given that only abrogative referenda are enshrined in Maltese legislation.

It now remains to be seen whether the anti-spring hunting referendum will be successful, which, in this case, would result in a substantive achievement for the environmental movement in Malta. Such an achievement would be taking place in spite of the Europeanization of politics on hunting, and would, therefore, signify the importance of the national contextualization of antagonisms within the Maltese social formation. Yet, it would then have to be seen if such a substantive achievement would be hegemonic – for example, through policy shifts amongst Malta's main political parties – or whether it would merely result in the continuation of antagonisms that were taking place in previous years.

NOTES

1. It is pertinent to note that the 2013 general elections resulted in a massive vote-swing from the outgoing Nationalist Party to the victorious Labour Party, which could raise doubts as to the persistence of such a tendency. This chapter, however, covers an earlier period.
2. This will also be referred to as the Hunters' Federation.
3. Trapping is also a contentious issue, which has been politicized in relation to Malta's EU accession, but is not being analyzed in this chapter.
4. One of Malta's national feasts celebrates the victory against the "great siege" of the Ottomans in 1565. At the time, Malta was ruled by the Sovereign Order of Saint John of Jerusalem of Rhodes and of Malta.

REFERENCES

Althusser, L. (1990). *Philosophy and the spontaneous philosophy of the scientists and other essays*. London: Verso.

Baldacchino, G. (2002). The nationless state? – Malta, national identity and the European Union. *West European Politics*, *25*(4), 191–206.

Boissevain, J. (1993). *Saints and fireworks – Religion and politics in rural Malta*. Malta: Progress Press.

Boissevain, J. (2000). The environment: Prospects for the millennium. In C. C. Vella (Ed.), *The Maltese islands on the move: A mosaic of contributions marking Malta's entry into the 21st century* (pp. 1–13). Malta: Central Office of Statistics.

Börzel, T. (2008). Environmental policy. In P. Graziano & M. P. Vink (Eds.), *Europeanization – New research agendas* (pp. 226–238). Hampshire: Palgrave Macmillan.

Börzel, T., & Buzogany, A. (2010). Environmental organizations and the Europeanization of public policy in Central and Eastern Europe: The case of biodiversity governance. *Environmental Politics*, *19*(5), 708–735.

Briguglio, M. (2001). *Ideological and strategic shifts from old labour to new labour in Malta*. M.A. dissertation, Department of Sociology, University of Malta.

Briguglio, M. (2010). Malta's labour party and the politics of hegemony. *Socialism and Democracy*, *24*(2), 213–226.

Briguglio, M., & Brown, M. (2009). Global and European Institutional Perspectives of the Green Economy, and Civil Society NGOs perspective. In M. Kennet & N. Baster (Eds.), *Proceedings: Preparing for Copenhagen COP15, Climate change and economics, the battleground of the growth/development paradox*. Green Economics Conference (pp. 52–66), 28 November 2009 at Oxford University Club. Oxford: Green Economics Institute.

Briguglio, M., & Pace, R. (2013). Malta. In J.-M. de Waele, F. Escalona, & M. Vieira (Eds.), *The palgrave handbook of social democracy in the European Union* (pp. 269–297). Hampshire: Palgrave Macmillan.

Busuttil, C. (2007). Afternoon hunting banned 'to accommodate foreigners. *The Times*, September 18.

Camilleri, G. (2010). *EU policy on bird hunting and the polemics surrounding bird hunting*. The case of Malta. B.A. (Hons.) dissertation, European Documentation and Research Centre, University of Malta.

Camilleri, I. (2006a). Bird Life welcomes changes but wants more. *The Times*, April 6.

Camilleri, I. (2006b). Commission to take legal action against Malta. *The Times*, June 29.

Camilleri, I. (2006c). Other EU states in trouble over hunting. *The Times*, December 20.

Camilleri, I. (2007a). Hunters happy with government's stance. *The Times*, January 18.

Camilleri, I. (2007b). Brussels expects no hunting this spring. *The Times*, January 22.

Camilleri, I. (2007c). Brussels condemns hunting decision. *The Times*, February 28.

Camilleri, I. (2007d). 'No Change' after top-level meeting. *The Times*, April 4.

Camilleri, I. (2007e). Malta replies to Brussels. *The Times*, April 10.

Camilleri, I. (2007f). Brussels to step up pressure on Malta. *The Times*, March 14.

Camilleri, I. (2007g). Malta given a month to file reply. *The Times*, March 22.

Camilleri, I. (2007h). Cyprus reprimanded over spring hunting. *The Times*, August 29.

Camilleri, I. (2007i). EU Commissioner calls spring hunting madness. *The Times*, November 8.

Camilleri, I. (2008a). Brussels shoots down hunters' proposal. *The Times*, January 27.

Camilleri, I. (2008b). Brussels agrees to send issue to European Court. *The Times*, January 31.

Camilleri, I. (2008c). Brussels asks court to ban spring hunting. *The Times*, February 27.

Camilleri, I. (2008d). Brussels expects spring hunting to stop. *The Times*, April 26.

Camilleri, I. (2008e). Brussels expected to rap Cyprus over spring hunting. *The Times*, May 6.

Camilleri, I. (2009a). European Court rules on spring hunting today. *The Times*, September 10.

Camilleri, I. (2009b). ECJ decides against Malta in spring hunting case. *The Times*, September 10.

Carter, N. (2001). *The politics of the environment – Ideas, activism, policy*. Cambridge: Cambridge University Press.

Debono, J. (2009). From power brokers to embarrassment. *Malta Today*, March 25.

Delanty, G., & Rumford, C. (2005). *Rethinking Europe – Social theory and the implications of Europeanization*. London: Routledge.

Dimitrakopoulos, D., & Richardson, J. (2001). Implementing EU public policy. In J. Richardson (Ed.), *European Union – Power and policy-making* (2nd ed., pp. 335–356). London: Routledge.

Edwards, M. (2009). *Civil society* (2nd ed.). Cambridge: Polity.

Euractiv.com. (2005, July 20). High noon for EU's environment policies. Retrieved from http://www.euractiv.com/en/innovation/high-noon-eu-environment-policies/article-142732. Accessed on February 16, 2008.

European Commission. (2009). European Court of Justice Case C-76/08 (Commission vs Malta) – Environment and Consumers. Retrieved from http://ec.europa.eu/avservices/services/showShotlist.do?out=PDF&lg=En&filmRef=65155. Accessed on February 16, 2011.

Eurostat. (2010). *Europe in figures. Eurostat yearbook 2010*. Luxembourg: Publications Office of the European Union.

Falter, C., & Scheuer, S. (2005). Nature. In S. Scheuer (Ed.), *EU environmental policy handbook: A critical analysis of EU environmental legislation* (pp. 36–45). Brussels: European Environmental Bureau.

Falzon, M.-A. (2008). Flights of passion. Hunting, ecology and politics in Malta and the Mediterranean. *Anthropology Today*, *24*(1), 15–20.

Fenech, N. (2007a). EU Commission invites hunters, BirdLife to Brussels meeting. *The Times*, February 9.

Fenech, N. (2007b). No hunting after 3 p.m. for two weeks. *The Times*, August 30.

Fenech, N. (2008). Hunters' federation looking for derogation on short spring hunting. *The Times*, January 21.

Giugni, M. G. (1995). Outcomes of new social movements. In H. Kriesi, R. Koopmans, J. W. Dvendak, & M. G. Giugni (Eds.), *New social movements in Western Europe* (pp. 207–237). London: UCL Press.

Government of Malta. (2003). *Europe. Do you agree that Malta should become a member of the European Union in the enlargement that is to take place on 1st May 2004?* Malta: Government of Malta.

Grech, H. (2007a). Stop the delaying tactics. *The Times*, February 14.

Grech, H. (2007b). Hunters suspend ultimatum. *The Times*, February 16.

Grech, H. (2007c). Malta deifies EU spring hunting. *The Times*, February 27.

Grech, H. (2007d). Bird migration report challenges government claims. *The Times*, May 19.

Grech, H. (2007e). Ornithologists witness over 200 incidents of illegal hunting. *The Times*, September 29.

Grech, H. (2008). Spring hunting ban possible before EU court ruling. *The Sunday Times*, January 20.

Griggs, S. (2005). Problematizing the mobilization of hospital directors. In D. Howarth & J. Torfing (Eds.), *Discourse theory in European politics – Identity, policy and governance* (pp. 117–138). Hampshire: Palgrave Macmillan.

Harwood, M. (2009). *The Europeanization of the maltese national administration: Mapping the changes in a Mediterranean member state*. PhD dissertation, European Documentation and Research Centre, University of Malta.

Hirczy, W. (1995). Explaining near-universal turnout: The case of Malta. *European Journal of Political Research*, *27*, 255–272.

Howarth, D. (2000). *Discourse*. Milton Keynes: Open University Press.

Howarth, D. (2004). Hegemony, political subjectivity and radical democracy. In S. Critchley & O. Marchart (Eds.), *Laclau: A critical reader* (pp. 256–276). London: Routledge.

Howarth, D. (2005). Applying discourse theory: The method of articulation. In D. Howarth & J. Torfing (Eds.), *Discourse theory in European politics – Identity, policy and governance* (pp. 316–349). Hampshire: Palgrave Macmillan.

Howarth, D., & Stavrakakis, Y. (2000). Introducing discourse theory and political analysis. In D. Howarth, A. J. Norval, & Y. Stavrakakis (Eds.), *Discourse theory and political analysis: Identities, hegemonies and social change* (pp. 1–23). Manchester: Manchester University Press.

Jessop, B. (2004). *The European Union and recent transformations in statehood*. Lancaster: Department of Sociology, University of Lancaster. Retrieved from http://www.lancs.ac.uk/fass/sociology/papers/jessop-eu-transformations-statehood.pdf. Accessed on September 1, 2008.

Johnston, H. (2011). *States and social movements*. Cambridge: Polity.

Laclau, E. (1996). *Emancipation(s) (phronesis S.)*. London: Verso.

Laclau, E. (2005). *On populist reason*. London: Verso.

Laclau, E., & Mouffe, C. (1985). *Hegemony and socialist strategy: Towards a radical democratic politics*. London: Verso.

Lane, J. C. (2000). A survey of elections in malta, 1921–1998. In C. C. Vella (Ed.), *The maltese islands on the move: A mosaic of contributions marking Malta's entry into the 21st century* (pp. 207–222). Malta: Central Office of Statistics.

Lenschow, A. (2006). Europeanization of public policy. In J. Richardson (Ed.), *European Union – Power and policy making* (3rd ed., pp. 55–72). Oxford: Routledge.

Mallia, S. (2006). Malta adopts birds directive. *The Times*, March 30.

Malta Environment and Planning Authority (MEPA). (2010). *Public attitudes survey 2008 – Analysis of results*. Malta: Ernst and Young.

Mangion, C. (2014, March 28). Almost 45,000 signatures for referendum to ban spring hunting. *Malta Today*. Retrieved from http://www.maltatoday.com.mt/lifestyle/environment/37428/almost_45000_signatures_for_referendum_to_ban_spring_hunting#.U00c4TOKDIU. Accessed on April 14, 2014.

Martin, I. (2014, April 12). Minors have signed FKNK Hunt Petition. *The Times*. Retrieved from http://www.timesofmalta.com/articles/view/20140412/local/Minors-have-signed-FKNK-hunt-petition.514593. Accessed on April 14, 2014.

Massa, A. (2008). Labour will respect European Court's ruling. *The Times*, February 13.

Mazey, S., & Richardson, J. (2006). Interest groups and EU policy-making: Organisational logic and venue shopping. In J. Richardson (Ed.), *European Union – Power and policy-making* (3rd ed, pp. 247–268). London: Routledge.

Micallef, M. (2007a). Birdwatchers condemn raptors slaughter. *The Times*, September 17.
Micallef, M. (2007b). Birdwatchers hit by lead pellets in hunting accident. *The Times*, September 20.
Micallef, M. (2008). Court bans spring hunting this year. *The Times*, April 25.
Mouffe, C. (2000). *The democratic paradox*. London: Verso.
Mouffe, C. (2005). *On the political*. London: Routledge.
Peregin, C. (2009). Ornis Committee advises ban on finch trapping. *The Times*, July 30.
Pollacco, C. (2004). *European integration: The Maltese experience*. Malta: Agenda.
Pullicino, G. (2007, March 2). Striking a balance. *The Times*. Retrieved from http://www.timesofmalta.com/core/print_article.php?id=253785. Accessed on March 2, 2007.
Reuters. (2007). Songbird killings in Cyprus rise: conservation group. December 20. Retrieved from http://www.reuters.com/article/2007/12/20/environment-cyprus-birds-dc-idUSHAR05781020071220. Accessed on April 24, 2012.
Roche, M. (2010). *Exploring the sociology of Europe*. London: Sage.
Rootes, C. (2007a). Environmental movements. In D. A. Snow, S. A. Soule, & H. Kriesi (Eds.), *The Blackwell companion to social movements* (pp. 608–640). Malden: Blackwell Publishing.
Rootes, C. (2007b). Conclusion: Environmental protest transformed? In C. Rootes (Ed.), *Environmental protest in Western Europe* (pp. 234–257). Oxford: Oxford University Press.
Steunenberg, B. (2006). Turning swift policy-making into deadlock and delay – National policy coordination and the transposition of EU directives. *European Union Politics*, *7*(3), 293–320.
Steunenberg, B., & Rhinard, M. (2010). The transposition of European law in EU member states: Between process and politics. *European Political Science Review*, *2*(3), 495–520.
The Sunday Times. (2006). Over our dead bodies, hunters warn. *The Sunday Times*, July 9.
The Sunday Times. (2008). EU federation to assist Maltese hunters in bid for derogation. *The Sunday Times*, January 20.
The Sunday Times. (2009). Hunters: We have won the war. *The Sunday Times*, September 13.
The Times. (2007a). Hunting in spring cannot be allowed. *The Times*, January 10.
The Times. (2007b). Hunters and trappers. *The Times*, February 8.
The Times. (2007c). MEPs approved diluted resolution. *The Times*, March 16.
The Times. (2007d). 53 ornithologists for BirdLife's monitoring camp. *The Times*, September 3.
The Times. (2007e). Federation warns wayward hunters. *The Times*, November 22.
The Times. (2008). Government to argue for spring hunting at the European Court. *The Times*, January 31.
The Times. (2009a). New Spring hunting ban would contradict Malta stand in European court – FKNK. *The Times*, March 11.
The Times. (2009b). FKNK insists spring hunting season should open. *The Times*, March 21.
The Times. (2009c). Majority of hunters respect spring hunting ban – CABS. *The Times*, May 5.
The Times. (2009d). Spring hunting decision tomorrow. *The Times*, September 9.
The Times. (2009e). BirdLife welcomes hunting decision. *The Times*, September 10.
The Times. (2009f). ECJ decision leaves hunting derogation possibility. *The Times*, September 10.
Torfing, J. (2005). Discourse theory: Achievements, arguments, and challenges. In D. Howarth & J. Torfing (Eds.), *Discourse theory in European politics – Identity, policy and governance* (pp. 1–32). Hampshire: Palgrave Macmillan.

Touraine, A. (1981). *The voice and the eye: An analysis of social movements.* Cambridge: Cambridge University Press.

Van der Heijden, H.-A. (1999). Environmental movements, ecological modernization and political opportunity structures. In C. Rootes (Ed.), *Environmental movements: Local, national and global* (pp. 199–221). London: Frank Cass Publishers.

Van der Heijden, H.-A. (2010). *Social movements, public spheres and the European politics of the environment. Green power Europe?* Hampshire: Palgrave Macmillan.

Vassallo, R. (2009). Yes, together spring hunting is possible, Gonzi tells hunters. *Malta Today*, October 7.

Vella, M. (2007). I'm not interested in votes'. *Malta Today*, March.

Vink, M. P., & Graziano, P. (2008). Challenges of a new research agenda. In P. Graziano & M. P. Vink (Eds.), *Europeanization – New research agendas* (pp. 3–20). Hampshire: Palgrave Macmillan.

Wallace, H., Wallace, W., & Pollack, M. A. (2005). *Policy-making in the European Union* (5th ed), Oxford: Oxford University Press.

Webster, R. (2000). What drives interest group collaboration at the EU level? Evidence from the European environment interest groups. *European Integration online Papers*, *4*(17), 1–22. Retrieved from http://eiop.or.at/eiop/texte/2000-017a.htm. Accessed on November 9, 2004.

Xuereb, M. (2008). No spring hunting this year, government says. *The Times*, April 26.

ABOUT THE CONTRIBUTORS

Saleh Ahmed is a graduate student in Sociology at Utah State University, majoring in environmental and community studies. He worked previously with World Bank Project in Bangladesh and with the International Labour Organization (ILO) Headquarter in Geneva, Switzerland.

Keith Bettinger received his PhD from the Department of Geography at the University of Hawai'i. He is currently conducting research on adaptation to climate change in Southeast Asia.

Michael Briguglio has a PhD in Sociology, and he lectures at the University of Malta. His main sociological research interests are politics, social movements, environment, culture and social policy, and he has published various academic papers.

Micah Fisher is a PhD student in the University of Hawai'i's Department of Geography. His research interests in Indonesia center around the dynamics of land-use change and efforts at collaborative governance.

Benny Furst is a geographer and environmental planner, holder of BA, MA and doctoral degrees from the Hebrew University of Jerusalem in Political Sciences, Geography and Urban planning. His PhD dissertation (2013) is about the effectiveness of environmental campaigns in Israel. Dr. Furst has published a number of studies about GreenPeace Israel, the wild flowers protection campaign, and lately about patterns of activism among environmental NGOs in Israel. Dr. Furst works at the planning division of the Israeli Ministry of Environmental Protection, and as a teaching fellow at the Inter-Disciplinary Center in Herzlia.

Nan Marie Greer is a professor of Environmental Anthropology at the University of Hawai'i-Kaua'i Community College. She works on indigenous rights issues, including consultancy research on land title and natural resource rights, in the Hawaiian Islands where she has worked for over 13 years, in Nicaragua where she has worked for over 20 years, and now in California as well.

Sya Buryn Kedzior is an Assistant Professor of Geography and Environmental Planning at Towson University in Maryland. Her work examines how people make sense of their natural environment, and how that understanding shapes and is shaped by struggles over resource use and management. She works primarily in northern India and North America.

Liam Leonard is the Senior Editor of the Advances in Sustainability and Environmental Justice Series. He is an Adjunct Associate Professor with the Department of Sociology and Anthropology at West Virginia University and a Research Fellow with WVU's Centre for Study of Violence. Dr. Leonard is also Founder and Editor of CRIMSOC: The Journal of Social Criminology, and a former President of the Sociological Association of Ireland.

Charles Z. Levkoe is a postdoctoral fellow at Wilfrid Laurier University. He is an Executive Committee member of the Canadian Association for Food Studies and works closely with food movement organizations across Canada. His teaching and research bridge the fields of sustainable agricultural, political and social geography, urban policy and environmental justice.

Robert J. Mason is a Professor of Geography and Urban Studies at Temple University. He is currently interested in broad-scale urban and regional approaches to furthering environmental sustainability, especially trends in land-use management in the United States, metropolitan sprawl in China, management of the Shirakami Sanchi World Heritage Area in Japan, and the emerging role of Japan's citizen environmental organizations.

Darren McCauley is Lecturer and Director for Postgraduate Studies in Sustainable Development in the School of Geography and Geosciences at the University of St. Andrews. He has held previous full time lectureships at Trinity College Dublin, Queens University Belfast and the University of Stirling.

Wendy Miles is a PhD candidate in the University of Hawai'i's Department of Geography and an Environmental Protection Agency 'Science to Achieve Results' Fellow. She was a Fulbright fellow to Indonesia in 2012, during which time she conducted dissertation field research on the potential impacts of the international carbon market on local communities and the environment in Central Kalimantan.

Milos Milicevic is a PhD Candidate at the Central European University's Department of Environmental Sciences and Policy, researching environmental and wider social effects of a proposed oil shale mining project in southern Serbia. His interests include issues of environmental and social justice, how these are articulated and enacted through social movement activity and civic action, and tackled by policy solutions.

Rebecca Schiff is an Assistant Professor in the Department of Health Sciences at Lakehead University. Rebecca has a long history of research and advocacy for sustainable, health food systems, having facilitated a civil society submission on Canadian food issues at the request of the U.N. Special Rapporteur on the Right to Food and served as an active member on the steering committee of Food Secure Canada/Sécurité alimentaire Canada (FSC/SAC) and the Executive Committee of Canadian Association for Food Studies.

Leslie E. Sponsel, PhD at Cornell University, is Professor Emeritus of Anthropology at the University of Hawai'i. He is founder and director of the independent Research Institute for Spiritual Ecology (RISE): http://spiritualecology.info.

Tamara Steger is an Associate Professor in the Department of Environmental Sciences and Policy at Central European University in Budapest, Hungary and founder of the Environmental and Social Justice Action Research Group (ACTJUST). Her current research interests include the anti-fracking movement and environmental justice in Central and Eastern Europe.